130
Advances in Polymer Science

Springer-Verlag Berlin Heidelberg GmbH

Thermoreversible Networks

Viscoelastic Properties and Structure of Gels

By K. te Nijenhuis

Springer

Author

Dr. K. te Nijenhuis
Laboratory of Polymer Science and Technology
Faculty of Chemical Technology and Materials Science
Delft University of Technolgy
Julianalaan 136
2628 BL Delft / The Netherlands

This series presents critical reviews of the present and future trends in polymer and biopolymer science including chemistry, physical chemistry, physics and material science. It is addressed to all scientists at universities and in industry who wish to help abreast of advances in the topics covered.

As a rule, contributions are specially commissioned. The editors and publishers will, however, always be pleased to receive suggestions and supplementary information. Papers are accepted for „Advances in Polmyer Science" in English.

In references Advances in Polymer Science is abbreviated Adv. Polym. Sci. and is cited as a journal.

Springer WWW home page: http://www.springer.de

ISSN 0065-3195
ISBN 978-3-662-14795-5   ISBN 978-3-540-70731-8 (eBook)
DOI 10.1007/978-3-540-70731-8

Library of Congress Catalog Card Number 61642

Originally published by Springer-Verlag Berlin Heidelberg New York in 1997
Softcover reprint of the hardcover 1st edition 1997

Typesetting: Macmillan India Ltd., Bangalore-25
SPIN: 10548385   02/3020 - 5 4 3 2 1 0 - Printed on acid-free paper

Editors

Prof. J. E. McGrath
Polymer Materials and Interfaces Lab.
Virginia Polytechnic and State University
Blacksburg
Virginia 24061, USA

Prof. Lucien Monnerie
Ecole Superieure de Physique et de Chimie Industrielles
Laboratoire de Physico-Chimie Structurale et Macromoléculaire
10, rue Vauquelin
75231 Paris Cedex 05, France

Prof. Seizo Okamura
No. 24, Minamigoshi-Machi Okazaki,
Sakyo-Ku, Kyoto 606, Japan

Prof. Charles G. Overberger
Department of Chemistry
The University of Michigan
Ann Arbor, Michigan 48109, USA

Prof. Helmut Ringsdorf
Institut für Organische Chemie
Johannes-Gutenberg-Universität
J.-J.-Becher Weg 18-20
D-55128 Mainz, FRG

Prof. Takeo Saegusa
KRI International
Inc. Kyoto Research Park 17
Chudoji Minamima-chi
Shimogyo-ku Kyoto 600, Japan

Prof. J. C. Salamone
University of Lowell
Department of Chemistry
College of Pure and Applied Science
One University Avenue
Lowell, MA 01854, USA

Prof. John L. Schrag
University of Wisconsin
Department of Chemistry
1101 University Avenue
Madison, Wisconsin 53706, USA

Prof. G. Wegner
Max-Planck-Institut für Polymerforschung
Ackermannweg 10
Postfach 3148
D-55128 Mainz, FRG

Foreword

Macromolecular science and engineering as a separate discipline has existed for no more than seventy-five years, and has undoubtedly been one of the major scientific growth areas in the 20th century. We now appreciate that the term represents the study of all aspects of high molecular weight species, from the synthesis and characterisation of plastics and rubbers, to the interpretation of the structure and function of biological material from an understanding of molecular physics and chemistry.

For many years most physical and chemical studies on polymeric materials had to be carried out on biopolymers, since no synthetic polymers existed, except as a 'mistake' in the bottom of an organic chemist's flask! Indeed, the first synthetic resins were manufactured some fifty years before there was any real understanding of their polymeric nature. Nowadays, it is taken almost for granted that the mechanical properties of a wide range of polymeric materials are a natural consequence of their long-chain structure.

Much work is still concerned with essentially linear polymers, from their synthesis to the characterisation as melts or glasses. However, also polymer gels and networks have always been of major interest and technological importance. Historically, at least, polymer networks can be divided into two main classes, chemically cross-linked materials (including bulk elastomers) and 'entanglement systems'. The latter are formed by the topological interaction of polymer chains, either in the melt or in solution when the product of concentration and molecular weight becomes greater than some critical value. In this case they behave as 'pseudo gels' at frequencies higher (timescales shorter) than the lifetime of the topological entanglements. Covalently cross-linked materials, on the other hand, are formed by a variety of routes including cross-linking high molecular weight linear chains, either chemically or by radiation, by end-linking reactant chains with a branching unit, or by step-addition polymerisation of oligomeric multifunctional precursors. They are true macromolecules, where the molecular weight is nominally infinite and they therefore possess an infinite relaxation time and an equilibrium modulus.

Rheological discrimination between these two classes of networks can be made by the technique of small deformation oscillatory shear mechanical spectroscopy. At very low frequencies, in the 'terminal zone', entangled melts flow as high viscosity liquids, while cross-linked networks show no such effect. For a rubbery gel

(i. e. above its glass transition) G', the storage modulus or real part of the complexmodulus, is largely frequency insensitive. A further discrimination can be made by adding excess solvent to the bulk 'gel'. Entanglement network systems will dissolve to form a more dilute polymer solution, whereas cross-linked networks will swell, but not dissolve. Indeed, it is common for swollen networks to be referred to simply as gels.

However, there are many systems which involve non-covalently cross-linked network systems formed from both synthetic and biological macromolecules. The term 'physical gel' has been coined to describe these systems, to distinguish them from 'chemical gels' which involve covalent cross-links. A major , and for technological reasons, perhaps the most important class of physical gels are those which are thermoreversible. That is they form (usually) on cooling, and can be melted by heating, only to reform on cooling again. A very few examples (cellulose ethers and the polysaccharide curdlan) systems are inversely thermoreversible, usually indicative of UCST-type phase behaviour. There are several other examples of thermoreversible gels from biological sources such as seaweeds, and from synthetic polymers, including PVC and polystyrene. However, many workers in synthetic polymer gels appear largely ignorant of directly analogous work on biopolymer gels and vice versa. At the same time, there have been several rheology papers on thermoreversible gels and their viscoelasticity which have completely neglected the important macromolecular detail, so application is almost futile.

The term, physical gel, is simple and attractive, and the word gel (attributed to Thomas Graham) originated from gelatin, a proteinaceous material derived by hydrolytic degradation of collagen. This is the principal protein component of white fibrous connective tissue. Its thermoreversible gelling properties must have been known since at least neolithic times, since when connective tissue (meat) is cooked and allowed to cool a (gelatin) gel is often seen.

It is very important at the outset to clarify that the range of systems which can form thermoreversible gels is very diverse, and there may be as many differences in the properties of these materials as there are similarities. However, as will be seen in Dr. te Nijenhuis' review, they all possess at least one property which can stand as the operational definition of a gel, they possess a plateau extending over an appreciable window of frequencies, i. e. they are, or can be coaxed under appropriate conditions to be, viscoelastic solids.

Technological use of thermoreversible gels is widespread, including the textile, paper, packaging, cosmetic, health care (including the pharmaceutical sector), oil field, photographic and food industries. More recent interest, particularly in Japan, has been shown in thermoreversible gels as examples of smart materials for switches, actuators and sensors. The photographic industry, in particular has, after 100 years, still not found a better material than gelatin for the film emulsion base.

The review itself is one of the first attempts to compare and contrast the rheological behaviour of different thermoreversible gelling systems. Dr. te Nijenhuis is, of course, a real authority in this area, and his work on both gelatin and PVC gels remains classic work in the field. He has two major advantages in writing this review, firstly his expertise can cover both theory and experiment, secondly he has published on both sides of the (bio)polymer fence and this allows him to draw together work on disparate systems. His review is certainly not just an uncritical list of systems and experiments. He has managed to draw together many examples of thermoreversible gel systems, and I am pleased to see included recent data on the new microbial gelling agent known as gellan gum. It is my pleasure to have been invited to provide this preface, and to introduce what I feel will be a much cited review article.

Division of Live Sciences
King's College London, January 1997

S.B. Ross-Murphy

Table of Contents

Advances in Polymer Science, Vol. 130

Abbreviations

a) Chemicals

CR	congo red
DBP	dibutyl phthalate
DMA	dimethylacetamide
DMSO	dimethyl sulfoxide
DMF	dimethylformamide
DOP	dioctyl phthalate = di-(2-ethylhexyl) phthalate
DPhE	diphenyl ether
EG	ethylene glycol
EO	ethylene oxide
PC	propylene carbonate
PO	propylene oxide

b) Amino acid residues

ALA	alanine
ARG	arginine
ASP	aspartic acid
GLU	glutamic acid
GLY	glycine
HYP	hydroxyproline
LYS	lysine
PRO	proline

c) Polymers

COSP	crystalline organosilicic polymer
EVA	ethylene-vinyl acetate copolymer
HDPE	high density polyethylene
LCP	liquid crystalline polymer
LDPE	low density polyethylene
LLDPE	linear low density polyethylene
MCLCP	main chain liquid crystalline polymer
PAN	polyacrylonitrile

PBLG	poly (γ-benzyl-α,L-glutamate)
PDMS	polydimethyl siloxane
PE	polyethylene
PEO	polyethylene oxide
PMMA	polymethyl methacrylate
a-PMMA	atactic polymethyl methacrylate
i-PMMA	isotactic polymethyl methacrylate
s-PMMA	syndiotactic polymethyl methacrylate
PPO	polypropylene oxide
PS	polystyrene
a-PS	atactic polystyrene
PVA	polyvinyl alcohol
PVC	polyvinyl chloride
SBS	styrene-butadiene-styrene triblock copolymer
SCLCP	side chain liquid crystalline polymer
SIS	styrene-isoprene-styrene triblock copolymer
UHMWPE	ultra high molecular weight polyethylene

d) Techniques

DSC	differential scanning calorimetry
GPC	gel permeation chromatography
IR	infra red spectroscopy
LAM	longitudinal acoustic mode
LS	light scattering
NMR	nuclear magnetic resonance
SANS	small angle neutron scattering
SAXS	small angle X-ray scattering
SEM	scanning electron microscopy
TEM	transmission electron microscopy
UV-vis	ultra violet-visible light spectroscopy
WAXS	wide angle X-ray scattering

e) Miscellaneous

DS	degree of saponification (polyvinyl alcohol)
EANC	elastically effective network chain
L-L	liquid-liquid (demixing)
L-S	liquid-solid (demixing)
TT	trans-trans (conformation)
WLF	Williams Landel Ferry (WLF-equation)

Symbols

a	exponent in Mark-Houwink equation	–
	unit cell parameter	nm
a_T	horizontal shift factor in WLF-equation	–
b	borax concentration	mol/m^3
	unit cell parameter	nm
	shift factor: G^o/G	–
b_o	concentration of borax added	mol/m^3
c	concentration	kg/m^3
	unit cell parameter	nm
c_1^o	parameter in WLF-equation (Eq. 23) at T_o	–
c_1^g	id at T_g	–
c_2^o	parameter in WLF-equation (Eq. 23) at T_o	K
c_2^g	id at T_g	K
c_{gel}	minimum polymer concentration at the gel point	kg/m^3
c_o	minimum polymer concentration at the gel point	kg/m^3
f	functionality of crosslinks	–
[f*]	compression modulus	N/mm^2
f_g	free volume fraction at T_g	–
k	Boltzmann constant	J/K
	reaction rate constant	variable
l_d	long distance	nm
m	slope in $\log(\sigma)$ vs. $\log(t)$ plot	–
n	relaxation rate exponent (Eq. 1)	–
	number of polymer molecules joined in a crosslink (i.e. 1/2 f)	–
n'	exponent in Eq. (50)	–
n_{cl}	number of f-functional crosslinks	m^{-3}
n_1	number of potential crosslinks still present	m^{-3}
$n_{4\text{-hyp}/1000}$	4-hyp content per 1000 amino acid residues present	–
$n_{(pro+hyp)/1000}$	imino acid content per 1000 amino acid residues present	–
p	reaction rate exponent	–
	fraction of associated monomeric units	–
	exponent in Eq. (49)	–
q	reaction rate exponent	–
	minimum sequence length	–
	exponent in Eqs. (48), (49) and (50)	–
s	fraction of syndiotactic diads	–
t	time	s, min, h

t_a	ageing time	s, min, h
t_c	critical gelation time	s, min, h
t_{ind}	induction time	s, min, h
v_2	volume fraction of polymer	–
w(M)	normalized weight distribution function of molecular weights	mol/kg
w_f	fraction of dangling ends	–
w_m	fraction of ideal network	–
w_n	fraction of network	–
w_s	sol fraction	–
z	parameter (Eqs. 15, 16, 17)	–
A	parameter (Eqs. 14, 16)	–
B	parameter (Eqs. 14, 16)	–
C_T	total concentration of alkali ions	eq/l
D	dispersion index = $\bar{M}_w/\bar{M}_n$	–
E(t)	time dependent Young's modulus	N/m^2
E′	storage Young's modulus	N/m^2
E″	loss Young's modulus	N/m^2
ΔG^o	standard molar Gibbs energy of reaction	J/mol
G(t)	time dependent shear modulus	N/m^2
$G'(\omega)$	storage modulus	N/m^2
$G''(\omega)$	loss modulus	N/m^2
G^*	complex shear modulus = $G' + iG''$	N/m^2
G_e	equilibrium shear modulus	N/m^2
G_r	reduced shear modulus: $G \cdot \rho_o T_o/(\rho T)$	N/m^2
G_R	reduced shear modulus: $G \cdot b(\rho/c^2)$	N/m^2
ΔH^o	standard molar enthalpy of reaction	J/mol
ΔH_m	melting enthalpy	J/mol
$\Delta H_{res/1000}$	average transition enthalpy per 1000 amino acid residues present	J/mol
K	constant in Mark-Houwink equation	m^3/kg
	equilibrium constant	variable
K^o	pre-exponential factor in equilibrium constant	variable
K_T	temperature dependent constant in Eq. (48)	s, min, h
M	molecular weight	kg/mol
M_c	molecular weight between crosslinks	kg/mol
M_e	molecular weight between entanglements	kg/mol
$\bar{M}_n$	number average molecular weight	kg/mol
$\bar{M}_v$	viscosity average molecular weight	kg/mol
$\bar{M}_w$	weight average molecular weight	kg/mol
N_1	first normal stress difference	N/m^2
P	degree of polymerisation	–
$\bar{P}_n$	number average degree of polymerisation	–
$\bar{P}_w$	weight average degree of polymerisation	–
R	gas constant	J/(mol·K)

S	gel strength	Ns^n/m^2
	entropy	J/(mol·K)
	shearing stress	N/m^2
ΔS^o	standard molar entropy of reaction	J/(mol·K)
T	temperature	°C or K
T_a	ageing temperature	°C or K
T_c	crystallization temperature	°C or K
	critical gelation temperature	°C or K
T_g	glass transition temperature	°C or K
T_{gel}	gelation temperature	°C or K
T_m	melting temperature	°C or K
T_o	reference temperature (WLF eq.)	°C or K
T_R	reduced temperature $= T/T_{gel}$	–
α	parameter in cumulative Flory distribution (Eq. 11)	–
	optical rotation	°
α_c	optical rotation at the gel point	°
α_f	expansion coefficient of free volume	K^{-1}
$[\alpha]$	specific optical rotation	°(ml/g) dm^{-1}
γ	amount of shear	–
	crosslinking index	–
γ_{gel}	crosslinking index at the gel point	
$\bar{\gamma}_n$	number average crosslinking index	–
$\bar{\gamma}_w$	weight average crosslinking index	–
$\bar{\gamma}_{w,g}$	weight average crosslinking index at the gel point	–
$\bar{\gamma}''_n$	number average crosslinking index in the network fraction	–
$\bar{\gamma}''_{n,g}$	number average crosslinking index in the network fraction at the gel point	–
$\bar{\gamma}_{max}$	maximum crosslinking index	–
$\bar{\gamma}_{w,max}$	maximum weight average crosslinking index	–
δ	loss angle	rad
	solubility parameter	$(MJ/m^3)^{1/2}$
η	viscosity	Ns/m^2 or Pas
η_∞	plastic viscosity	Ns/m^2 or Pas
$[\eta]$	intrinsic viscosity	m^3/kg
η^*	complex viscosity $= G^*/\omega$	Ns/m^2 or Pas
λ	parameter in cumulative Flory distribution	mol/kg
	wave length	nm
μ	parameter in cumulative Flory distribution	mol/kg
ν	number of elastically effective network chains per unit volume	m^{-3}
ρ	specific density	kg/m^3
ρ_o	specific density at the reference temperature T_o	kg/m^3

σ	shear stress	N/m^2
σ_o	yield stress	N/m^2
ς	number of monomeric units along the polymer chain in crystalline crosslink	–
τ	relaxation time	s
	time constant	s
	turbidity	–
ϕ	volume fraction of polymer	–
ϕ_2	volume fraction of polymer	–
ϕ_s	volume fraction of polymer domains	–
χ	Flory Huggins interaction parameter	–
	weight fraction of helices present in gelatin solutions or gels	–
ω	angular frequency	rad/s
ω_c	angular frequency at the cross-over point	rad/s
$\Gamma(n)$	Legendre gamma fraction	–

1 Introduction

1.1 Preamble

It is well-known that many polymers, synthetic and natural, form physical, thermoreversible aggregates in dilute solutions, whereas in moderately concentrated solutions gels can be formed. Examples are poly(vinyl chloride), polyacrylonitrile, poly(vinyl alcohol), atactic polystyrene, mixtures of syndiotactic and isotactic poly(methyl methacrylates), liquid crystalline polymers, gelatin, agarose, carrageenans etc.

Several reviews concerning physical gels have been produced in the recent past: e.g. in 1987 Clark and Ross-Murphy's excellent review "Structural and mechanical properties of biopolymer gels" [1] and in 1992 Guenet's extensive monograph "Thermoreversible gelation of polymers and biopolymers" [2]. The present review deals with the viscoelastic properties of thermoreversible gels, this topic being only partly covered by the reviews mentioned. It deals with small deformation viscoelasticity of thermoreversible gels (synthetic and natural) as well as with properties which deepen insight into the gel structure: e.g. small-angle X-ray scattering (poly(vinyl chloride) gels), optical rotation (gelatin, agarose and carrageenan gels), small-angle neutron scattering (poly(vinyl alcohol) and block copolymer gels), and differential scanning calorimetry (most of the considered gels).

Viscoelastic properties change dramatically during the gelation process of a polymer solution: the system is liquid-like before crosslinking starts and remains a liquid till the viscosity becomes infinite. At that moment, which is called the gel point, there is at least one molecule with an infinite molecular weight. After the gel point, an equilibrium shear modulus develops, at first for frequencies tending to zero. A schematic view of the development of the viscosity and the equilibrium shear modulus is shown in Fig. 1. In particular, the measurement of the storage modulus as a function of frequency is a powerful tool in the determination of the development of a network, as schematically shown in Fig. 2. The gradual formation of a network is clearly visible: at the very onset of the crosslinking process the system behaves liquid-like, the slope of the $\log G'$ vs $\log \omega$ curve being 2, whereas in the course of time a network is gradually developed: only after continued crosslinking is a rubber plateau present in the measured frequency window.

Various terms are used in the literature for the development of a network structure: ageing, annealing and maturation. In many cases network formation is clearly time dependent. In other cases network formation seems to be instantaneous and a time dependence cannot be observed. In physical networks this may be the result of a sudden temperature decrease, resulting in a spinodal phase separation. During network formation the increase in the storage modulus may be of the order of 10^4 or more. It is clear that a horizontal plateau

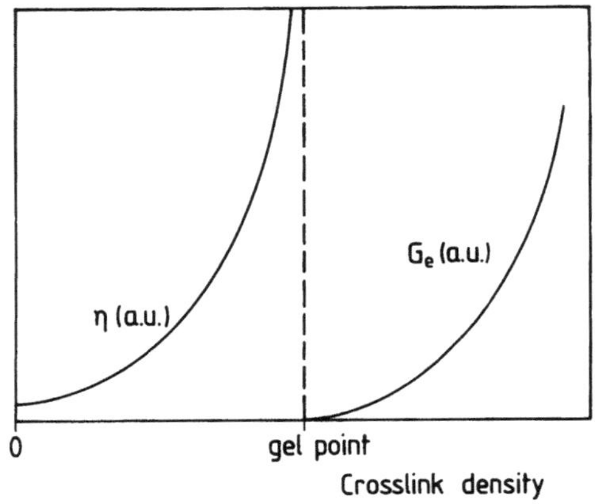

Fig. 1. Schematic view of the viscosity and equilibrium shear modulus before and after the gel point, respectively

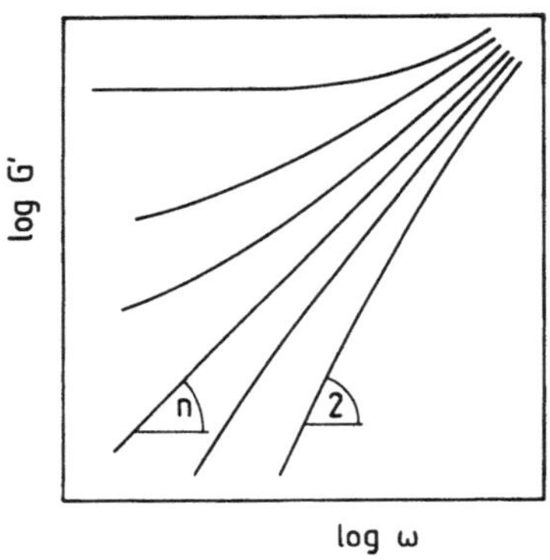

Fig. 2. Gradual formation of a polymer network, as depicted by the increase of the storage modulus as a function of angular frequency

is reached just after the gel point, and hence at earlier stages of network formation than shown in this figure. However, in order to determine this rubber plateau, the viscoelastic properties have to be measured at very low frequencies (in principle, at the gel point the horizontal plateau only exists at the limit for $\omega \rightarrow 0$). Against these low frequency measurements, two objections may be raised: first, instruments generally cannot measure at frequencies lower than 10^{-3} rad/s and second, during such a low frequency cycle the properties of the physically maturing system can change significantly. In the schematic in Fig. 2 the gel point is located somewhere between liquid-like and rubber-like behaviour. It is generally accepted that at this point the $\log G'$ vs $\log \omega$ curve is a straight line, as will be shown later in this section.

1.2 Definition of a Gel

Although gels must have been used since prehistoric times, the definition of a gel is still a matter of discussion [3]. In 1926 Jordan Lloyd [4] made her famous comment that:

The colloidal state, the gel, is one which is easier to recognize than to define.

In 1949 P.H. Hermans [5] defined a gel as:

A coherent system of at least two components, which exhibits mechanical properties characteristic of a solid, where both the dispersed component and the dispersion medium extend themselves continuously throughout the whole system.

In the three editions of his famous book "Viscoelastic Properties of Polymers", Ferry [6] defined a gel as:

A substantially diluted system which exhibits no steady state flow.

Kramer et al. [3] in the opening of the first issue of Polymer Gels and Networks in 1993, reviewed the term 'Gel' and gave their opinion of a phenomenological definition of a gel as:

(1) A gel is a soft, solid or liquid-like material of two or more components, one of which is a liquid, present in substantial quantity.
(2) Solid-like gels are characterized by the absence of an equilibrium modulus, by a storage modulus, $G'(\omega)$, which exhibits a pronounced plateau extending to times at least of the order of seconds, and by a loss modulus, $G''(\omega)$, which is considerably smaller than the storage modulus in the plateau region.

According to the present author the applicability of this definition depends on the patience of the experimentalist: if he allows himself more time, then he might conclude that a relaxation process starts at low frequencies (as depicted by the presence of a minimum in the loss modulus at low frequencies) or after a long time in creep experiments. Another, more philosophical definition (or a complaint) could be:

A gel is a gel, as long as one cannot prove that it is not a gel.

However, by making use of this definition, one has to conclude that many systems which look like a gel are in fact not covered by this definition: aqueous poly(vinyl alcohol)/borate systems, which are known to show liquid-like behaviour at low frequencies [7], solutions of phase separated atactic polystyrene at temperatures above the glass transition temperature of the swollen polystyrene crosslinks, solutions of ABA block copolymers above the glass transition temperature of the swollen A-blocks, and even gelatin, which also shows creep behaviour, as shown by Ross-Murphy et al. [8, 9] and by Kramer et al. (private communication), and a relaxation mechanism at extremely low frequencies, as is shown in Fig. 7 of the Section on gelatin, and possibly also poly(vinyl chloride) in plasticizers [10–13]. The advantage of the approach of Kramer et al. [3] is that these systems certainly are covered by their practical definition.

1.3 Classification of Gels

In classifying networks Flory [14] proposed to subdivide them into four types:

1. Well-ordered lamellar structures, including mesophases;
2. covalent polymer networks, completely disordered;
3. networks formed through physical aggregation, predominantly disordered, but with regions of local order;
4. particulate, disordered structures.

Examples of the first type are soap gels, phospholipids and clays; electrostatic and/or Van der Waals forces play an important role in the gel formation. The second category of gels consists of the well known chemical networks, formed by crosslinking of high molecular weight polymers or by crosslinking reactions in polymerisation or polycondensation processes. Thermoreversible gels belong in general to the third type; many polymer and biopolymer gels are thermoreversible and the crosslinks are formed by physical interactions (crystallization, helix formation, complex formation etc.). The fourth class of gels includes flocculent precipitates, which usually consist of particles of large geometric anisotropy (needles or fibrils); examples are V_2O_5 gels and globular and fibrillar protein gels.

The thermoreversible gels, which are the subject of the present review, belong in general to the third type of gels in Flory's classification. Schematic views of the various kinds of thermoreversible gels are shown in Fig. 3. In poly(vinyl chloride), poly(vinyl alcohol), polyacrylonitrile and polyethylene, micellar crystallites are responsible for the gel formation (case A), whereas helix formation (case B) is the origin of the biopolymer gels gelatin, agarose, carrageenans and gellan gum. Phase separation (cases C, D and E) causes network formation in atactic polystyrene solutions and in ABA block copolymer solutions, where the phase separated A-blocks are not soluble in the solvent used. In these cases (C and D) the crosslink regions are not ordered but relatively dense amorphous regions, unless the A-blocks are semicrystalline (case E). If instead the B-blocks are not soluble in the solvent used, other types of gels will develop, e.g. of type 1; an example is the gelation of ethylene oxide-propylene oxide-ethylene oxide triblock copolymers. Gelation can also find its origin in complex formation (case F), such as between syndiotactic and isotactic poly(methyl methacrylates), between poly(vinyl alcohol) and borate or congo red. In liquid crystalline polymer solutions gelation will be caused by interaction between mesogenic groups in the main chain and/or in the side chain (case G).

1.4 Determination of Gel Point

For the determination of the gel point several methods are used in the literature. A short review will be given in this section.

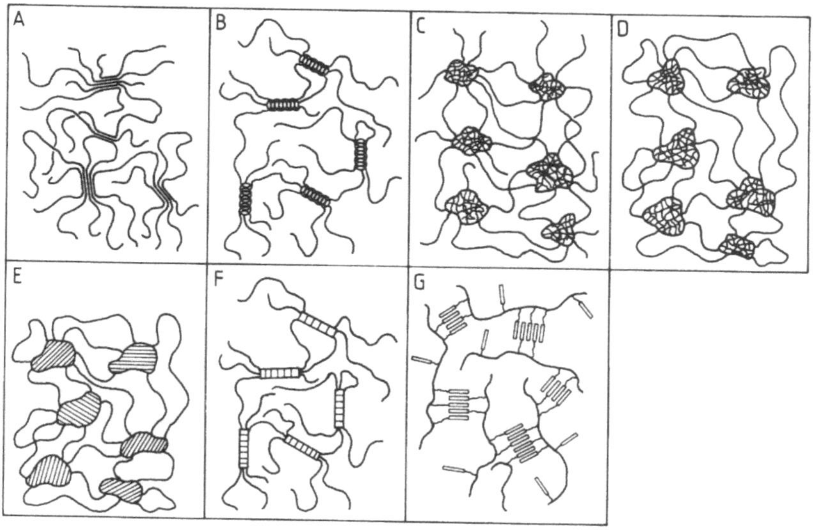

Fig. 3A–F. Schematic view of various kinds of Flory's type 3 gels: **A** PVC/plasticizer; **B** aqueous gelatin; **C** atactic PS in CS_2; **D** triblock copolymer SBS in tetradecane; **E** PO-EO-PO triblock copolymer in water; **F** s-PMMA and i-PMMA in toluene; **G** dissolved SCLCP

1.4.1 Solubility

As the gel point is the point where at least one molecule just has an infinite molecular weight, it is the point where the system is just not completely soluble any more. Consequently, extraction of the system is a method to determine the gel point. It will be clear that this method is in general not an appropriate one and definitely not suitable for physical networks.

1.4.2 Tilted Test Tube and Falling Ball Methods

The most simple methods are the "tilted test tube method" and the "falling ball method". In the tilted test tube method the polymer solution or gel is observed while rotating the tube during a temperature decrease or temperature increase or during a certian period at constant temperature. The gel point of the solution or the melting temperature of the gel is defined as the point where the system stops or starts to flow, respectively. For relatively concentrated systems it is difficult to determine whether the viscosity is infinite or just very high. The same holds for the falling ball method, where a ball of a small diameter is placed on the surface of the gel. During a temperature increase the ball starts to move through the system when the gel melting temperature is reached. For gels that cannot withstand relatively large forces the falling ball method might be destructive.

A sophisticated improvement is the sphere magnetorheometer (see e.g. [15–22]) that makes use of a small magnetic sphere that moves through or oscillates in the crosslinking system under the action of a magnetic force. In this way the viscosity and/or shear modulus can also be measured.

1.4.3 Measurement of Mechanical Properties

One of the methods to determine the gel point is the determination of properties such as the storage modulus at a certain frequency as a function of time at a certain temperature or as a function of temperature, when the network formation is apparently instantaneous. The instant or temperature where the storage modulus rises from immeasurably low values is a measure of the gel point – the gel time or the gel temperature – respectively. Another way is to determine the point where the storage and loss moduli cross each other as a function of time or temperature. Te Nijenhuis [11, 23, 24] made use of the first method to determine the gel point and the maximum gelation temperature of gelatin solutions. Ross-Murphy [9] analysed both methods for the determination of the gel point of aqueous gelatin solutions and concluded that they are rather inaccurate.

1.4.4 The Method of Winter and Chambon

Winter and Chambon [25–31] found experimentally that the mechanical behaviour of a chemically crosslinking system could be described at the gel point by a power law relaxation shear modulus:

$$G(t) = St^{-n}. \tag{1}$$

The gel strength S and the relaxation exponent n characterize the critical gel. For stoichiometric balanced endlinking networks, $n = 0.5$, while for imbalanced networks, $n = 0.5$ for excess of crosslinker and $n > 0.5$ for lack of crosslinker. Hence, upon plotting the relaxation modulus at the gel point vs time on double logarithmic scales, straight lines are obtained with slope $-n$, while lines with a convex and concave curvature with respect to time are obtained before and after the gel point, respectively (see Fig. 4).

According to the theory of linear viscoelasticity of polymers, the dynamic moduli G′ and G″ are given by

$$G'(\omega) = \omega \int_0^\infty G(t) \sin(\omega t)\, dt \tag{2}$$

$$G''(\omega) = \omega \int_0^\infty G(t) \cos(\omega t)\, dt. \tag{3}$$

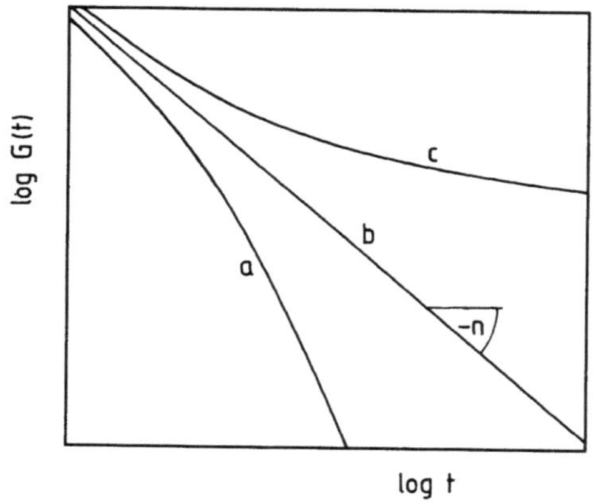

Fig. 4. Double logarithmic plot of the relaxation shear modulus vs time for a crosslinking system; *curve a* – before the gel point; *curve b* – at the gel point; *curve c* – after the gel point

Substitution of Eq. (1) into Eqs. (2) and (3) leads to

$$G'(\omega) = \frac{S\pi\omega^n}{2\Gamma(n)\sin(\frac{1}{2}\pi n)} \tag{4}$$

$$G''(\omega) = \frac{S\pi\omega^n}{2\Gamma(n)\cos(\frac{1}{2}\pi n)} \tag{5}$$

$$\frac{G''}{G'} = \tan(\delta) = \tan(\tfrac{1}{2}\pi n) \tag{6}$$

where $\Gamma(n)$ is the Legendre gamma function.

It appears from Eqs. (4) and (5) that parallel, straight lines with slopes n are obtained at the gel point upon plotting both dynamic moduli on double logarithmic scales vs ω. This is schematically shown in Fig. 5; their slopes and their mutual distance are correlated by

$$n = \frac{2}{\pi}\arctan\left(\frac{G''}{G'}\right). \tag{7}$$

Although this method has been developed for covalently crosslinked systems, it was also successfully applied to physical systems by Te Nijenhuis and Winter [31]. From then on many investigations were carried out to determine the gel point of physically crosslinking systems with the aid of the Winter-Chambon method. One thing has to be emphasized, however. Winter and Chambon developed their experimental model for covalently crosslinked systems, for which they were able to prove by dissolution experiments that the gel point was attended with power law behaviour of the dynamic moduli. This is not possible for physically crosslinking systems, because physical gels are destroyed by dissolution experiments. Nevertheless, during the crosslinking process of physically crosslinked systems a situation arises that closely resembles the critical state of covalently crosslinked systems.

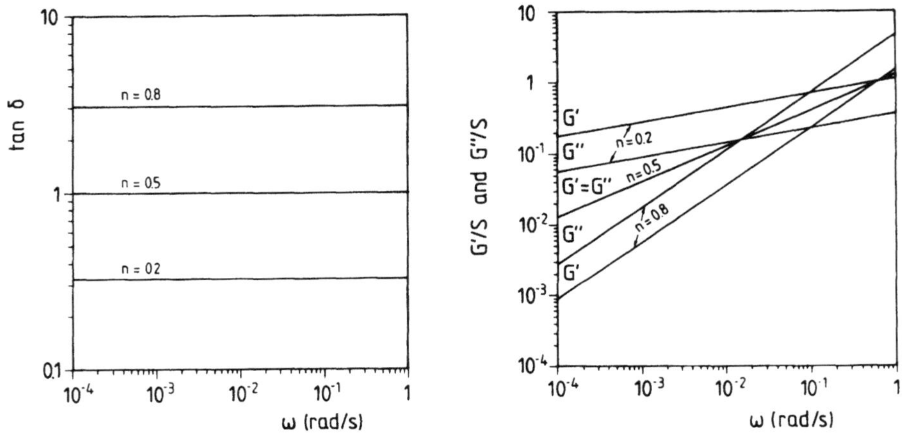

Fig. 5. Loss tangent, tan δ, and the dynamic moduli, G′ and G″, at the gel point of various systems ($n < 0.5$, $n = 0.5$ and $n > 0.5$, respectively) plotted vs angular frequency

A fifth method to determine the gel point follows from the relationships between the equilibrium shear modulus and the network structure, as discussed in the next section.

1.5 Relationships Between G_e and the Network Structure

1.5.1 Introduction

Networks obtained during the crosslinking process of high molecular weight polymers are in general far from ideal, because only a (small) fraction of the polymer present contributes to the rubber modulus. At the very beginning of the crosslinking process many polymer molecules are present that do not belong to the network (i.e. sol weight fraction w_s). Moreover, many polymer molecules or parts of them form dangling or free ends (weight fraction w_f); only the ideal part of network (weight fraction of ideal network w_m) contributes to the rubber modulus. The network fraction w_n consists of free ends and elastically active chains (see Fig. 6). Hence, we have

$$w_n = w_f + w_m \qquad w_s + w_n = 1. \tag{8}$$

A definition of dangling ends and other substructures in non-ideal networks has been given by Dušek (see e.g. [32]).

1.5.2 The Flory-Stockmayer Model

In order to calculate the network parameters both Flory [33–35] and Stockmayer [36, 37] had already presented in the 1940s a statistical theory concerning

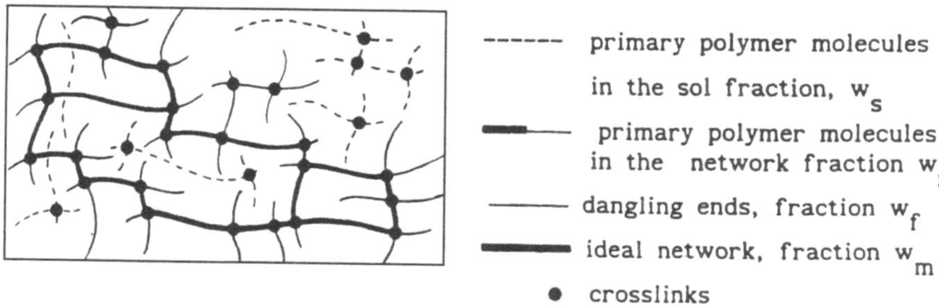

Fig. 6. Schematic representation of a gel network

the crosslinking process of primary polymer molecules, e.g. the vulcanization of natural rubber. They defined a crosslinking index, $\bar{\gamma}$, as the average number of crosslinks per primary polymer molecule in the system as a whole (sol fraction + network fraction). The relationship between $\bar{\gamma}$ and the sol fraction, w_s, that they found for network formation in monodisperse polymer, which is crosslinked with tetrafunctional crosslinks, reads

$$\bar{\gamma} = \frac{-\ln w_s}{1 - w_s}. \tag{9}$$

With the aid of this result Morton and Ferry [38] derived a relationship between the equilibrium shear modulus, G_e, and the sol fraction:

$$G_e = \frac{cRT}{M}\left[\bar{\gamma}(1 - w_s^2) - 2\right](1 - w_s)$$

$$= \frac{cRT}{M}\left[-(1 + w_s)\ln w_s - 2(1 - w_s)\right]. \tag{10}$$

Thus, from the value of G_e the sol fraction w_s can be calculated and, subsequently, the average crosslinking index $\bar{\gamma}$.

1.5.3 Extension of the Flory-Stockmayer Model

The results above are only valid for tetrafunctional crosslinking of monodisperse polymer. However, in many thermoreversible systems the crosslinks have functionalities that are much larger than four. Moreover, the polymers used are not monodisperse in general. In order to be able to calculate network parameters the present author [39–44] extended the Flory-Stockmayer model for polydisperse polymer which is crosslinked with f-functional crosslinks. It was possible to calculate network parameters for polymers of various molecular weight distributions (monodisperse polymer with $D \equiv \bar{M}_w/\bar{M}_n = 1$, a Schulz-Flory distribution with $D = 1.5$, a Flory distribution with $D = 2$, a cumulative

Schulz-Flory distribution with $D > 1.5$ and a cumulative Flory distribution with $D > 2$). The molecular weight distribution of many polymers may be approximated by a cumulative Flory distribution, which is given by

$$w(M)\,dM = \frac{\lambda}{\alpha}\left(e^{\lambda M} - e^{\mu M}\right) dM \tag{11}$$

where

$$\lambda = \frac{1}{\bar{M}_n}\,\frac{-\alpha}{\ln(1-\alpha)} \qquad \mu = \frac{\lambda}{1-\alpha} \tag{12}$$

and

$$\frac{\bar{M}_w}{\bar{M}_n} = \frac{(\alpha - 2)\ln(1-\alpha)}{\alpha}\,. \tag{13}$$

The relation between G_e, w_s, f and α reads

$$G_e = \frac{2cRT}{\bar{M}_n}\left[A\left(1 - w_s^{0.5f}\right)\frac{f-2}{f} - \frac{1-w_s}{B}\right]\cdot\frac{-\alpha}{\ln(1-\alpha)} \tag{14}$$

where

$$A = \frac{\alpha - 2 + \sqrt{\alpha^2 + \dfrac{4(1-\alpha)}{w_s}}}{2z(1-\alpha)} \tag{15}$$

$$B = \frac{\dfrac{Az}{1+Az} - \dfrac{Az(1-\alpha)^2}{1+Az(1-\alpha)}}{\ln\dfrac{1+Az}{1+Az(1-\alpha)}} \tag{16}$$

$$z = 1 - w_s^{0.5f-1}\,. \tag{17}$$

If $\bar{M}_w$ and $\bar{M}_n$ are known, then α can be calculated with the aid of Eq. (13); thereupon from the equilibrium shear modulus, G_e, the sol fraction, w_s, can be calculated with the aid of Eqs. (14)–(17) provided the crosslink functionality, f, is known. Subsequently the weight average crosslinking index in the system as a whole, $\bar{\gamma}_w$, and the number average crosslinking index in the network fraction, $\bar{\gamma}_n''$, can be calculated via

$$\bar{\gamma}_w = A(2-\alpha) \tag{18}$$

and

$$\bar{\gamma}_n'' = \frac{ABz}{1-w_s}\,. \tag{19}$$

If the crosslink functionality is unknown, then from additional, independent measurements, e.g. small angle X-ray scattering, the number of crosslinks might be calculated, and subsequently the weight average crosslinking index, $\bar{\gamma}_w$. Thereupon the functionality of the crosslinks can be calculated with the aid of Eqs. (14)–(17).

Other models, presented in the past, for the calculation of network parameters during a crosslinking process of high molecular weight polymers, have been described, e.g. by Charlesby and Pinner [45, 46], Langley et al. [47–49] and Graessley et al. [50–52]. There is much agreement between the results of those models and the described model. An important discrepancy is the presence of entrapped entanglements in the models of Langley and Graessley, which is neglected in the described model. However, in relatively low concentrated gels entanglements do not play an important role.

Although this model is only one of the models presented in the literature, it has been used throughout this review for the calculation of the network parameters of various systems. In fact, it was developed during the preparation of this review in order to be able to calculate all kinds of network parameters. The advantage of the relatively simple calculation of network parameters (for one value of the equilibrium shear modulus only a fraction of a second is needed) is the direct correlation between the progress of the ageing or maturing process and $\bar{\gamma}_w$ and/or f, whereas the equilibrium shear modulus is, beyond doubt, not directly proportional to the progress of the reaction.

At the gel point these crosslinking indices are simplified to

$$\lim_{w_s \uparrow 1} \bar{\gamma}_w = \bar{\gamma}_{w,g} = \frac{2}{f-2} \tag{20}$$

$$\lim_{w_s \uparrow 1} \bar{\gamma}''_n = \bar{\gamma}''_{n,g} = \frac{f}{f-2} \tag{21}$$

independent of the molecular weight distribution of the primary polymer. Hence, from measurement of the equilibrium shear modulus as a function of time or temperature, the weight average crosslinking index, $\bar{\gamma}_w$, can be calculated as a function of time or temperature; subsequent extrapolation of $\bar{\gamma}_w$ to $2/(f-2)$ yields the gel point. This is shown schematically in Fig. 7.

It should be emphasized that just after the gel point the equilibrium shear modulus scales to the third power of $\Delta\bar{\gamma}_w \equiv \bar{\gamma}_w - \bar{\gamma}_{w,g}$, independent of the molecular weight distribution of the primary polymer:

$$\lim_{w_s \uparrow 1} G_e \propto (\Delta\bar{\gamma}_w)^3 = (\bar{\gamma}_w - \bar{\gamma}_{w,g})^3 \,. \tag{22}$$

1.6 Scope of the Review

The present review deals with the viscoelastic properties of thermoreversible gels. Every section starts with an introduction of the discussed polymer and

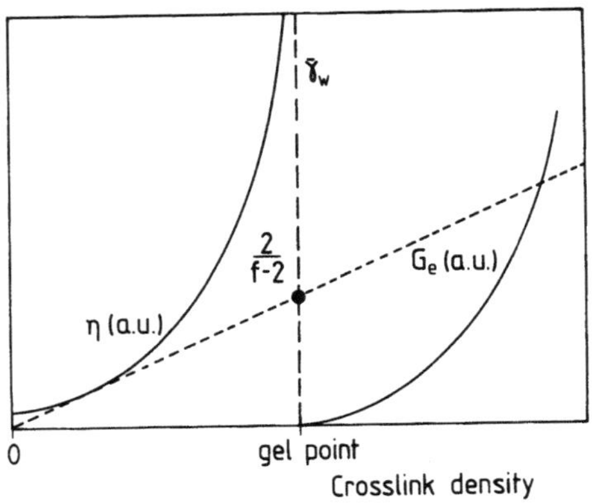

Fig. 7. Schematic views of the weight average crosslinking index, $\bar{\gamma}_w$ (– – – – –), and the viscosity, η, and equilibrium shear modulus, G_e, before and after the gel point, respectively

a more or less detailed description of the crosslink structures. The discussion of the viscoelastic properties will be compared with results obtained by other techniques in order to deepen the insight into the crosslinking process.

Sections 2–7 deal with thermoreversible gels of ordinary synthetic polymers (PVC, PVA, PMMA, PS, PAN and PE) and Sects. 8 and 9 with thermoreversible gels of tri- and multi-block copolymers and with gels of liquid crystalline polymers, respectively. Sections 10–13 concern thermoreversible biopolymer gels (gelatin, agarose, carrageenans and gellan gum).

2 Poly(vinyl chloride)

2.1 Introduction

As early as 1922 Plotnikow [53] published his investigations on the photo-polymerisation of vinyl chloride. Besides his measurements of the kinetics of the polymerisation reaction, he also paid attention to properties of the product and he stated that gels are obtained by dissolving the product in aniline (50–70%) or in tetraline (70–85%). In 1926 Flumiani [54] published his colloid chemical findings on the photo-polymerisation product of vinyl chloride. He reported that, for the determination of molecular weight, the temperature, the various heating periods, the ageing of the solution, the solvent itself and the method of preparation of different concentrations of the solution influenced the dispersion state of the polymer in the solution. In 1928 he also published his studies on the structure of gels [55]. One of his interesting statements was: if one realises that the photoproduct of vinyl chloride is a relatively simple system and that the gelation of that system is a rather difficult to understand phenomenon, it would be very important to study the gelation of systems in general.

In 1943 Leaderman [56] observed rubberlike behaviour of plasticized poly(vinyl chloride) in creep experiments. He assumed the existence of secondary forces between the molecules, which decrease in magnitude with increasing temperature and plasticizer content. Reed [57] noticed that ageing took place in plasticized poly(vinyl chloride). This held for homopolymers as well as for copolymers with up to 10% of vinyl acetate. Heating to 100 °C restored the original properties. Doty et al. [58] studied dilute solutions of poly(vinyl chloride) in dioxane. Aggregates were formed after cooling a solution from 70 °C to room temperature. Sedimentation measurements showed the existence of a minimum aggregation number. From sedimentation measurements on solutions in butanone, Hengstenberg and Schuch [59] calculated this number to be 11 on average, with an upper limit of 25. From additional viscosity measurements, these authors derived an effective hydrodynamic diameter (12–18 nm) which appears to be of the same order of magnitude as that of the primary molecules. From dissymmetry of scattered light, however, Doty et al. [58] calculated an aggregation diameter of about 200 nm. As was pointed out by other authors, these large dimensions could be the consequence of microgel particles [60–63]. Apparently, these microgel particles must be formed by a large number of primary aggregates which are connected by macromolecules being entrapped in more than one of these aggregates. As reported by Hengstenberg and Schuch [59], Hendus showed, with the aid of small angle-X-ray scattering (SAXS), that the radius of the primary aggregates is indeed rather small, viz. 7–8 nm.

In the 1940s Alfrey et al. [64–66] proved by means of X-ray diffraction that the crosslinks occurring in poly(vinyl chloride)/plasticizer gels are formed by crystallites. According to Bonart [67], Hendus showed with the aid of SAXS that the small-angle spacing of the lamellar structures in poly(vinyl chloride) increases with increasing plasticizer content from 15 nm (0% plasticizer) to 30 nm (80% plasticizer) as a consequence of preferential collection of the plasticizer between the lamellae. For more syndiotactic poly(vinyl chloride) these values appear to be a little lower. Qualitatively, the same results are reported by Gezovich and Geil [63] and Singleton et al. [68]. Together with increasing syndiotacticity, the degrees of aggregation in solution and of crystallinity in the solid material also increase [69–77]: thus the more syndiotactic the poly(vinyl chloride) used, the more rapid the gelation process proceeds; these two dependencies on syndiotacticity affirm the idea that gelation is due to crystallinity. Moreover, the temperatures at which gels are converted into solutions are considerably higher than the temperatures at which solutions gelatinize [78, 79]. In general it was assumed that crystallites of poly(vinyl chloride) are build up merely of syndiotactic sequences. From X-ray diffraction, Hellwege et al. [80] showed that the minimum number of syndiotactic sequences in a crystallite should be 12. However, with this postulate the commonly assumed 10% crystallinity of nearly atactic commercial poly(vinyl chloride) cannot be explained. However, according to Crugnola [81], crystallization is also possible with one isotactic sequence between syndiotactic sequences. This, of course, can increase the degree of crystallinity tremendously. Juijn et al. [82–84] showed that crystallinity of commercial poly(vinyl chloride) would be only 0.045% if at least 12 syndiotactic sequences were built into a crystallite. From model investigations, however, they concluded that crystallization would also be possible if a larger but even number of isotactic sequences were present between syndiotactic sequences, despite the objections against the trans and gauche sequences (60° rotations of the all trans configuration in the isotactic sequence). In this way, they arrived at a crystallinity of 8.5% for commercial rigid poly(vinyl chloride). However, in their calculations these authors did not take into consideration the length (the weight) of the sequences. An improved calculation leads to a crystallinity of 28%, or if one starts at a commonly detected crystallinity of 8.5%, to a minimum sequence length of 19 monomeric units. This rather strong tendency to form crystallites could probably provide a sufficient number of crosslinks in poly(vinyl chloride)/plasticizer systems to make them behave so extremely rubberlike. In spite of all these and other investigations [85–87] Straff and Uhlmann [88] concluded from their SAXS measurements that there is an absolute absence of crystallinity. This might be the result of too short maturation times, however.

Keller et al. [86, 89, 90] proposed the existence of two families of X-ray effects in stretched commercial poly(vinyl chloride): A-crystals with an orientation of the a-axis (i.e. in the direction of the polymer backbone) in the stretch direction and B-crystals with an orientation of the c-axis (i.e. perpendicular to the polymer backbone) in the stretch direction. The B-crystals are smaller and are supposed to form the network junctions (crosslinks) of fringed micellar crystallites, whereas the large A-crystals are supposed to be well developed lamellar crystals, entrapped in the gel and with less connections with the network. Upon stretching, the A-crystals show unusually sharp reflections at 0.52 nm on the meridian, whereas under the same conditions the B-crystals cause two other reflections on the equator at 0.472 and 0.525–0.530 nm, but they are much broader and less sharp. The sharp reflection at 0.52 nm is only found in gels and disappears at increasing temperature. The reflections of the B-crystals are also present in melt pressed samples. In fact, it is strange that the better developed A-crystals melt more easily than the B-crystals. Normally it is assumed that, upon stretching, the polymer chains are oriented in the stretching direction and that the meridional reflections in an X-ray diffraction pattern are indicators for periodicity along the chain. However, by 1960 Asahina and Okada [91] had observed a very broad X-ray diffraction intensity distribution along the equatorial direction, and a sharp meridional diffraction spot. Ohta et al. [85] studied the isothermal crystallization behaviour of commercial poly(vinyl chloride) and concluded that the exponent n in the Avrami equation has values between 1 and 2. They explain this behaviour as heterogeneous lineal ($n = 1$) or homogeneous lineal ($n = 2$) growth of the crystals. Their main conclusions are:

a) gelation is caused by crystallites; this was already proven by Alfrey et al. [64–66] and by Hendus (see [59] and [67]) with the aid of X-ray diffraction and confirmed by Dorrestijn et al. [92];
b) gelation takes place, if the chain composition is heterogeneous, just as for a copolymer consisting of stereoirregular homopolymer; in that case segments are long enough to be able to crystallize into fringed micelles, but not long enough to be able to fold;
c) gels are formed by supercooling if the stereoregular segments are long enough.

With respect to the measured crystallinity, one has to realise that on quenching the maximum possible crystallinity will not be obtained. It is generally assumed that the molten state is a purely amorphous state which is maintained on quenching a sample [93]. However, there really exists a state of "intermediate order". The normally measured crystallinity of about 10% might be too low because the amorphous background might contain a relatively high crystallinity. According to Mammi and Naida [94] and Baker et al. [95] the transition from the crystalline to the amorphous state consists of a kind of nematic structure, although birefringence was not observed.

According to Keller [96] the line broadening of the 002 reflection shows that the B-crystals consist of at least 13 monomeric units; however Bernoullian statistics calculations show that the amount of syndiotactic sequences of 13 or more units is only 0.07%, whereas the crystallinity is about 10%. Explanations for this discrepancy might be that Bernoullian statistics are not applicable to vinyl chloride polymerisation or that the above-mentioned Juijn model [82, 84] is correct. In this model an even number of isotactic sequences can be built into a syndiotactic regular part of the chain. It is concluded that the possibility of additional links, different from those that could be formed between the syndio or "near synido" sequences, should be seriously considered.

Yang and Geil [97] claim that the gelation of poly(vinyl chloride) starts with the formation of intermolecular hydrogen bridges: a molten gel which is quenched to room temperature will gel immediately. In a DSC scan, however, it does not show as endothermic immediately: hence, at the beginning of the gelation process, crystallization would not occur unless, by the formation of hydrogen bridges, a gel is formed. On the other hand, Te Nijenhuis [40] demonstrates from model calculations combined with small-angle X-ray scattering that the crystalline crosslinks are very small early in the gelation process, and are hard to determine using DSC. From their study of thermoreversible gelation of block copolymers He et al. [98] conclude that poly(vinyl chloride) gel moduli can be compared with these block copolymer gels only if the copolymer contains at least 50% of crystallizable sequences. The solvent also plays an important role, but the effect of the solubility parameter is not yet clear: in this respect bromobenzene and 1,1,2,2-tetrachloroethane are strong gelation solvents for poly(vinyl chloride) ($\delta = 17.6$ and 20.1 respectively $(MJ/m^3)^{1/2}$), whereas di-(2-ethylhexyl) phthalate and butanone ($\delta = 16.2$ and 19.1 respectively) are relatively weak gelation solvents (NB $\delta_{PVC} = 19.7$). In this respect the work of Najeh et al. [99] has to be mentioned. These authors investigated the solvent dependence of a large homologous series of monoesters and diesters with the aid of rheological techniques, differential scanning calorimetry and light scattering. The monoesters act only as plasticizers for poly(vinyl chloride), whereas in diester solutions the solvent molecules participate in the formation of additional links, as shown schematically in Fig. 8. This, of course, gives rise to an increased elasticity of PVC gels.

From their measurements of electrical birefringence among other things, Guenet et al. [100–102] claimed that crystallization really does occur during gelation of poly(vinyl chloride) solutions. Upon quenching, dilute poly(vinyl chloride) solutions crosslinked aggregates are formed. A combination of results from elastic light scattering, quasi elastic light scattering, and optical and electrical birefringence yielded information on the pregels and the subsequently formed gels. The authors confirmed the existence of two types of physical links in poly(vinyl chloride) aggregates: first, weak links which gradually disappear with increasing temperature and which completely disappear at 62 °C and, second, strong links which are not destroyed at that temperature but which will melt only at temperatures between 250 and 300 °C. At very low concentrations ($c \leq 5$ kg/m^3) the weak links are absent; at higher concentrations

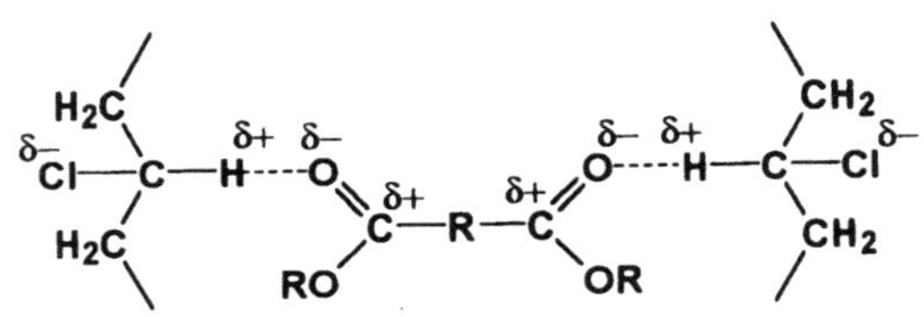

Fig. 8. Formation of a link between two PVC molecules by a diester

($c > 10\ kg/m^3$), however, they appear. It is assumed that the weak links are responsible for the physical ageing process in thermoreversible poly(vinyl chloride) gels. By using NMR spectroscopy, Spěváček and Schneider [103, 104] also observed two types of aggregates in *O*-dichlorobenzene: weaker aggregates are completely decomposed at 80–100 °C, whereas strong aggregates start to decompose at temperatures above 160 °C. This is in agreement with investigations by Te Nijenhuis and Dijkstra [10–12] who observed that the maximum gelation temperature is approximately 100 °C, depending on the kind of solvent, whereas complete rejuvenation of the gels was possible at 160 °C. Weak aggregates do not exist in nitrobenzene, whereas strong aggregates do not appear in grafted copolymers poly([ethylene-*co*-vinyl acetate]-*g*-vinyl chloride). The presence of weak and strong aggregates closely resembles the A-crystals and B-crystals mentioned by Keller et al. [86, 89, 90]. From their small angle neutron scattering experiments, Guenet et al. [105] conclude that the short range molecular structures of the pregels (aggregates formed in dilute solutions) and gels are fiberlike structures. This is an agreement with the electron microscopy observations of Yang and Geil [97]. In this respect the solvent plays an important role, especially with respect to the swelling behaviour of the amorphous regions between the crystalline crosslinks.

In a careful study of the thermal properties of poly(vinyl chloride) gels in various solvents, Mutin and Guenet [105, 106] show the existence of two endotherms. The first is caused by melting of the gel; this endotherm is no longer present after the first heating run. The second, on the other hand, fixed at 50 °C (independent of concentration), is only present in aged gels and is still present in the second run. However, this second run was taken only after 24 h of renewed ageing at room temperature. This is different from the measurements of Yang and Geil [97], which were taken almost instantaneously after cooling to room temperature, i.e. after a much shorter time than the measurements by Mutin and Guenet. Based on these DSC measurements, Mutin and Guenet [106] produced the phase diagram of poly(vinyl chloride)-diethyl malonate shown in Fig. 9.

Inoue et al. [107, 108] produced a phase diagram of PVC/γ-butyrolactone on the basis of light scattering experiments that yielded the spinodal curve, and the tilted-test-tube method that yielded the sol-gel transition curve. The result is shown in Fig. 10, where four regions can be seen. Above the spinodal curve they found two regions: (i) a homogeneous solution and (iv) a region where gelation occurred. Below the spinodal curve, phase separation took place by spinodal decomposition in the regions (ii) and (iii); in region (iii) the liquid-liquid phase separation is accompanied by gelation. The authors conclude that the spinodal decomposition is not a necessary condition for gelation.

Jackson et al. [109] studied the ageing behaviour of 20–35 wt% cyclohexanone solutions of PVC with Raman spectroscopy. After about 24 h of ageing the gels became more homogeneous and the overall rate of crystallization seemed to slow down, and was essentially completed after about two weeks. Similar results were obtained by Dorrestijn et al. [92] from SAXS measurements on 10 wt% solutions in di-(2-ethylhexyl phthalate). According to Jackson et al.,

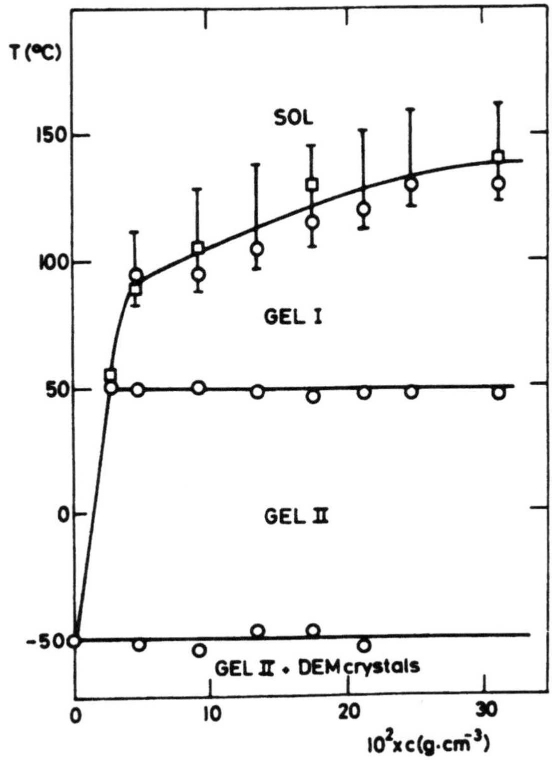

Fig. 9. Temperature-concentration phase diagram of PVC-diethyl malonate gels aged for 24 h at 20 °C; (○) temperatures determined from the first heating by DSC (scan speed 20 °C/min: bars indicate the width of the melting endotherm) and (□) melting temperatures determined visually by the ball-drop method (heating rate 2 °C/min). Reproduced from Macromolecules [Ref. 106] by the courtesy of the authors and The American Chemical Society

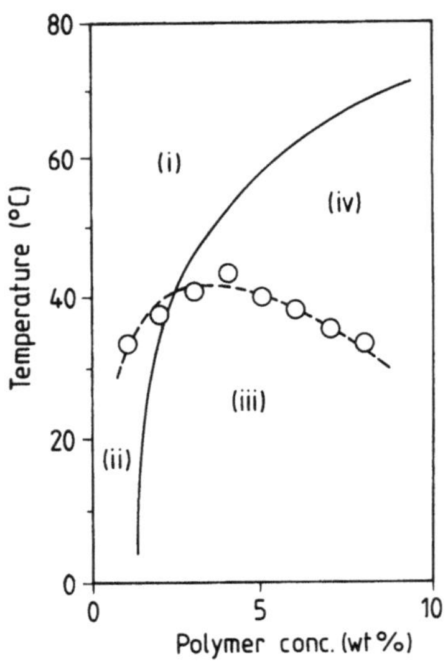

Fig. 10. Sol-gel transition curve (*solid line*) and the spinodal curve (*broken line*) of PVC/γ-butyrolactone. Reproduced from Polym J [Ref. 107] by the courtesy of the authors and The Society of Polymer Science, Japan

from then on the increase in modulus would not be the result of progressing crystallization but the consequence of the formation of hydrogen bonds between cyclohexanone and poly(vinyl chloride).

2.2 *Viscoelasticity*

2.2.1 *Introduction*

By means of tensile measurements, Walter [110] made the following observations on gels of poly(vinyl chloride) in di-(2-ethylhexyl) phthalate.

a) In the concentration range up to 20% of polymer the gels behaved ideally in the sense of the classical theory of rubber elasticity. From 20 to 50% a small energetic term was perceptible. For more concentrated systems there is a rapid increase in the contribution of internal energy to the elasticity with concentration; however, one has to realise that the highly concentrated plasticized polymers are in the glassy state at the temperatures of the measurement of the Young modulus [111, 112].
b) The ageing speed depends strongly on the concentration: at low concentrations an equilibrium value was reached within a few months during ageing at 30 °C, whereas the concentrated systems, with more than 40% of polymer, did not even reach an equilibrium value within two years.

Oscillatory mechanical measurements were taken by Ferry et al. [38, 113–116] on dilute and moderately concentrated solutions of poly(vinyl chloride) in plasticizers. They were not able to make use of the time temperature superposition principle [117] in the normal way. This is as expected, as the internal equilibrium structure of the system changes with temperature [118, 119]. Moreover, the thermal history, which at the same time appears to be so extremely important, was not well defined in their systems. Teplov et al. [120] measured the compression modulus of a 62.5 wt% solution of poly(vinyl chloride) in various plasticizers and they also found that this modulus increased linearly with the logarithm of ageing time. The rate of gelation depends strongly on the kind of solvent used, being faster in the series dibutyl phthalate, di-(2-ethylhexyl) phthalate, tricresyl phosphate, didecyl phthalate and dioctyl sebaccate. Juijn et al. [82, 83] also reported that the increase of the modulus of PVC/plasticizer systems (70–100% PVC) is linear with log ageing time over more than a year. They also concluded that the increase of the modulus of elasticity is accompanied by an increase of an endotherm. This also confirms the statement that the gelation process is caused by crystallization. In a careful study of the time and temperature dependence of the viscoelastic properties (i.e. G' and G'') of a 9.9 wt% solution of poly(vinyl chloride) in di-(2-ethylhexyl) phthalate), Te Nijenhuis and Dijkstra [10] also found that the storage modulus was a linear function of log ageing time, its slope being strongly dependent on the ageing temperature.

2.2.2 *Time Dependence*

An example of the time dependence of the storage modulus of a 9.9 wt% solution of poly(vinyl chloride) (Solvic 239) in di-(2-ethylhexyl) phthalate (DOP), after quenching the solution from 160 to 90 °C, is shown in Fig. 11 [10–12]. The gradual formation of a network is shown: after short ageing times (0.2 h) the system behaves like a liquid, the storage modulus being strongly dependent on the frequency, whereas, as the ageing proceeds, the low frequency storage modulus becomes nearly independent of frequency. The curve at 0.75 h is a linear one with a slope of 0.77. Later on it will be shown that this curve apparently corresponds to the curve at the gel point.

In Figs. 12 and 13, the storage moduli are plotted vs log t_a at temperatures of 79 °C and 37 °C, respectively. At high temperature the curves become linear after some ageing time. The ageing process is faster at lower temperatures: linear curves are obtained immediately; in that case the increase of the modulus slows down. The slopes are independent of frequency, but strongly dependent on temperature. An analogous effect was reported by Juijn et al. [82, 83] who measured the annealing effect of unplasticized commercial poly(vinyl chloride) (as a function of time): in Fig. 14 the endothermic effect, as measured by DSC experiments, is plotted vs log annealing time. Again a linear dependence is shown. In order to determine the gel point of the 9.9 wt% PVC/DOP system at 90 °C in Fig. 15, log tan δ is plotted vs log ageing time for angular frequencies varying from 0.39 to 31.6 rad/s [31]. According to the Winter and Chambon method the gel time would be 0.75 h, i.e. the cross point of the different curves. At this point $\tan\delta = \tan(n\pi/2) = 2.9$ or $n = 0.79$. In Fig. 16 log G′ and log G″

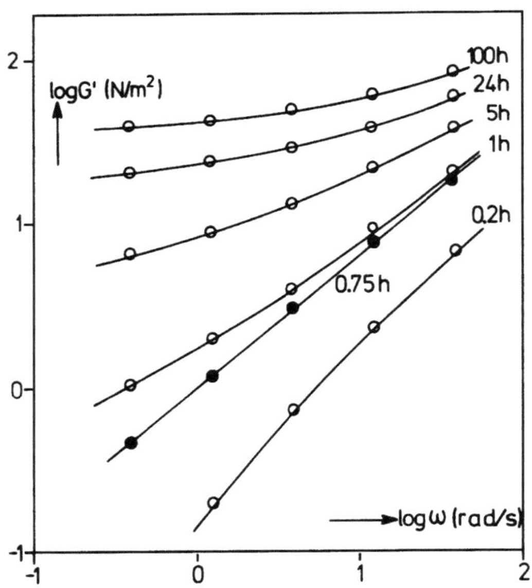

Fig. 11. Storage modulus of 9.9 wt% PVC/DOP plotted vs frequency for several ageing times; ageing temperature 90 °C. Reproduced from Physical Networks, Polymers and Gels [Ref. 12] by the courtesy of Chapman & Hall

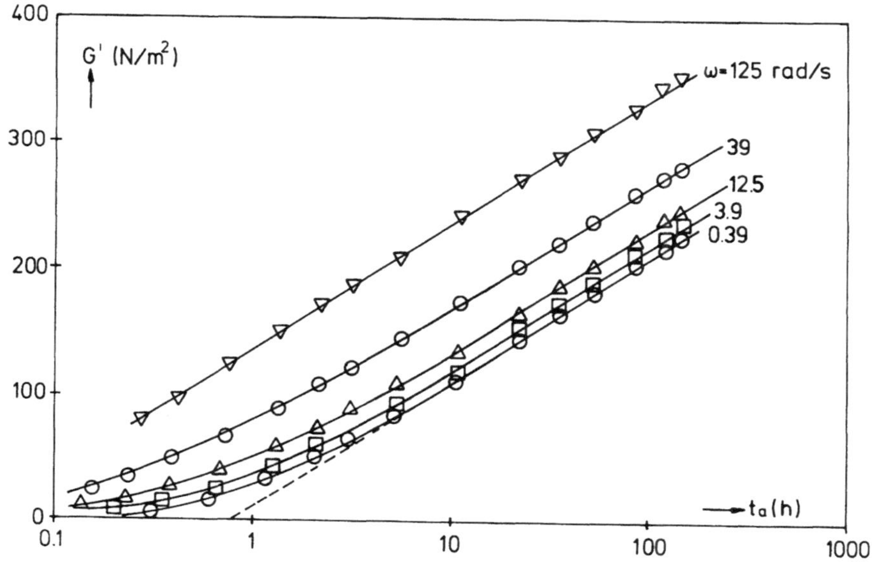

Fig. 12. Storage modulus of 9.9 wt% PVC/DOP plotted vs ageing time for several angular frequencies; ageing temperature 79 °C. Reproduced from Rheol Acta [Ref. 10] by the courtesy of Steinkopff Publishers Darmstadt, FRG

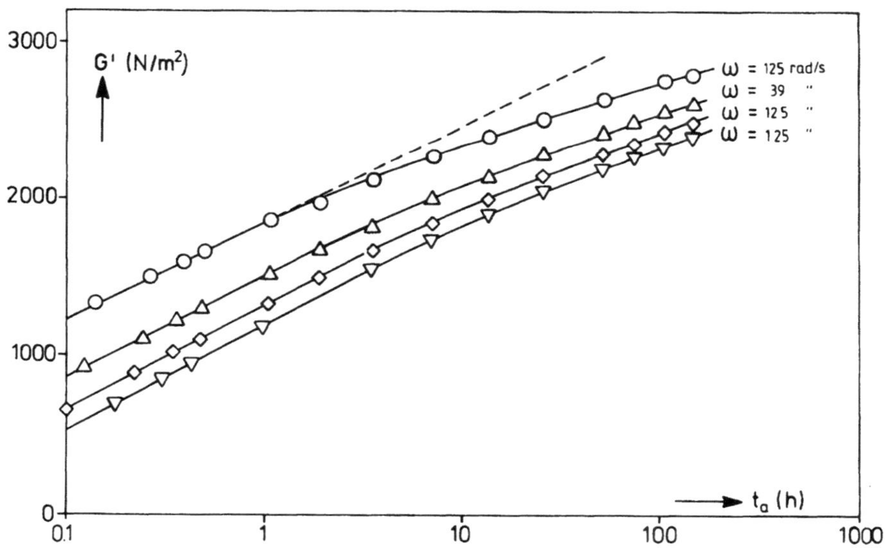

Fig. 13. Storage modulus of 9.9 wt% PVC/DOP plotted vs ageing time for several angular frequencies; ageing temperature 37 °C. Reproduced from Rheol Acta [Ref. 10] by the courtesy of Steinkopff Publishers Darmstadt, FRG

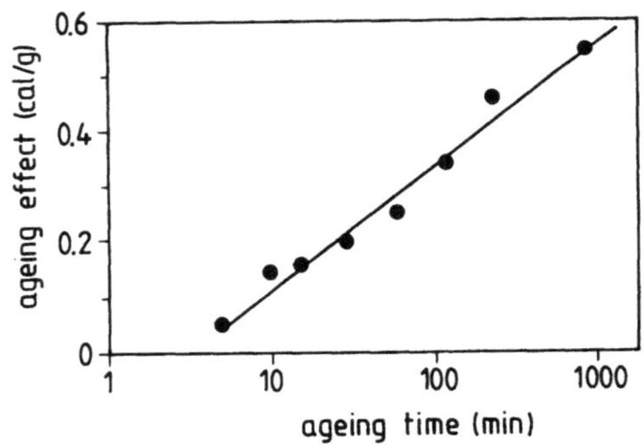

Fig. 14. Influence of ageing time on the magnitude of the ageing effect (ageing temperature 100 °C; preheating temperature 165 °C). Reproduced from Kolloid z u Z Polymere [Ref. 83] by the courtesy of the authors and of Steinkopff Verlag Darmstadt, FRG

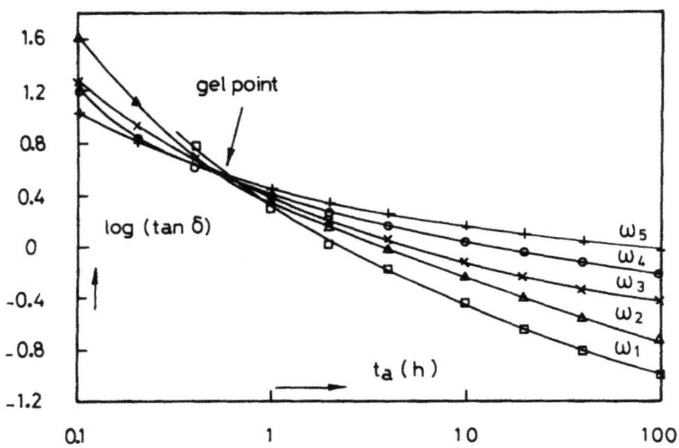

Fig. 15. Loss tangent of 9.9 wt% PVC/DOP plotted vs ageing time for several frequencies. Ageing temperature 90 °C; $\omega_i = 0.39, 1.26, 3.9, 12.6, 31.6$ rad/s. Reproduced from Macromolecules [Ref. 31] by the courtesy of The American Chemical Society

are plotted against log ω for the ageing time of 0.75 h [12]. The curves are linear and parallel with a slope $n = 0.77$, nearly equal to n calculated from tan δ at the cross point. It is interesting that there is a situation where the PVC/DOP system behaves as a chemical system in the critical gel state with a deficiency of crosslinker.

2.2.3 Concentration Dependence

A linear dependence on the logarithm of the ageing time is also shown for higher concentrations of poly(vinyl chloride) in DOP [82, 84, 110, 120]. The most systematic study has been done by Walter [110]. Results are shown in Fig. 17.

According to Walter the systems behave elastically as rubber up to 20% of polymer in DOP; from then on, up to 65% of polymer, small, and above 70%

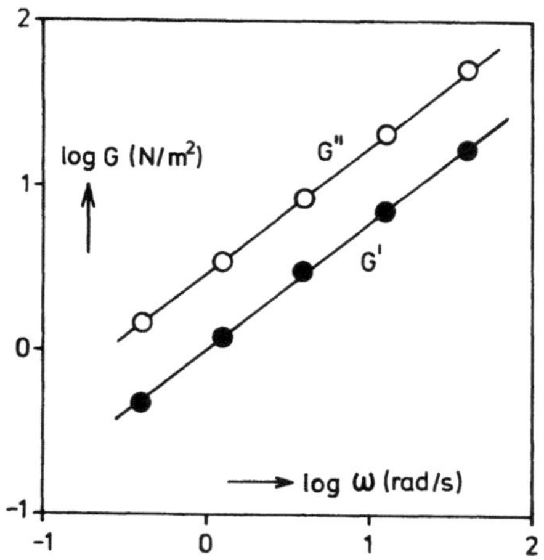

Fig. 16. Double logarithmic plot of the dynamic moduli of 9.9 wt% PVC/DOP vs angular frequency; ageing time 0.75 h; ageing temperature 90 °C. Reproduced from Physical Networks, Polymers and Gels [Ref. 12] by the courtesy of Chapman & Hall

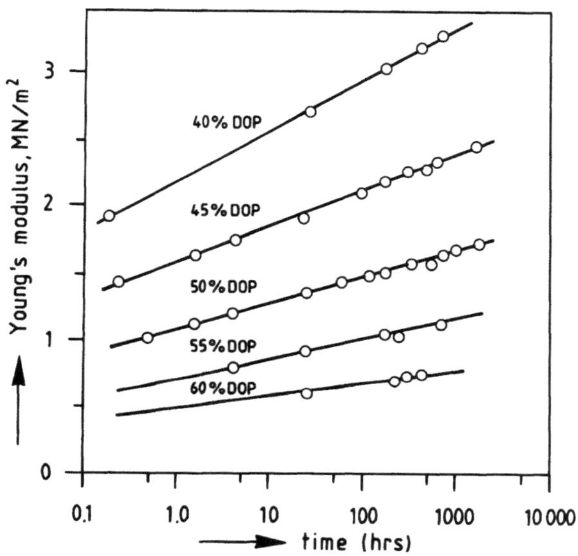

Fig. 17. Change in the 60-s modulus at 30 °C after molding of several PVC/DOP compounds. Reproduced from J Polym Sci [Ref. 110] by the courtesy of John Wiley & Sons, Inc

strong energy contributions are perceptible. However, that strong energy contribution is mainly due to the approach or even passing through the glass-rubber-transition. The gel melting temperature as a function of concentration is shown as a log c vs T_m^{-1} plot in Fig. 18. It appears that over the whole concentration range (3.5–65%) a straight line is obtained, from which the gelation enthalpy can be calculated to be -32.8 kJ/mol. The amount of crystallinity of the

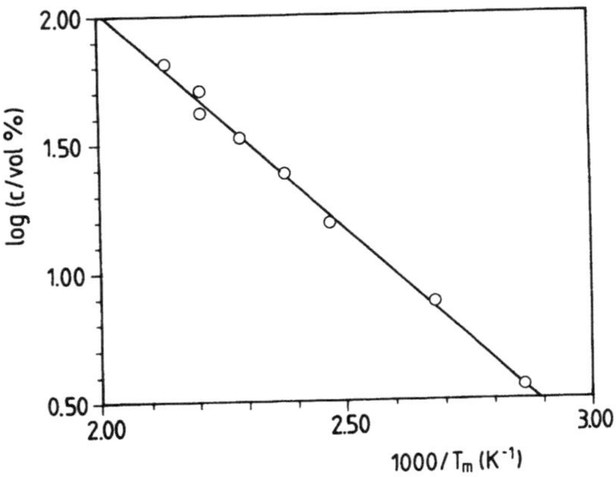

Fig. 18. log c vs $1/T_m$ plot for PVC/DOP after one month of ageing at room temperature (data collected from [110])

poly(vinyl chloride) used can be calculated from Δh_u as given by Anagnostopoulos et al. [121] and by Lyngaee-Jørgensen [122], being 2.51 and 2.75 kJ/mol, respectively. This leads to a crystallinity in the poly(vinyl chloride) used of 13.1 and 11.9% respectively, quite reasonable values for a commercial poly(vinyl chloride) (see e.g. [71]). Extrapolation to c = 100% yields a melting temperature of the PVC used of 223 °C.

2.2.4 Solvent Dependence

A systematic study of the gelation properties of various solvents and plasticizers has been reported by Walter [110], Juijn et al. [82, 84], Teplov et al. [120] and Te Nijenhuis et al. [10, 11, 13, 31]. As an example, results are shown in Fig. 19 for a 9.9 wt% solution of Solvic 239 in three different plasticizers [11], namely DOP, dibutyl phthalate (DBP) and Reomol ATM (the trimellitic acid ester of alphanol 79, a mixture of branched alcohols). It appears that DOP and especially DBP are low-ageing plasticizers. This conclusion is also evident from the other studies mentioned [82–84, 110, 120].

2.2.5 Temperature Dependence

Temperature is also an important parameter in the thermoreversible ageing of PVC solutions. This can be established from Figs. 13 and 14. Unfortunately, the temperature or temperature history is not very well defined in most studies. Careful studies were reported by Juijn et al. [82, 83] and Te Nijenhuis et al. [10–12, 92] where the temperature and its history were defined very well. For 9.9 wt% solutions of Solvic 239 in various plasticizers the slope ($dG'/d\log t_a$) of curves as shown in Figs. 13 and 14 is plotted against temperature in Fig. 20. The curves closely resemble crystallization rate curves with maxima half way

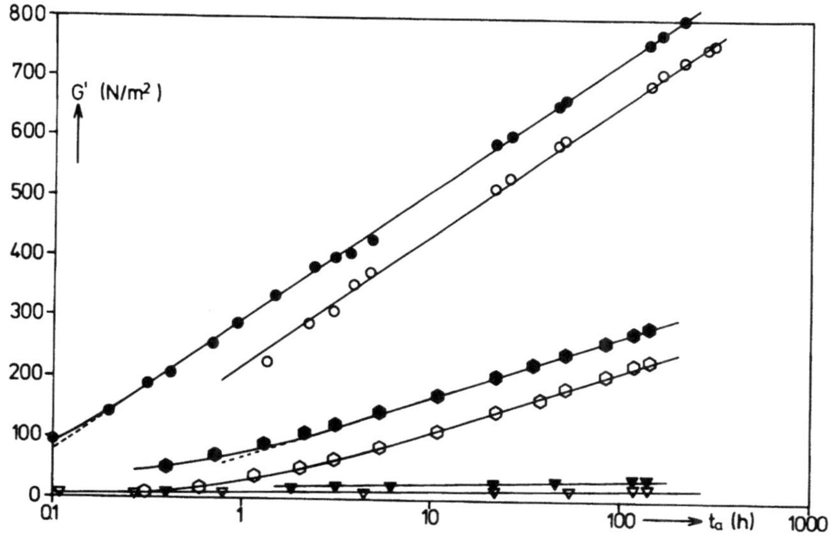

Fig. 19. Storage modulus of 9.9 wt% PVC in various plasticizers plotted vs log ageing time for two angular frequencies; ageing temperature 80 °C; *open symbols* ω = 0.39 rad/s, *filled symbols* ω = 39 rad/s; (○) PVC/Reomol; (⬡) PVC/DBP; (▽) PVC/DOP. Reproduced from Macromolecules [Ref. 31] by the courtesy of The American Chemical Society

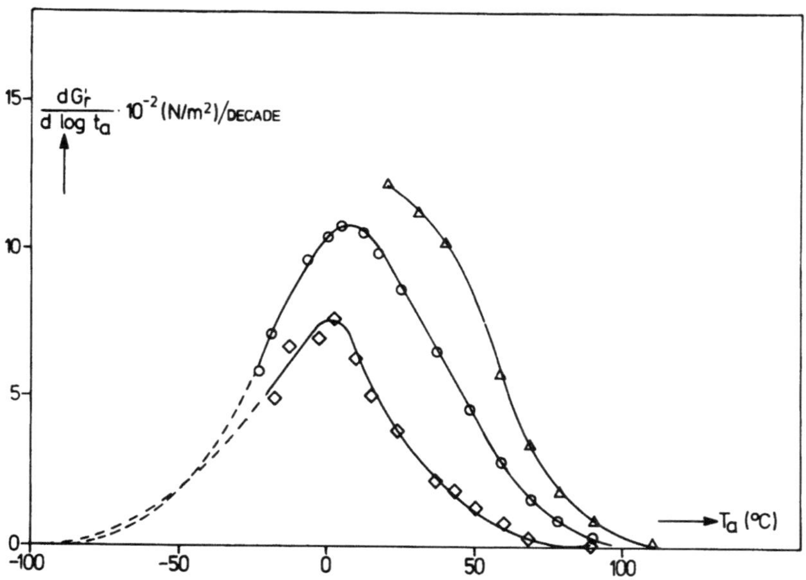

Fig. 20. Influence of temperature on the reduced rate of ageing, $dG'_r/d\log t_a$, of 9.9 wt% PVC/plasticizer systems: (△) PVC/Reomol; (○) PVC/DOP; (◇) PVC/DBP. Reproduced from Physical Networks, Polymers and Gels [Ref. 12] by the courtesy of Chapman & Hall

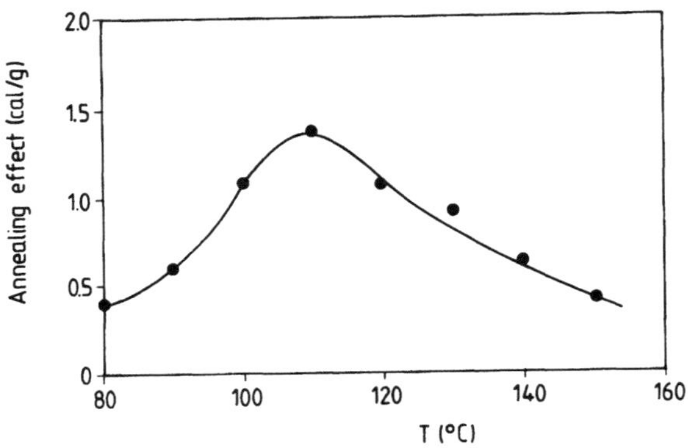

Fig. 21. Influence of the ageing temperature on the magnitude of the ageing effect of commercial poly(vinyl chloride) (ageing time 16 h). Reproduced, with permission of the author, from [82]

between the glass transition temperature (~ −90 °C) and their melting temperature (~ 120 °C). An analogous effect was reported by Juijn et al. [82] as shown in Fig. 21, where the annealing effect, shown as an endothermic effect in DSC measurements, at various temperatures after 16 h of ageing is plotted vs temperature for commercial, unplasticized poly(vinyl chloride) which was preheated at 180 °C. Again a maximum effect is shown half way between T_g (~ 85 °C) and T_m (~ 160 °C).

2.2.6 Molecular Weight Dependence

The next important parameter in the tendency of polymer molecules to form gels in moderately concentrated solutions is the molecular weight. If the polymer molecular weight is higher, less crosslinks are needed per unit volume to form networks. Hence one would expect that the tendency of polymers to form networks is higher when their molecular weight is higher. This was proven by Dorrestijn and Te Nijenhuis [13]. They showed that for a series of poly(vinyl chlorides) the tendency to form gels increases tremendously with increasing molecular weight from 36 to 64 kg/mol. However, the authors also attribute this increased tendency to the increased syndiotacticity. As a matter of fact, the range of increasing molecular weights was obtained by polymerisation at decreasing polymerisation temperatures. It is well-known that by decreasing the polymerisation temperature the syndiotacticity of the synthesized poly(vinyl chloride) also increases. Thus, by decreasing the polymerisation temperature, the molecular weight as well as the syndiotacticity increase, and both properties have the same effect on the tendency to form gels in moderately concentrated solutions. For that reason, the authors also synthesized poly(vinyl chlorides) at one temperature (40 °C) but with various amounts of chain transfer agents in order to control the molecular weight. One would expect that these polymers of molecular weights 40–75 kg/mol have the same degree of syndiotacticity. In Fig. 22, results are shown for the storage modulus of 9.9 wt% solutions in DOP,

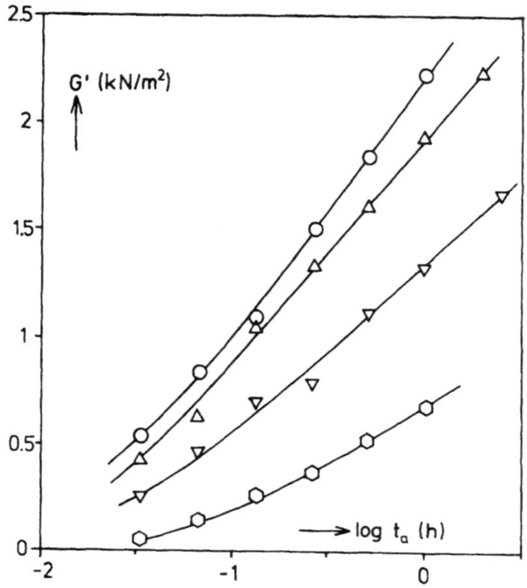

Fig. 22. Storage modulus, measured at $\omega = 2\pi$ rad/s, of relatively low temperature (40 °C) polymerised poly(vinyl chlorides) plotted vs log ageing time; ageing temperature 26 °C; number average molecular weights in kg/mol: (○) 75, (△) 63, (▽) 53, (⬡) 40. Reproduced from Colloid Polym Sci [Ref. 13] by the courtesy of Steinkopff Verlag Darmstadt, FRG

measured at $\omega = 6.28$ rad/s and at 26 °C. It appears that increasing polymer molecular weights causes an increasing tendency to form gels.

2.2.7 Chlorination of Poly(vinyl chloride)

Chlorination of poly(vinyl chloride) destroys the regularity of the tacticity. Dorrestijn et al. [123] executed chlorination in two different ways. First, poly(vinyl chloride) was chlorinated in solution so that no preference existed for the kind of hydrogen atom which was replaced by chlorine atoms: the tacticity of the polymer molecules is destroyed more or less completely. Second, the polymer was chlorinated in powder form; in this case there is a large preference for the amorphous parts, whereas the crystalline parts are much less available for chlorination and remain more or less intact. Even if the polymer is attacked preferentially at the isotactic sequences, the tendency to form gels will decrease, as the presence of an even number of isotactic sequences in between syndiotactic sequences causes an enormous increase in the crystallinity of poly(vinyl chloride) [82, 84]. Dorrestijn et al. showed that chlorination had a dramatic influence on the tendency to form gels in the case of solution chlorination, whereas powder chlorination had only a minor effect despite much higher chlorine contents.

2.2.8 Extraction of Commercial Poly(vinyl chloride)

According to Keller [96], a preponderence of syndiotacticity is obtained during the early stages of the polymerisation process of vinyl chloride. Thus, extraction of poly(vinyl chloride) with a solvent in which the more atactic part is dissolved would separate the polymer into a fraction with a higher tendency (the residue) and the other one with a lower tendency (the extract) to form gels. This was confirmed by Dorrestijn and Te Nijenhuis [13]. Extraction of Marvylan S7102 (DSM) ($\bar{M}_n = 56$ kg/mol) with acetone resulted in a residue (mass fraction 0.79, $\bar{M}_n = 65$ kg/mol) and a polymer extract (mass fraction 0.21, $\bar{M}_n = 37$ kg/mol). Results of the measurements of the tendency to form gels are shown in Fig. 23, where the storage modulus, measured at 26 °C and at an angular frequency of $\omega = 6.28$ rad/s, of a 9.9 wt% solution in DOP of both the residue and the extracted polymer is plotted against ageing time. The differences are dramatic: the residual polymer has a much larger tendency to form gels than the original polymer. This tendency nearly disappears for the extracted polymer. The authors, however, remark that the residual polymer has a higher and the extracted polymer a lower molecular weight, compared to the original one, and, as is shown in Sect. 2.2.6, the molecular weight is an important parameter in the gelation tendency. They have also shown results of the same measurements of original polymers with molecular weights close to the residual and the extracted polymer, respectively. This is a better way to compare the residual and extracted polymers. This comparison shows that the residue polymer is really more

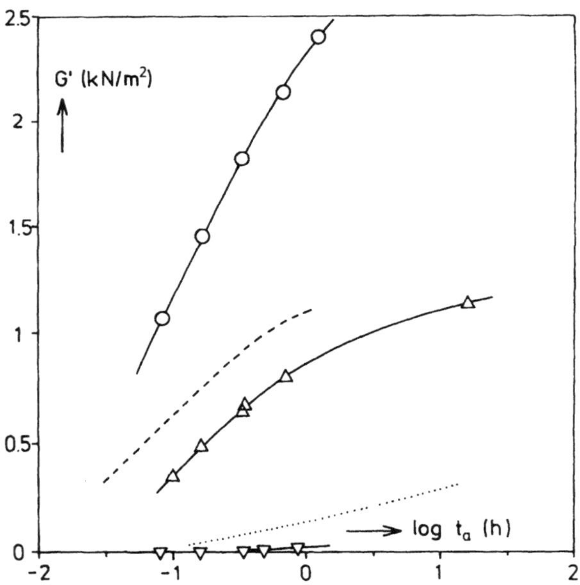

Fig. 23. Results of extraction of PVC (Marvylan S7102). Storage modulus, measured at $\omega = 2\pi$ rad/s, plotted vs log ageing time; ageing temperature 26 °C. (△) Original Marvylan ($\bar{M}_n = 56$ kg/mol); (○) residue sample ($\bar{M}_n = 65$ kg/mol); (▽) extract sample ($\bar{M}_n = 37$ kg/mol); (----) Marvylan S7502 ($\bar{M}_n = 64$ kg/mol); (····) Marvylan S5702 ($\bar{M}_n = 36$ kg/mol). Reproduced from Colloid Polym Sci [Ref. 13] by the courtesy of Steinkopff Verlag Darmstadt, FRG

crystalline and (so) more syndiotactic than the original polymers and the extract polymer less crystalline and (so) less syndiotactic.

2.2.9 Time Temperature Superposition

In order to increase the frequency window of the dynamic moduli, normally use is made of the time-temperature superposition principle. However, this principle can only be applied to systems in which the structure does not depend on temperature and, consequently, not to ageing (gelating) systems. Te Nijenhuis et al. [10–12], however, were able, by making use of an extrapolation procedure, to extend the procedure to ageing systems in the following way. After having aged the system for a certain time at a fixed temperature T_a, the moduli were measured over a frequency range as large as possible. Subsequently the gel is cooled very quickly to a new temperature T_o and the excess ageing at this temperature is measured at several frequencies. Extrapolation of these measurements to the moment of the temperature quench yields the storage and loss moduli at temperature T_o of the gel, with the same internal structure as that at T_a just before the quench. With these extrapolated values, the curves of $\log G'$ and $\log G''$ against $\log \omega$ are constructed. These curves can be used for the application of the time temperature superposition (see Fig. 24 for a schematic application). An example of such a master curve, as reported by Te Nijenhuis [11], is given in Fig. 25 for a 9.9 wt% solution of Solvic 239 in DOP for an ageing temperature of 49 °C and a reduction temperature of 25.7 °C. The temperature dependence of the horizontal shift factor $\log a_T$ has to be measured with

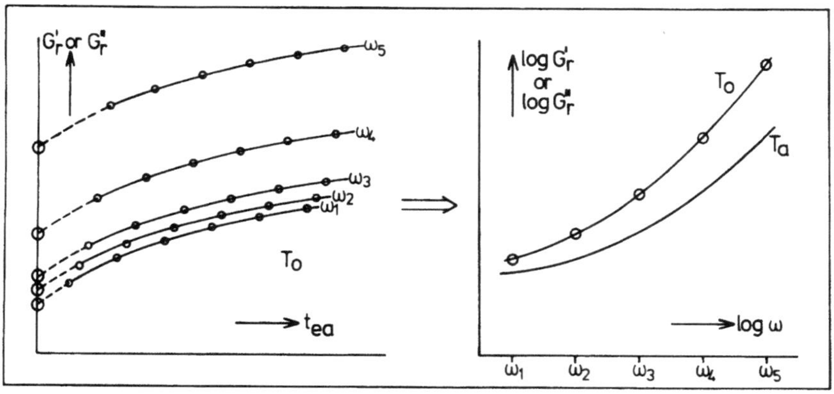

Fig. 24. Schematic view of the application of the time-temperature superposition principle. *Left hand side*: excess-ageing plotted vs excess-ageing time. *Right hand side*: original reduced ageing curve (T_a °C) and the corresponding curve at the reference temperature (T_o °C) obtained by extrapolation of the excess-ageing to zero excess-ageing time. Reproduced from Physical Networks, Polymers and Gels [Ref. 12] by the courtesy of Chapman & Hall

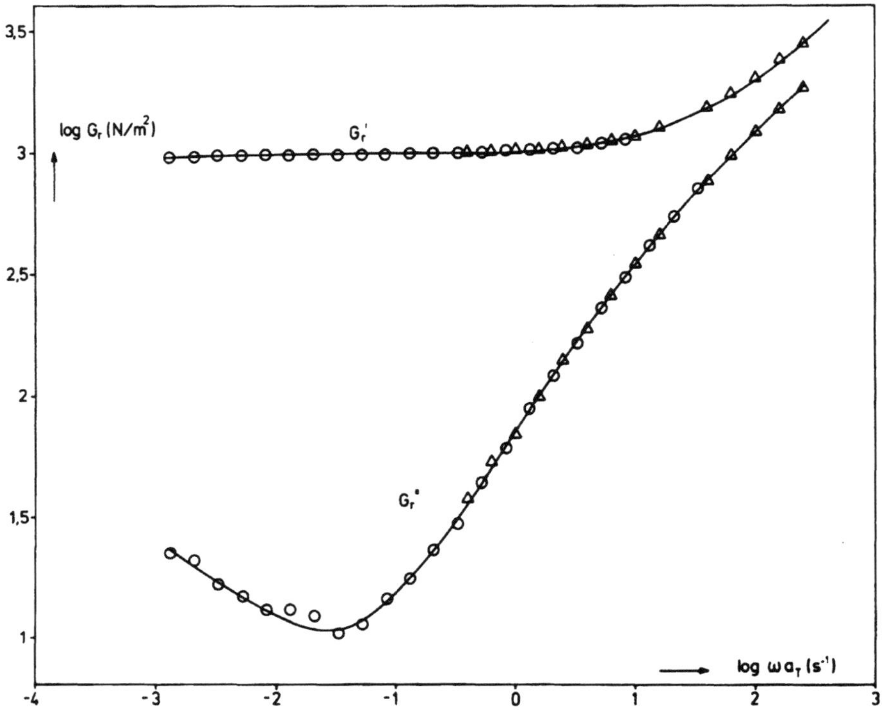

Fig. 25. Master curve of the storage and loss moduli of 9.9 wt% PVC/DOP; ageing temperature 49 °C; ageing time 3 h; (○) 49 °C; (△) 25.7 °C (= reference temperature). Reproduced from Rheol Acta [Ref. 10] by the courtesy of Steinkopff Publishers Darmstadt, FRG

many different samples. It appeared that the WLF equation [117] worked very well:

$$\log a_T = \frac{-c_1^o (T - T_o)}{c_2^o + T - T_o} \tag{23}$$

and the values of c_1^g, c_2^g, f_g and α_f of PVC/DOP and PVC/DBP at their respective glass transition temperatures were in good agreement with values normally obtained for polymers [10–12]. An additional demonstration of the correctness of the method is the application of the time temperature superposition principle to a 9.9 wt% solution in DOP which was aged for 16 months at room temperature [11]. In this case, the excess ageing at lower temperatures is imperceptibly slow, so that the normal way of determining the shift factor could be used as long as only a temperature decrease is applied. Results of $\log a_T$ measured in this way are compared with the above-mentioned results in Fig. 26. The full line was calculated with the parameters c_1^g and c_2^g. The agreement was surprisingly good.

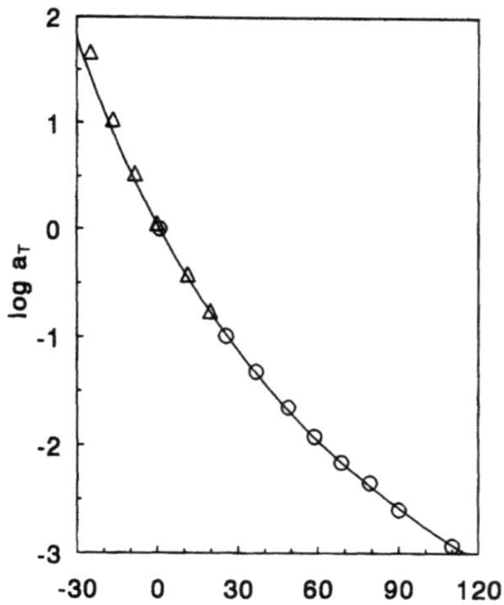

Fig. 26. Influence of the temperature on log a_T of 9.9 wt% PVC/DOP. (○) Measurements corresponding with the extrapolation procedure; (△) measurements corresponding with the normal application of the time-temperature superposition principle (see Fig. 24). The *full line* corresponds with the WLF-equation (reference temperature 1 °C). Reproduced from Physical Networks, Polymers and Gels [Ref. 12] by the courtesy of Chapman & Hall

2.2.10 Modulus Curves

By the application of the time-temperature superposition principle described in Sect. 2.2.9, moduli were obtained over a frequency range up to seven decades. In Fig. 27, master curves of 9.9 wt% PVC/DOP are shown after ageing over 6 days at different temperatures. From this figure the following can be concluded:

a) the span of the rubber plateau is considerable;
b) the height of the rubber plateau increases with decreasing temperature;
c) the loss moduli become independent of temperature at high frequencies (note that all curves were reduced to a reference temperature of 25.7 °C);
d) the loss moduli have minimum values at low frequencies for ageing temperatures below 60 °C.

The presence of a minimum in the loss modulus points to a relaxation mechanism at still lower frequencies, which will be accompanied by a maximum in the loss modulus and a decrease in the storage modulus, i.e. to the existence of a non-infinite rubber plateau. However, assuming that the crosslinks in the poly(vinyl chloride) gel, formed by micellar crystallites, cause infinite relaxation times, then after a maximum in the loss modulus and the corresponding decrease in the storage modulus a new rubber plateau should be reached at lower frequencies and at a lower level than the original one. Such behaviour can be studied with the aid of a special thermal history: an example is given in Fig. 28 for gels which were aged at room temperature and subsequently at a higher temperature [10–12]. From a careful analysis of the time dependence of the moduli at higher excess ageing temperatures, Te Nijenhuis and Dijkstra [10, 11] conclude that an increase in temperature causes melting of unstable crystallites, followed by a recrystallization process in which entrapped entanglements are formed. An important conclusion of the previous Section is that, as

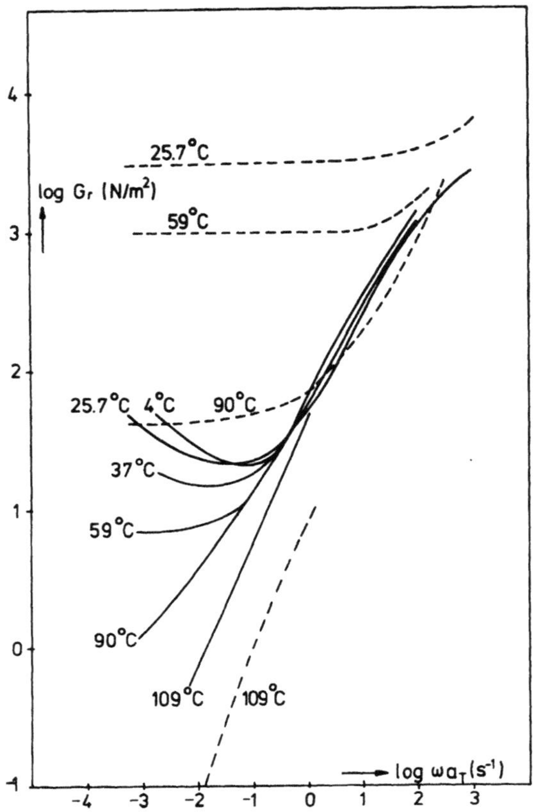

Fig. 27. Master curves of the storage and loss moduli of 9.9 wt% PVC/DOP for several ageing temperatures; ageing time about 144 h; reference temperature 25.7 °C; *full lines*: loss modulus; *broken lines*: storage modulus. Reproduced from Rheol Acta [Ref. 10] by the courtesy of Steinkopff Publishers Darmstadt, FRG

a consequence of quick melting of the microcrystallites at an increase of temperature, a decrease of temperature is needed for the described method of the application of the time-temperature superposition.

2.2.11 Excess Ageing after a Temperature Quench Following Previous Ageing at Higher Temperatures

By means of the procedure which was developed for the application of the time-temperature superposition, the excess ageing at a lower temperature has to be followed as a function of time. The excess ageing appears to be strongly dependent on the amount of pre-ageing at the higher temperature. Apparently, any ageing process depends on the thermal history. Such a dependence is shown in Fig. 29 where, for one frequency ($\omega = 0.393$ rad/s), the excess ageing is given as a function of time t_{ea} elapsed at 25.7 °C after previous ageing at 69 °C for 0, 0.27, 1, 23, 146 and 647 h respectively. If previous ageing is extended over a long time, then the excess ageing process is very slow; the crosslinking density is so

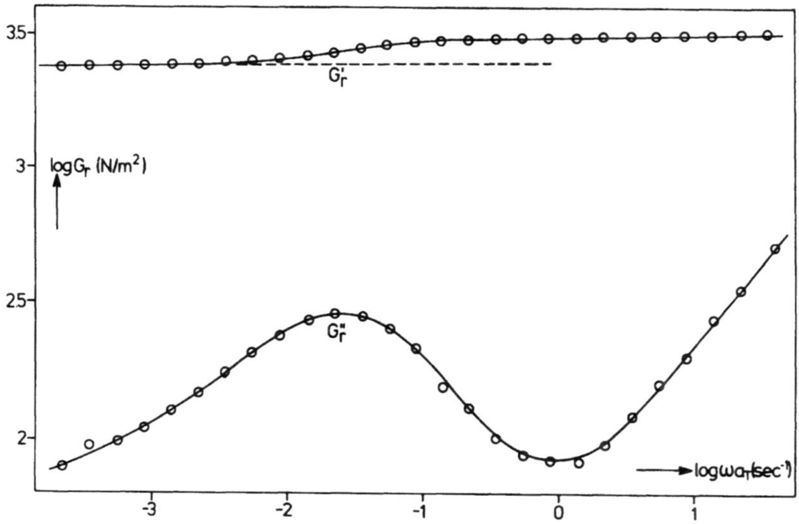

Fig. 28. Dispersion region attributed to slip in entanglements in 9.9 wt% PVC/DOP; measurements after 313 h of excess-ageing at 49 °C preceeded by 23 h of ageing at 25.7 °C; reference temperature 25.7 °C. Reproduced from Rheol Acta [Ref. 10] by the courtesy of Steinkopff Publishers Darmstadt, FRG

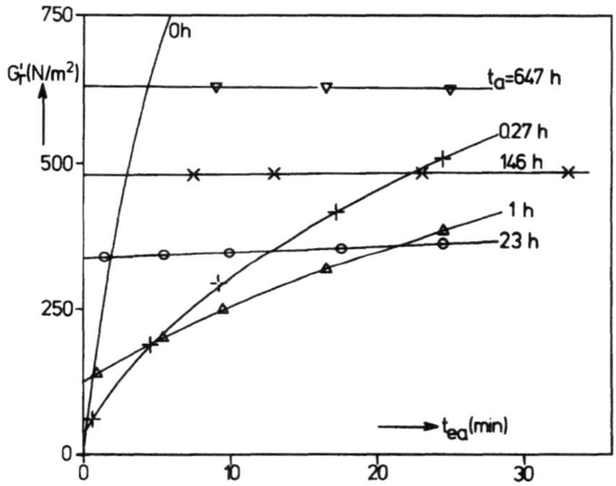

Fig. 29. Excess-ageing of 9.9 wt% PVC/DOP at 25.7 °C as indicated by the storage modulus, after previous ageing at 69 °C for several pre-ageing times; angular frequency $\omega = 0.393$ rad/s. Reproduced from Rheol Acta [Ref. 10] by the courtesy of Steinkopff Publishers Darmstadt, FRG

high that the mobility of chains necessary for further growth of existing crystallites has become too small. On the other hand, at short pre-ageing times the crosslinking density is still very low, so that the mobility of the chains is high enough to enable rapid additional growth of crystallites. Moreover, new nuclei which are stable at the lower temperatures, might be formed. A high rate of excess ageing is the consequence; the curves even cross those of longer pre-ageing times because the surface area of growing crystallites is higher, so that the increase of the crosslinking density is faster. By kinetic calculations Te Nijenhuis

[11] was able to discriminate between the growth of already existing crystallites and the growth of newly formed crystallites: at short pre-ageing times the newly formed crystallites contribute (much) more to the ageing process than the already existing crystallites, whereas at long pre-ageing times the amount of newly formed crystallites is negligible.

2.2.12 Temperature Increase

2.2.12.1 Sudden Temperature Increase

It has already been mentioned in Sect. 2.2.10 that a sudden temperature increase causes melting of unstable crystallites, which is followed by a recrystallization process; in addition to this process trapped entanglements are formed.

2.2.12.2 Gradual Temperature Increase

Recently Garcia et al. [124] reported measurements on poly(vinyl chloride) gels (12–36 wt%) which were aged at room temperature and subsequently heated at a rate of 2 °C/min. If the gel was aged for a long time (7 days) at room temperature, then two plateaus in the storage modulus as a function of temperature were present, one with a stability up to about 45 °C, the other, lower one, with a stability up to about 90 °C. In gels which were aged for only 15 min, such behaviour was not shown. The same kind of measurements were reported by Gallego et al. [125]. This behaviour is in agreement with the experiences of Mutin and Guenet [106], who report the existence of an endotherm at 50 °C, exclusively in aged gels, independent of concentration and an endotherm at the gel melting temperature. According to Kawanishi et al. [126] the formation of an aged gel takes place in two steps: first the presence of cyrstallites forming a network structure and second an interconnected structure of polymer rich and polymer poor regions. The first is the origin of the high temperature modulus plateau (the lowest), whereas the second, together with the first, is the origin of the low temperature modulus plateau (the highest), which is gradually formed.

2.2.13 Small-Angle X-ray Scattering Combined with Viscoelasticity

In the work of Dorrestijn et al. [92], the ageing process of 9.9 wt% PVC gels was followed with dynamic mechanical analysis as well as with small-angle X-ray scattering (SAXS). First a secondary maximum in the scattering intensity as a function of the scattering angle was observed at a scattering angle which appeared to be independent of the amount of ageing (i.e. the ageing time) and of the kind of plasticizer (see Fig. 30). From this scattering angle, the authors calculated (with the aid of Bragg's law) a long distance of 30 nm. This distance

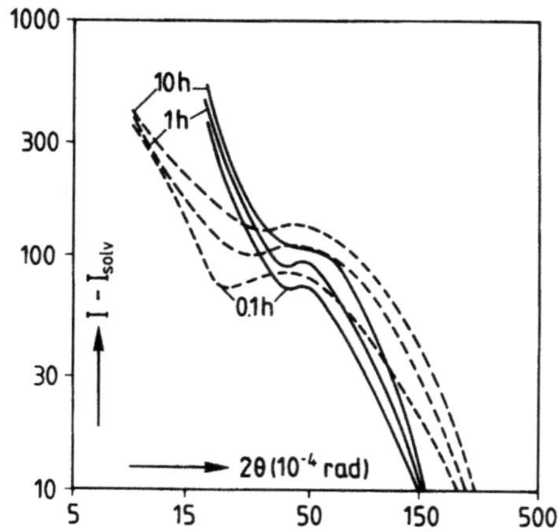

Fig. 30. Intensity of small-angle X-ray scattering of two 9.9 wt% PVC gels plotted vs the scattering angle for various ageing times; *full lines* PVC/DOP; *broken lines* PVC/Reomol. Reproduced from Makromol Chem, Macromol Symp [Ref. 40] by the courtesy of Hüthig & Wepf Verlag Publishers, Zug, Switzerland

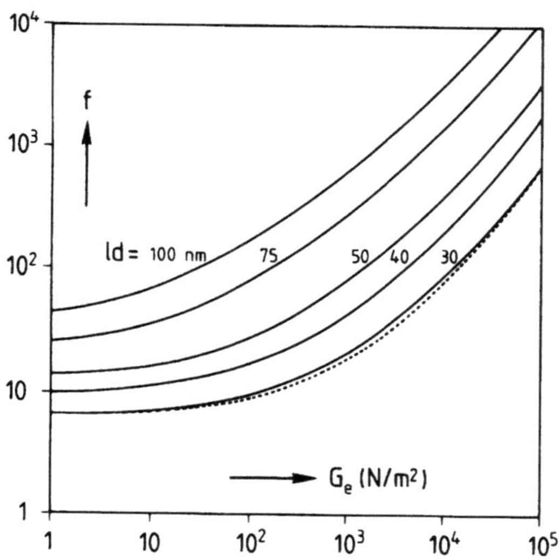

Fig. 31. Influence of the rubber equilibrium shear modulus on the functionality of the crosslinks in a poly(vinyl chloride) gel for various values of the distance between crosslinks. *Full lines* calculated for a Flory distribution of the molecular weights; *dashed line* calculated for a uniform polymer ($M = \bar{M}_w$); the 30 nm curve for the Schulz-Flory distribution lies between the corresponding other two curves. Reproduced from Makromol Chem, Macromol Symp [Ref. 40] by the courtesy of Hüthig & Wepf Verlag Publishers, Zug, Switzerland

was considered to be the average mutual distance of the microcrystalline crosslinks. Ohta et al. [85] also arrived at the conclusion that the formation of nuclei only occurs at the start of the crystallization process. An increase of the long distance with increasing plasticizer content was found by Gezovich and Geil [63,64], by Hendus (see [67]) and by Dorrestijn (1982, personal communication). From results obtained by Hendus one can conclude that there is a linear relationship between $1_d^{-1/3}$ and the plasticizer content from 0 to 80% of plasticizer. Second, between the invariant (i.e. the integral of the scattering intensity times the scattering angle over the scattering angle), which is a measure of the volume fraction of the dispersed phase, and the storage equilibrium modulus a linear relationship exists. It was assumed that the microcystallites are ordered on average by a simple cubic packing of the microcrystallites. In that case, from the long distance of 30 nm the number of microcrystallites in the above-mentioned gels is calculated to be $3.7 \times 10^{22}/m^3$. Te Nijenhuis [40] calculated the number of polymer molecules that are incorporated in such microcrystallites. This number increases from 10 to 45 and 360, plateau moduli increasing from 10^3 to 10^4 and 10^5 N/m^2, respectively (see Fig. 31). Very recently Te Nijenhuis [127] showed that the crosslink functionality of the system mentioned in Fig. 30 increased from 16 to 36 over a period from 0.1 to 1000 h. During this period the equilibrium shear modulus increased from 500 to 2800 N/m^2.

2.3 Conclusions

Many studies have dealt with the gelation and ageing process of poly(vinyl chloride) solutions. It is generally accepted from X-ray diffraction and model calculations that the crosslinks in these gels are tiny crystallites, although according to several authors hydrogen bonding also plays an important role, especially in the presence of diesters. Many examples of the viscoelastic behaviour of these gels could be explained by the presence and growth of those crystalline crosslinks. The dependence of the viscoelastic behaviour on time, concentration, type of solvent, temperature, molecular weight, degree of syndiotacticity, chlorination and extraction of the polymer and the temperature history has been examined relatively thoroughly. It even appeared possible to make use of the WLF-equation through a special application of the time-temperature superposition. In a combined study of SAXS and viscoelasticity of 10 wt% gels the functionality of the crystalline crosslinks could be calculated: for the system studied it increased in time from 16 to 36. A systematic study of the dependence of the crosslink functionality on temperature, temperature history, concentration and degree of syndiotacticity would be of great value. It can be seen that poly(vinyl chloride) gels have been the subject of many studies and rather a good qualitative picture of the gelation process has been obtained.

3 Poly(vinyl alcohol)

3.1 Introduction

Poly(vinyl alcohol) is unique as a synthetic polymer in that it has a large number of hydroxyl groups that can react with many kinds of functional groups. It has many applications, varying from thickening and gelation agent and dispersion stabiliser to solution spun fibres. Hydrogen bonds related to the large number of hydroxyl groups play an important role in the physical behaviour of poly(vinyl alcohol) [128]. It is produced by saponification or hydrolysis of poly(vinyl esters), generally poly(vinyl acetate). Its properties depend, among other things, on the degree of hydrolysis of the mother polymer, in general higher than 90 mol%.

The crystal structure of poly(vinyl alcohol) is similar to that of polyethylene: the CHOH group is small enough to fit easily into the CH_2 space in polyethylene. Pairs of chains are linked into sheets as far as the stereochemical irregularity allows [129]. Gelation instead of crystallization is attributed to the copolymeric nature of the polymer, which is the result of incomplete hydrolysis of the mother polymer [7]. In this respect, the stereoregularity also plays an important role: isotactic poly(vinyl alcohol) is dissolved easily in cold water, but syndiotactic poly(vinyl alcohol) does not dissolve even in boiling water. If the stereoregularity is not high, gels may be formed in aqueous solutions, where tiny crystallites act as multifunctional crosslinks. It many cases more or less turbid gels are formed.

Gelation of poly(vinyl alcohol) solutions, whether or not thermoreversible, can take place in a number of ways:

- in water: gelation is accompanied by blue coloration in the presence of I_2/KI; this coloration points to the formation of helical structures which are formed by syndiotactic sequences in the poly(vinyl alcohol) molecules; hydrogen bonding also seems to play an important role; this is concluded from the fact that no gelation takes place in the presence of rhodanide ions or urea;
- in aqueous solutions of inorganic anions like borates, titanates, antimonates, chromate etc., gels are formed immediately on mixing solutions of poly(vinyl alcohol) and the anions [130–133]; these gels are non-permanent in that they are viscid in nature, i.e. self healing; apparently the crosslinks are non-permanent;
- in aqueous solutions of specific hydrogen bonding reagents: well-known in this sense are resorcinol and congo red; the gelation mechanism is dependent on the reagent: congo red forms helical structures as revealed by blue coloration in the presence of I_2/KI; this is not the case in resorcinol catalysed gelation;
- in non-aqueous solutions, where e.g. ethylene glycol or dimethyl sulfoxide (whether or not mixed with water) are used as solvents.

3.2 Aqueous Solutions and Gels

3.2.1 Introduction

In aqueous solutions with a polymer concentration of more than 1%, entangled aggregates of hydrogen bonded poly(vinyl alcohol) molecules are formed. This is the consequence of the time dependent formation of crystal nuclei and (micro-) crystalline regions, as found by Morawetz [134]. From WAXS measurements, it was concluded that the gels are partially crystalline systems [135–137]. The microcrystallites in poly(vinyl alcohol) gels have the same atomic distances as in solid crystals of poly(vinyl alcohol), and thus it might be concluded that the crystals are not entered by solvents and that they do not change during gelation [135–137]. The tendency of poly(vinyl alcohol) to form microcrystalline junction zones, by which aggregates or gels are formed, depends, among other things, on the stereo-regularity of the poly(vinyl alcohol) used, on its degree of hydrolysis and on the temperature history.

From pioneering work by Prins et al. [138–140] it was proposed that spinodal liquid-liquid phase separation, followed by crystallization of parts of the PVA chains, is the origin of gel formation in poly(vinyl alcohol) solutions in water and in ethylene glycol/water mixtures. Komatsu et al. [108] investigated the dynamics of the sol-gel transition of aqueous poly(vinyl alcohol) solutions by light scattering. The sol-gel transition of aqueous poly(vinyl alcohol) solutions was determined with the tilted-test-tube method, whereas spinodal and binodal curves were obtained from light scattering studies. The result is shown in Fig. 32. It appears that the spinodal curve crosses the sol-gel transition curve (which has its origin in crystallization behaviour) and that gelation occurs both above (spinodal decomposition) and below this curve (liquid-liquid demixing). In this respect it is worth-while mentioning the work of Wu et al. [141, 142] who explained the results of their SANS studies on the melting behaviour of aqueous poly(vinyl alcohol) gels with a two-phase model: polymer rich domains are surrounded and interconnected by a polymer depleted phase.

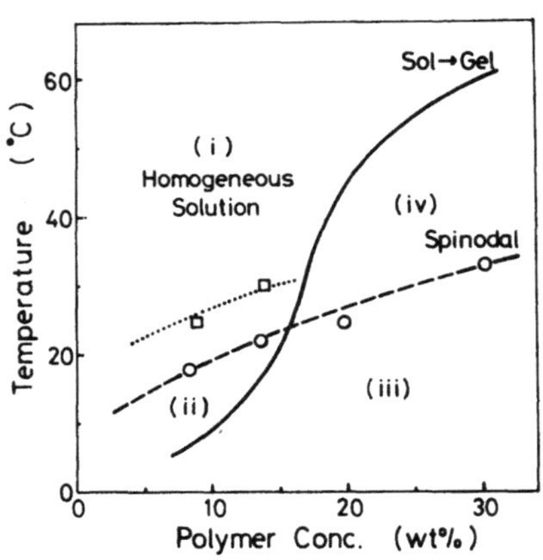

Fig. 32. Binodal and spinodal curves and sol-gel transition of aqueous poly(vinyl alcohol) solutions. Reproduced from J Polym Sci, Polym Phys Ed [Ref. 108] by the courtesy of the authors and of John Wiley & Sons, Inc

Gelation occurs when a solution, formed at high temperatures ($> 80\,°C$) is cooled. The gel temperature depends strongly on the cooling rate, but reproducibility is reached for cooling rates smaller than 1 °C/hour [143]. The gels obtained are relatively brittle and generally not clear due to crystallinity of the relatively large crosslinks. For high molecular weight poly(vinyl alcohol) solutions the melting point of the gels increases from 14 to 28.5 °C for concentrations of 100–212 kg/m^3. By plotting ln c vs $1/T_m$, a straight line is obtained [144] from which the melting enthalpy was calculated to be $\Delta H_m^o = -37.5 \pm 0.7$ kJ/mol of crosslinks.

In a series of papers, Prokopová et al. [145–150] reported their studies on rheology, light scattering, calorimetry and infrared spectroscopy of moderately concentrated aqueous poly(vinyl alcohol) solutions. Their main conclusions are as follows.

a) Aqueous poly(vinyl alcohol) solutions undergo the process of ageing, reflected predominantly in a rise of viscosity and first normal stress difference N_1 with time (see Figs. 33a and 33b), which in concentrated solutions is followed by the formation of aggregated supermolecular structures.
b) The temperature of dissolution is very important; the viscosity and first normal stress difference of solutions prepared at 80 °C increased more slowly with time than those prepared at 90 °C. An extrapolation procedure shows that a temperature of at least 92 °C is needed to dissolve the polymer "completely" [145–147]. These observations might be explained by assuming that many microcrystallites are still present at 80 °C, which means that only relatively few potential junction zones are still present. Ageing proceeds relatively slowly. When the polymer is dissolved at 90 °C, most of the

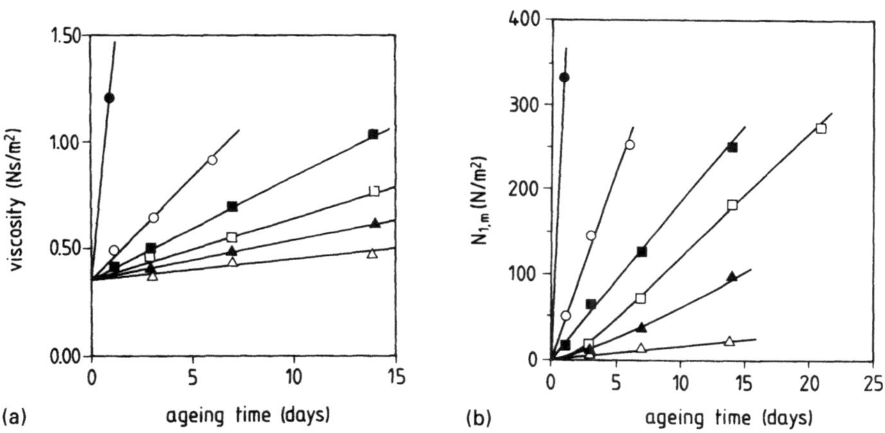

Fig. 33a, b. Time dependence of: **a** viscosity; **b** first normal stress difference for 16 wt% aqueous solutions of PVA99 ($M_w = 22.3$ kg/mol) for several ageing temperatures; dissolution of poly(vinyl alcohol) took place at 80 °C over 5 h; shear rate 71.2 s^{-1}; temperature in °C: (●) 10; (○) 13; (■) 15; (□) 18; (▲) 30; (△) 50. Reproduced from Colloid Polym Sci [Ref. 146] by the courtesy of the authors and of Steinkopff Verlag Darmstadt, FRG

microcrystallites have disappeared, so that many more potential junction zones will be present. This enables a rapid ageing process at lower temperatures. Dissolving the polymer at still higher temperature results in a slow ageing process again. However, in this case the number of potential junction zones is very high, and the nuclei are destroyed so that the ageing process becomes fast only if new nuclei are formed; apparently this takes a long time.

c) It was demonstrated by light scattering measurements [148] that ageing is a process in which supermolecular formations arise, irrespective of the concentration of the solution, where the predominant part of the polymer remains in the non-aggregated form. The amount of supermolecular formations, which is dependent on the concentration of the ageing solutions, increases with time. If parts of the polymer molecule interpenetrate different crystallites, those crystallites become linked. This gives rise to crosslinked systems (gels). The bonds between the macromolecules are of a physical nature and their existence is probably related to the considerable crystallization ability of poly(vinyl alcohol) (precrystallites [143] or fringed micelles [151]). Similar results were obtained recently by Wu et al. [141, 142] with the aid of small angle neutron scattering.

d) During swelling of poly(vinyl alcohol) in aqueous solution, the original paracrystalline regions do not disappear [152] even at temperatures above 100 °C and act as crosslinks; during the swelling process ordered structures are formed, as revealed by infrared spectroscopy and differential scanning calorimetry [149, 150]. In the absence of water the ordered structures are formed less readily and in smaller amounts; probably the decrease of T_g in the presence of water plays an important role in this respect.

3.2.2 Stereoregularity

The gelling properties of aqueous solutions of poly(vinyl alcohol) are strongly affected by the amount of syndiotacticity. Isotactic sequences in poly(vinyl alcohol) are believed to form intramolecular hydrogen bonds, which do not play an important role in the aggregation properties of poly(vinyl alcohol) [153] (it is known that isotactic poly(vinyl alcohol) dissolves easily in cold water). On the other hand, syndiotactic sequences in poly(vinyl alcohol) are believed to form intermolecular junctions [153] if their lengths are larger than a critical value of n monomeric units. These junctions form the crosslinks that are responsible for the aggregation behaviour of aqueous poly(vinyl alcohol) solutions (syndiotactic poly(vinyl alcohol) does not dissolve even in boiling water). Critical sequence lengths reported in the literature vary from five to ten monomeric units depending on temperature and syndiotacticity [154]. Ogasawara et al. [155–159] have also shown that syndiotacticity of the polymer plays an important role in the ageing process of poly(vinyl alcohol): an increase of syndiotacticity is accompanied not only by an increase in the tendency to form gels, but also by an increase in crystallinity.

Interesting results are shown in Figs. 34 and 35 where the elastic modulus E of gels that were matured at room temperature, is plotted vs temperature for solutions of two kinds of poly(vinyl alcohol) (1: M = 220 kg/mol, c = 8.4 wt%, s = 0.584; 2: M = 88 kg/mol, c = 6.4 wt%, s = 0.643). It appears that the

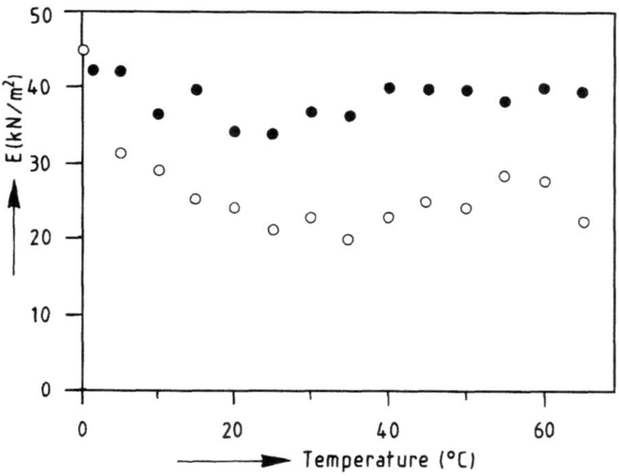

Fig. 34. Effect of syndiotacticity on Young's modulus of aqueous PVA gels chilled at room temperature; (○) M = 220 kg/mol; c = 8.4–10.2 wt %; s = 0.584; (●) M = 88 kg/mol; c = 6.4–7.2 wt %; s = 0.643 (data collected from [157])

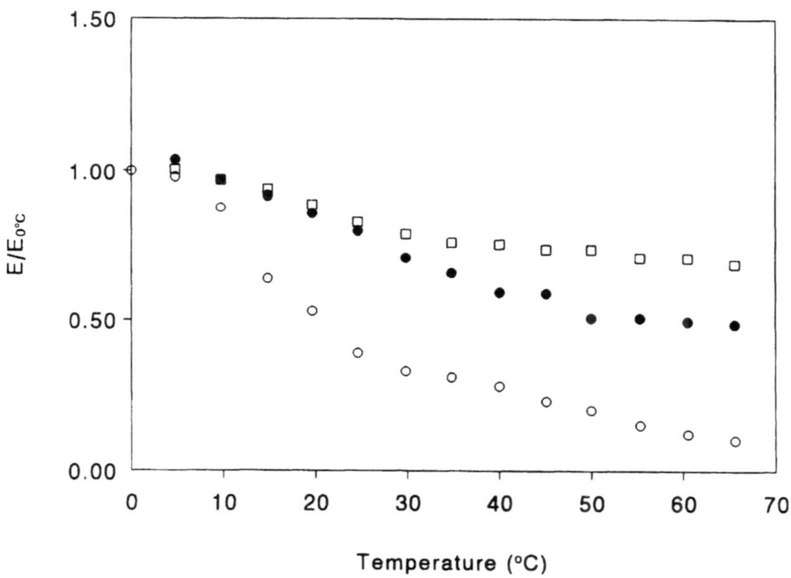

Fig. 35. Relationship between E/E_o and the temperature rise for PVA gels of different molecular weight and syndiotacticities, chilled at room temperature (E_o = Young's modulus at 0 °C); (○) M = 220 kg/mol, s = 0.584; (●) M = 240 kg/mol, s = 0.609; (□) M = 88 kg/mol, s = 0.643. Reproduced from Colloid Polym Sci [Ref. 157] by the courtesy of the authors and of Steinkopff Verlag Darmstadt, FRG

Table 1. Melting enthalpies of atactic PVA of various origins [157]

Origin	ΔH_m^o (kJ/mol)	Corresp. number of H-bonds
Poly(vinyl acetate)	− 37	1.8
Poly(vinyl formate)	− 46	2.2
Poly(vinyl trifluoroacetate)	− 84	4.0

syndiotacticity of poly(vinyl alcohol) plays a predominant role in the gelation properties of this polymer (molecular weight and concentration are less important as shown in Figs. 34 and 35). The stability of gels formed with poly(vinyl alcohol) of higher syndiotacticity is also better, as becomes clear from Fig. 35: the more syndiotactic the polymer, the higher the ordered structures melt and the slower the modulus decreases with temperature.

The tendency to form gels is strongly determined by the origin of the poly(vinyl alcohol). The method of Eldridge and Ferry [160] to calculate the gel melting enthalpy, ΔH_m^o, a method against which Ogasawara et al. [155] and Yamaura et al. [161] and also the present author have objections, is often used in the literature to calculate the enthalpy of crosslink formation. For poly(vinyl alcohol) of various origins, results are shown in Table 1. According to these authors, the rate of the ageing process shows similar dependence. For atactic poly(vinyl alcohol) it is believed that hydrogen bonds are the origin of the gelation process. ΔH of a hydrogen bond is about −21 kJ/mol, so that the number of hydrogen bonds would be only from two to four, whereas the number of monomeric units taking part in a crosslink, is about seven per polymer chain [154] as mentioned before. From x-ray crystalline diffraction patterns it was shown that, for a filament gel spun from very concentrated solutions, the crosslink loci in the poly(vinyl alcohol) gels were syndiotactic sequences. This means that hydrogen bonding does not play a predominant role. It is to be noted here that Ogasawara et al. [155] tried to synthesize poly(vinyl alcohols) of the same molecular weight, but differing in s-content. They only partly succeeded. For that reason, they fractionated their polymers to approximately constant molecular weights. The s-content was determined with the aid of infrared spectroscopy before the fractionation. As a matter of fact, fractionation on molecular weights might also affect the s-content. For gels of poly(vinyl alcohol) with s = 0.65 but of different molecular weights, ln c vs $1/T_m$ plots yield straight lines with slopes that are almost independent of molecular weight. The enthalpy of crosslink formation is calculated to be approximately $\Delta H_m^o = -168$ kJ/mol, which is about five times as high as for atactic poly(vinyl alcohol), where it was found that $\Delta H_m^o = -37$ kJ/mol. The melting temperatures are also much higher: for 5% gels up to 117°C for poly(vinyl alcohol) with s = 0.643. Unfortunately, the authors did not report on measurements of viscoelastic properties like G′ and G″.

3.2.3 *Temperature History*

From the early 1980s on, Watase and Nishinari [162–168] reported their extensive studies on aqueous poly(vinyl alcohol) gels on ways to obtain high modulus gels. Their method consists of freeze drying and subsequent immersion in water of aqueous gels, or by application of freezing-thawing cycles with a special time-temperature history to aqueous gels. It seems that the upper temperature during the thawing cycle should not exceed 0 °C. An example of the increase of the modulus by such a treatment is shown in Fig. 36, where the molecular weight dependence of the Young storage modulus measured at 2 Hz is presented for various freezing times of aqueous poly(vinyl alcohol) gels, and in Fig. 37, where E′ and tan δ are shown for various concentrations. The high concentration gels show rubber elastic behaviour up to 45 °C; at higher temperatures the moduli start to decrease and the loss tangents to increase, and hence a transition to liquid-like behaviour occurs at temperatures above 45 °C. The authors also show that the evacuation time at low temperatures has a dramatic effect on the modulus.

3.2.4 *Degree of Saponification*

Poly(vinyl alcohol) is obtained from a mother polymer, in general poly(vinyl acetate), by saponification or hydrolysis of the ester group. The degree of saponification or hydrolysis determines the gelation properties: in general a degree of at least 90 mol% is needed for poly(vinyl alcohol) to be able to form aqueous gels. Its copolymeric character depends on the conversion of the hydrolysis reaction. Beltman [7] showed that an 8% aqueous solution of PVA88 (the number following PVA means the conversion of the hydrolytic reaction in mol%) is stable at 5 °C but that the viscosity of an 8% solution of

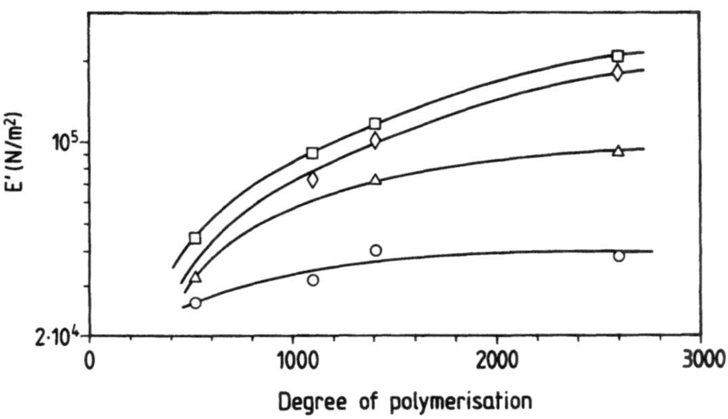

Fig. 36. Dependence of the 2 Hz storage Young modulus of aqueous poly(vinyl alcohol) gels on the degree of polymerisation for various freezing periods: (○) 48 h; (△) 72 h; (◇) 96 h; (□) 120 h. Reproduced, with permission of the authors and publishers, from [164]

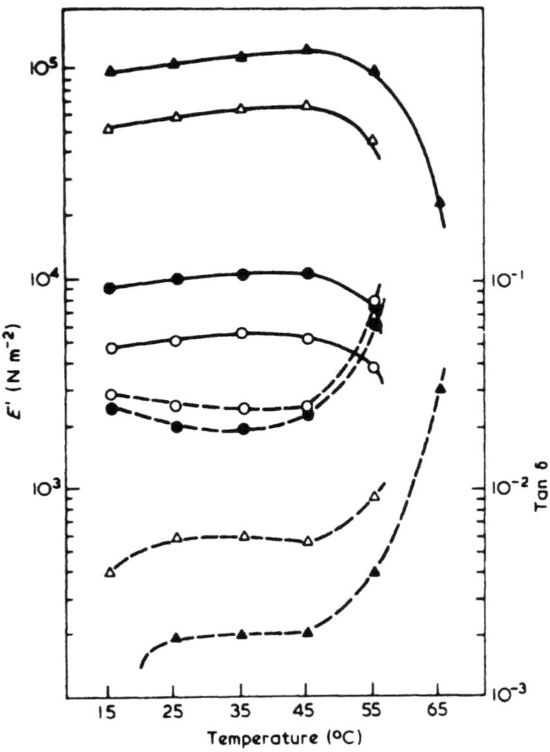

Fig. 37. Temperature dependence of E′ and tan δ, measured at 2 Hz, of aqueous poly(vinyl alcohol) gels of various concentrations, frozen at temperatures between −20 and −23 °C for 12 h and immersed in distilled water for 3 days at 25 °C: (○) 8 wt%; (●) 10 wt%; (△) 15 wt%; (▲) 20 wt; E′: *solid lines*; tan δ: *broken lines.* Reproduced from Polym Commun [Ref. 162] by the courtesy of the authors and Elsevier Science Ltd, The Boulevard, Langford Lane, Kidlington 0X5 1GB, UK

PVA98.5 increases fourfold in 100 h. The conclusion is that an increased hydrolysis conversion has a spectacular influence on the aggregation behaviour of poly(vinyl alcohol) solutions. Watase and Nishinari [168–170] demonstrated the influence of the degree of saponification from concentration dependent rheological and thermal properties. The effect of saponification of the mother polymer is shown in Fig. 38, where the storage Young modulus of 15 wt% PVA gels is plotted as a function of the degree of hydrolysis after several freezing-thawing cycles. It is clear that the degree of saponification strongly influences the tendency to gel formation. Differential scanning calorimetry experiments reveal very sharp endotherms at the melting point of the gels, as shown in Fig. 39. The concentration dependence of the melting temperature is shown in Fig. 40. From the slopes of the various straight lines, which depend on the degree of saponification, the melting enthalpy ΔH_m was calculated and plotted as a function of the degree of saponification in Fig. 41. From extrapolation of the straight line to zero melting enthalpy one could conclude that the minimum amount of hydrolysis for this polymer would be 90 mol%. This is in agreement with results obtained by Beltman [7], who showed that aqueous solutions of poly(vinyl alcohol) with a degree of saponification of 0.88 did not gel, although, of course, gelation was also affected by the polymer molecular weight.

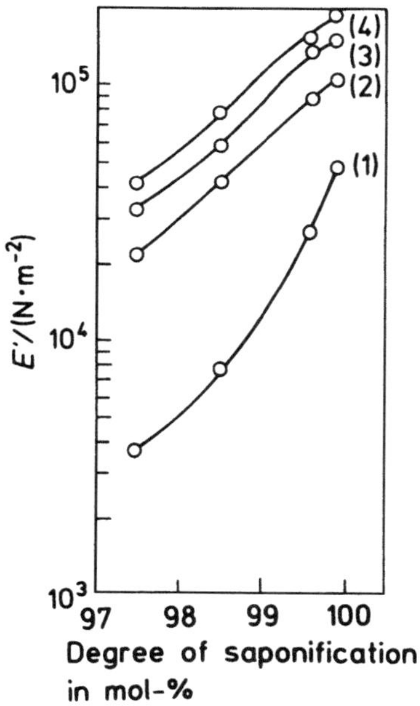

Fig. 38. Dependence of the storage Young modulus of 15 wt% aqueous poly(vinyl alcohol) gels for various numbers of freezing-thawing cycles, as indicated, on the degree of saponification. Reproduced from Makromol Chem [Ref. 166] by the courtesy of the authors and of Hüthig & Wepf Verlag Publishers, Zug, Switzerland

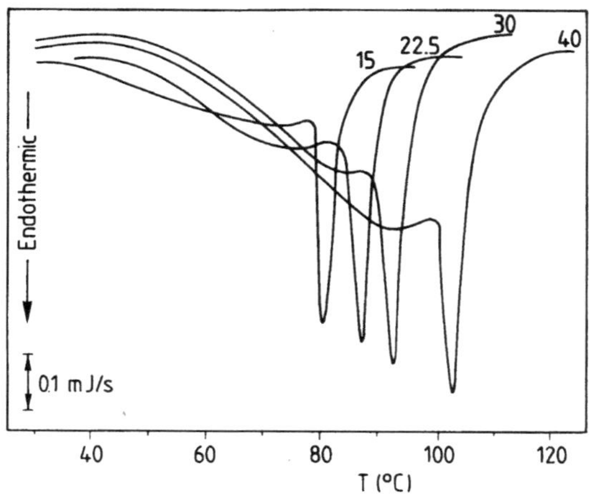

Fig. 39. Heating DSC curves for aqueous poly(vinyl alcohol) gels (M = 75 kg/mol, degree of saponification = 0.999) of various concentrations, as indicated at the curves in wt%. Reproduced from Makromol Chem [Ref. 168] by the courtesy of the authors and of Hüthig & Wepf Verlag Publishers, Zug, Switzerland

3.3 Other Solvents

3.3.1 Introduction

In the 1950s and early 1960s observations on non-aqueous poly(vinyl alcohol) gels were already reported. Rehage [135] mentioned melting temperatures of 10% poly(vinyl alcohol) gels in water (14 °C), glycerol (64 °C), ethylene glycol

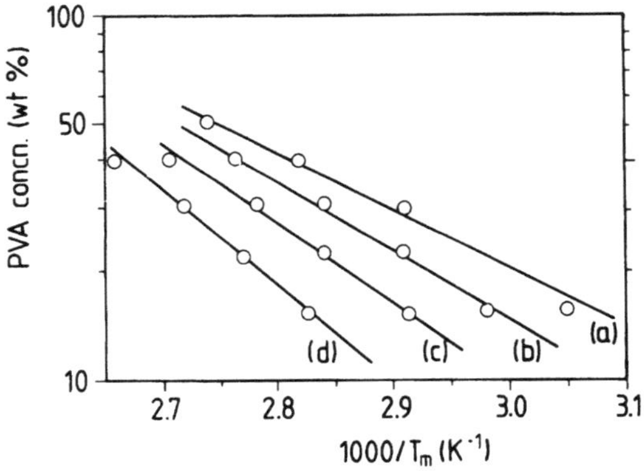

Fig. 40. Eldridge-Ferry-plots for aqueous poly(vinyl alcohol) gels of various degrees of saponification DS: *line a* – DS = 96 mol%; *line b* – DS = 97.5 mol%; *line c* – DS = 98.5 mol%; *line d* – DS = 99.9 mol%. Reproduced from Makromol Chem [Ref. 168] by the courtesy of the authors and of Hüthig & Wepf Verlag Publishers, Zug, Switzerland

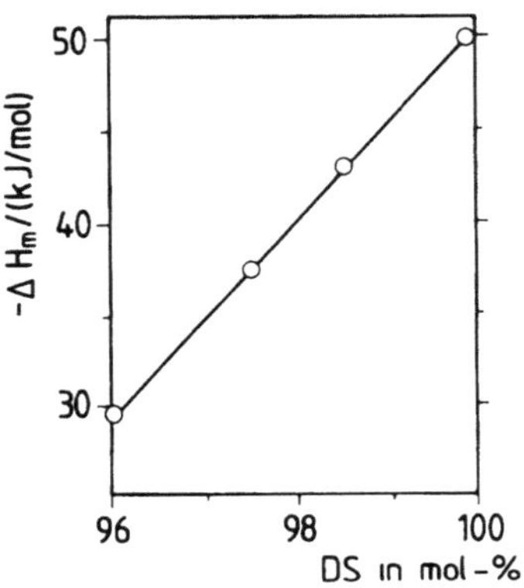

Fig. 41. Dependence of $-\Delta H_m$ for formation of network junctions in aqueous poly(vinyl alcohol) gels on degree of saponification DS. Reproduced from Makromol Chem [Ref. 168] by the courtesy of the authors and of Hüthig & Wepf Verlag Publishers, Zug, Switzerland

(112 °C, gel point 102 °C), 1,3-propanediol (130 °C), 1,4-butanediol (142 °C) and 1,6-hexanediol (183 °C). According to Pritchard [143] viscoelastic behaviour of poly(vinyl alcohol)/glycerol gels was investigated by Hirai et al. Results showed that the behaviour of these gels was almost entirely rubber-like.

3.3.2 Ethylene Glycol/Water Systems

The phase behaviour and thermoreversible gelation of poly(vinyl alcohol) in ethylene glycol was investigated by Berghmans et al. [171–174] and by Nishinari and Watase [175]. On the basis of a combination of rheological, optical, calorimetrical and X-ray scattering techniques, Berghmans et al. [171–174] demonstrated that the thermoreversible gelation of poly(vinyl alcohol) in ethylene glycol is associated with a combination of liquid-liquid demixing and a crystallization or liquid-solid demixing, which is influenced by molecular weight and solvent quality. They proposed two kinds of mechanisms for their gelation process:

a) gelation above 100 °C: slow direct crystallization to a dispersion of crystallites, which form the polyfunctional crosslinks in the network; the formation of a gel requires a few days;
b) gelation by cooling to room temperature: liquid-liquid demixing, followed by crystallization in the polymer rich regions; the gelation process is very fast: within a few minutes a gel is formed.

Results of dynamic mechanical measurements are shown in Fig. 42: none of the gels show a real equilibrium plateau for G′, so a real permanent network is not obtained. Moreover, the gels show thixotropic (two phase) behaviour, which follows from the observations that the gels are liable to a continuous stress relaxation upon applying a constant deformation and that the linearity of the dynamic moduli already disappears at strains of less than 1%; the non-linear response, however, is almost instantaneous and, by turning back into the linear region, the initial values are quickly regained. These phenomena are typical for two-phase systems and especially for thixotropic ones.

According to Nishinari and Watase [175] the elastic moduli of poly(vinyl alcohol)/ethylene glycol-water gels showed a maximum around an EG mole fraction of 0.35. The polymer concentration dependence was exponential: $E \sim c^n$, where the exponent was minimal at an EG mole fraction of 0.35. The

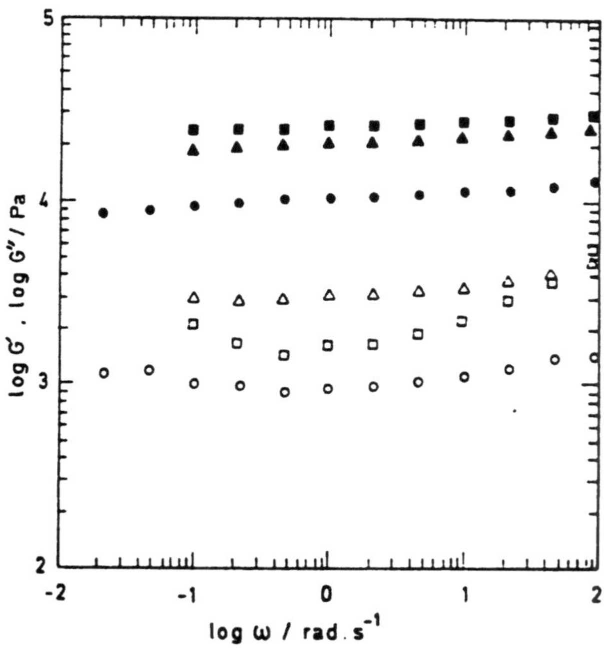

Fig. 42. Dynamic moduli vs angular frequency for poly(vinyl alcohol)/ethylene glycol-gels prepared by: (● ○) slow cooling to room temperature; (■ □) gelation at 100 °C; (▲ △) quenching to room temperature; *filled symbols* G′; *open symbols* G″. Reproduced from Brit Polym J [Ref. 173] by the courtesy of the authors and of John Wiley & Sons, Inc

endothermic gel melting enthalpy was calculated from Eldridge-Ferry plots [160], as shown in Fig. 43, and ΔH_m shown to increase with increasing EG content. This indicates that the junction zones in poly(vinyl alcohol) gels become more heat resistent with increasing ethylene glycol content. From their DSC curves the authors conclude that solvent properties are dramatically changed around an EG mole fraction of 0.3. This is shown in Fig. 44, where the melting temperature and the heat of gel fusion are plotted vs EG content. According to

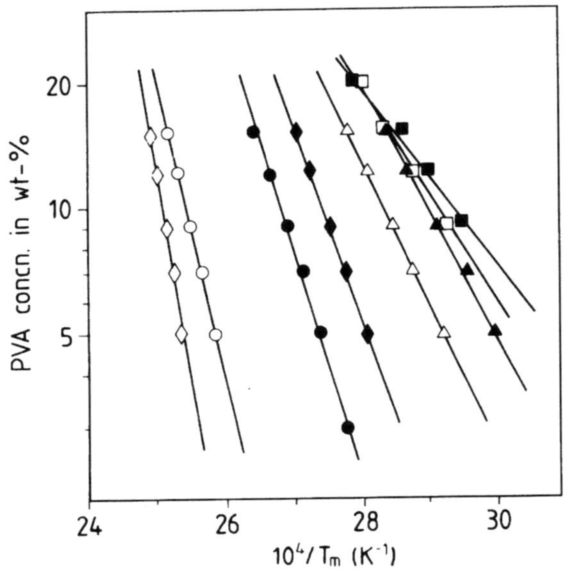

Fig. 43. Eldridge-Ferry plots for poly(vinyl alcohol)/ethylene glycol-water gels in mixtures of ethylene glycol and water of various ethylene glycol contents in mole fraction mf: (◇) 1 mf; (○) 0.794 mf; (●) 0.458 mf; (◆) 0.357 mf; (△) 0.222 mf; (▲) 0.147 mf; (□) 0.087 mf; (■) 0.04 mf. Reproduced from Polym J [Ref. 175] by the courtesy of the authors and The Society of Polymer Science, Japan

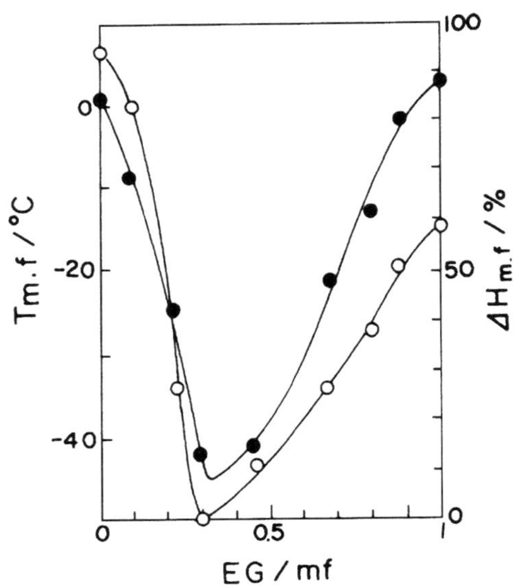

Fig. 44. Melting temperature T_{mf} and heat of fusion ΔH_{mf} as a function of ethylene glycol content for poly(vinyl alcohol) gels in mixtures of ethylene glycol and water. Reproduced from Polym J [Ref. 175] by the courtesy of the authors and The Society of Polymer Science, Japan

the authors the state of gels is determined by the balance between the solubility and the crystallinity. Below EG mole fractions of 0.3, free water solubilizes poly(vinyl alcohol), while above this EG content, free EG molecules solubilize poly(vinyl alcohol). The result is that the number of junction zones becomes maximal at the EG mole fraction of 0.3.

3.3.3 Dimethyl Sulfoxide/Water Systems

Hyon et al. [176, 177] obtained gels of excellent transparency in solutions of atactic poly(vinyl alcohol) in DMSO/water (80/20 v/v) mixtures. They showed that the elasticity was much higher than in pure water as a solvent. The same results were obtained by Ohkura et al. [178, 179]. Moreover, these authors showed that the gelation rate was also very fast compared with aqueous solutions. Sol-gel diagrams were obtained for solutions of a series of poly(vinyl alcohols) ($\bar{M}_n$ varying from 8.2 to 890 kg/mol) in DMSO/water (60/40 v/v) mixtures and the critical gelation concentration was dependent on molecular weight but independent of temperature between −20 and −60 °C, except for the lowest molecular weight sample, as shown in Fig. 45. The gels were transparent at temperatures below 0 °C, and turbid above 0 °C. It was pointed out that phase separation plays an important role above −20 °C, whilst below −20 °C phase separation did not occur [178–181]. Phase separation of PVA gels occurs at 23 °C in competition with network formation due to crystallisation: rates of L-L-demixing and crystallisation are of comparable rates. However, below −20 °C crystallisation is so fast that the network formation is completed before phase separation occurs, resulting in transparent gels. The structure of the gels was investigated by small angle neutron scattering [180, 181]. For gels with polymer concentrations of 50 kg/m^3 the distance between the crystalline cross-links was calculated to be 14.8, 16.0 and 20.8 nm for poly(vinyl alcohol) molecular weights of 8.2, 26.0 and 72 kg/mol, respectively.

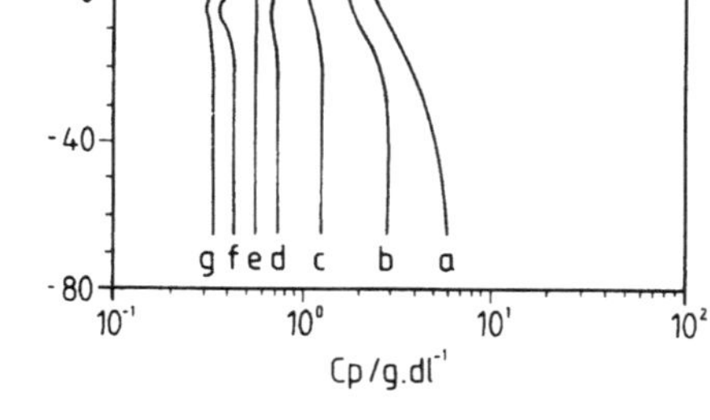

Fig. 45. Sol-gel transition diagrams of atactic poly(vinyl alcohol) in a mixture of DMSO and water (60/40 v/v) for samples of various number average molecular weights in kg/mol and values of D = $\bar{M}_w/\bar{M}_n$ (from right to left): *curve a* – 8.2 and 2.04; *curve b* – 26 and 1.61; *curve c* – 72.2 and 1.97; *curve d*-239 and 1.90; *curve e* – 471 and 1.99; *curve f* – 524 and 1.93; *curve g* – 889 and 2.17, respectively. Reproduced from Polymer [Ref. 178] by the courtesy of the authors and Elsevier Science Ltd, The Boulevard, Langford Lane, Kidlington 0X5 1GB, UK

The average number of polymer molecules involved in crosslinks might be calculated in the way used for poly(vinyl chloride) gels [40], provided elastic moduli of the gels are known. Unfortunately they were not reported for the PVA gels. On the other hand, although it was proven that the crosslinks of the PVA-gels are crystallites (with sizes of approximately 7 nm) most crystallites do not work as crosslinks.

Nishinari and Watase [169, 170] investigated the influence of the DMSO concentration in DMSO/water mixtures on the gelation behaviour of poly(vinyl alcohol) solutions with the aid of dynamic viscoelasticity and differential scanning calorimetry. In addition, the influence of the degree of saponification was taken into account. It appeared that in the concentration range of 5 to 25% of polymer [169] the gel-to-sol transition temperature, the endothermic enthalpy and the Young modulus showed maximum values at a mole fraction of 0.277 DMSO (i.e 60 vol.% DMSO), independent of degree of saponification. Results are shown in Fig. 46. This value was also obtained by Tanigami et al. [182] in their investigations of the syneresis phenomenon in these gels: ageing for up to 500 days had no effect on the mole fraction DMSO where the maximum in the storage Young modulus is present, although the poly(vinyl alcohol) concentration increased by syneresis from 10 to 28 wt%. However, it is not known whether the proportion of water and DMSO in the gel is the same after the syneresis.

For higher concentrations the maximum melting temperature was shifted to higher DMSO contents [169, 170], although the maximum melting enthalpy remained at the DMSO mole fraction of 0.277.

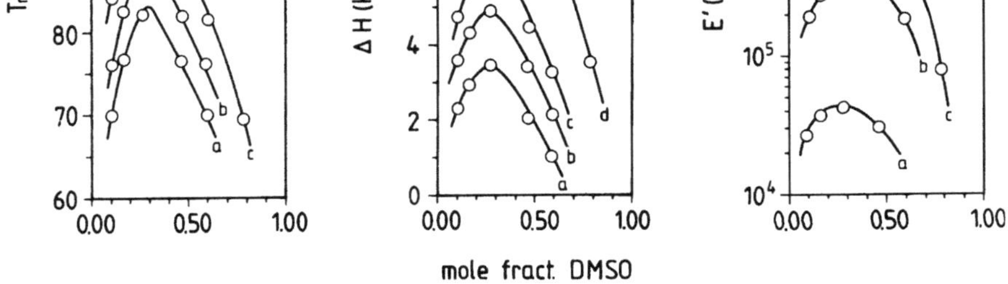

Fig. 46a–c. Dependence of: **a** melting point T_m; **b** endothermic enthalpy ΔH_m; **c** E′ of poly(vinyl alcohol) gels in mixtures of DMSO and water, on DSMO concentration in mole fraction: poly(vinyl alcohol) concentrations: *line a* – 5 wt%; *line b* – 10 wt%; *line c* – 15 wt%; *line d* – 25 wt%. Reproduced from Polym J [Ref. 169] by the courtesy of the authors and The Society of Polymer Science, Japan

The authors attribute these phenomena to the difference in solvent power of DMSO and water: DMSO is a better solvent than water. Okada et al. [183] showed that low concentration poly(vinyl alcohol) solutions at DMSO contents above mole fractions of 0.5 are transparent, while below mole fractions of 0.13 the system is non-homogeneous. For more concentrated systems, opaque gels are formed in the low DMSO content systems. Watase and Nishinari [170] attribute these phenomena to the complexation between DMSO and water: at low DMSO contents the free water content is higher, so that the gelling ability of poly(vinyl alcohol) is lower. Hydrated compounds of DMSO/water are reported as $(CH_3)_2SO \cdot 2H_2O$ and as $(CH_3)_2SO \cdot 3H_2O$, i.e. DMSO mole fractions of 0.33 and 0.25 respectively, just around 0.277. In this respect it is interesting to mention that a maximum in the Young modulus of agarose-DMSO-water systems was observed at a DMSO mole fraction of 0.277 [204].

3.4 Gelation by Addition of Borate Ions

3.4.1 Early Studies

Aqueous poly(vinyl alcohol) solutions can be gelled by inorganic anions like borate, titanate, antimonate, chromic anions etc. [130–133], which can condense with organic hydroxyl groups in aqueous solution. In most cases use is made of borates, and borax ($Na_2B_4O_7 \cdot 10H_2O$), sodium metaborate ($NaBO_2$) or sodium perborate ($Na_2B_2O_4$) is mostly used. The borate anions are converted into $B(OH)_4^-$ by protolysis. This might react with poly(vinyl alcohol) as shown in Fig. 47.

In 1949, Deuel and Neukom [131] suggested that the origin of crosslinks in aqueous poly(vinyl alcohol)/borate gels is the complexation shown in Fig. 48. This concept of didiol type complexation was adopted by authors in more recent times [130, 184–193]. Schultz and Myers [185] measured the dynamic moduli G′ and G″ of poly(vinyl alcohol)/borate gels (at small deformation amplitudes in order not to disturb the gel [185]). At low frequencies, the gels showed liquidlike behaviour; this means that the borate crosslinks are dynamic in nature. Such results were also reported by Beltman [7]. He measured the dynamic moduli of a gel formed in a 4 wt% PVA98.5 solution with 2.5 wt% of borax (= 0.0069 mol/l, which is equivalent to 0.0276 mol/l of borate) at various temperatures (15–75 °C). Results are shown in Fig. 49.

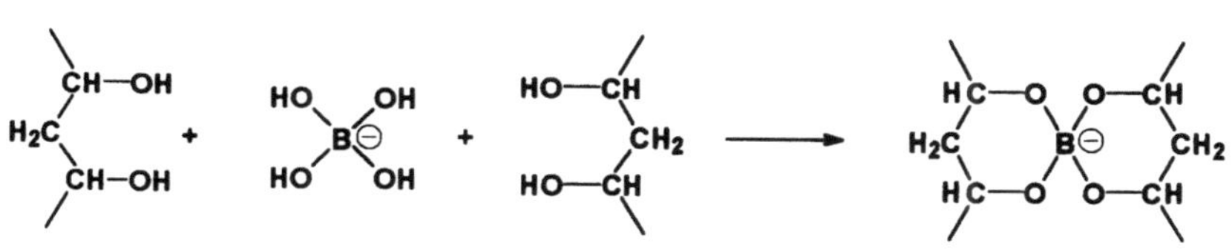

Fig. 47. Possible complex formation in aqueous poly(vinyl alcohol)/borate gels

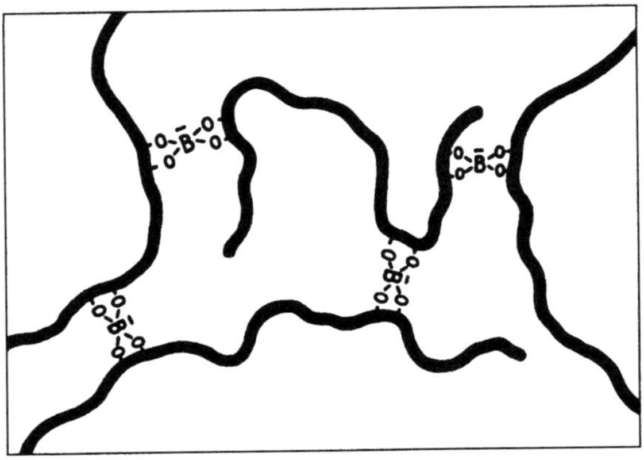

Fig. 48. Schematic view of a network formed by borate crosslinking of poly(vinyl alcohol). Reproduced from Makromol Chem [Ref. 131] by the courtesy of Hüthig & Wepf Verlag Publishers, Zug, Switzerland

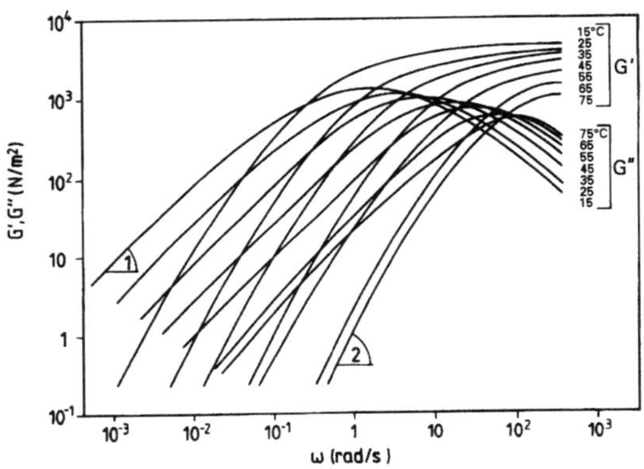

Fig. 49. Dynamic moduli for a PVA98.5/borax gel plotted vs angular frequency at different temperatures; PVA concentration 4%, borax concentration 2.5%. Reproduced, with permission of the author, from [7]

The high frequency range shows rubberlike behaviour, and the low frequency range liquidlike behaviour (slopes of 2 and 1 for G′ and G″, respectively). The height of the rubber plateau is strongly dependent on temperature and at 15 °C about five times as high as at 75 °C. Application of the time-temperature superposition principle (WLF-equation) in order to obtain a master-curve is not possible in the normal way. Nevertheless, Schultz and Myers [185] made use of this principle by introducing an extra vertical shift. However, a real mastercurve cannot be obtained in this way, because the structure of the polymer gel is temperature dependent. From the measurements reported by Beltman [7] and by Schultz and Myers [185], the crosslinking index was calculated with the aid of the Te Nijenhuis model [39–42] (for a crosslink functionality of 4) as a function of temperature (the crosslinking index yields the number of crosslinks in the system, which is equal to the number of borate molecules used to form the crosslinks). In this way, it is possible to calculate ΔH^o for the formation of crosslinks. For both series of measurements $\ln \bar{\gamma}_w$ is plotted vs 1/T and nearly parallel straight lines are obtained (see Fig. 50). The heights of the lines are quite different. This is caused by, among other things, a difference in

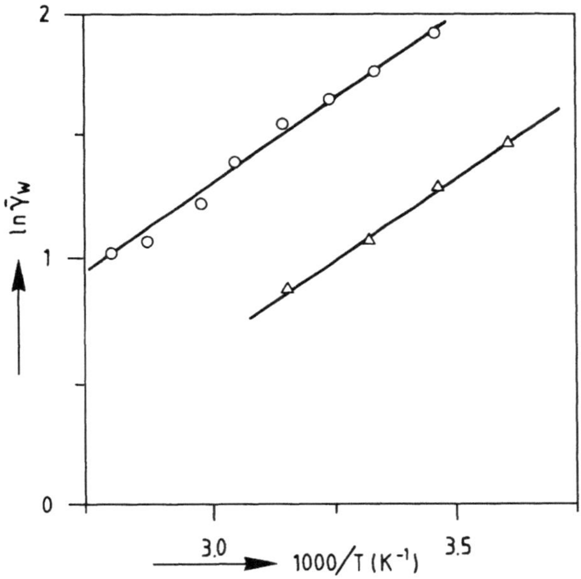

Fig. 50. Logarithm of weight average crosslinking index $\bar{\gamma}_w$, calculated from the rubber storage modulus of PVA/borate gels, plotted vs reciprocal temperature; ○ measurements by Beltman (PVA concentration 4 wt%; borate concentration 0.275 mol/l; [7]) △ measurements by Schultz and Myers (PVA concentration 4.4 wt%; borate concentration 8.7×10^{-3} mol/l) [185]

borate concentrations. In the Beltman case the borate concentration is so high that the amount used to form crosslinks is negligible. This is not the case with the results of Schultz and Myers [185] (for the highest values of G′ about 10% of added borate is used to form crosslinks). From Fig. 50 it appears that the crosslinking enthalpy ΔH° has nearly the same value in both cases: $\Delta H^\circ = -11.3 \pm 0.7$ kJ/mol (Beltman) and $\Delta H^\circ = -10.9 \pm 0.7$ kJ/mol (Schultz and Myers). These are rather low values, much lower than the values calculated for aqueous poly(vinyl alcohol) solutions (−37 to −168 kJ/mol, depending on the stereoregularity of poly(vinyl alcohol)) and also much lower than the values shown in Table 1 for atactic poly(vinyl alcohol). This means that the equilibrium of

$$\text{PVA} + \text{borate} \rightleftharpoons \text{crosslink} \tag{24}$$

is a dynamic equilibrium with a small heat of activation. This is the reason why the gels show liquidlike behaviour at low frequencies. Beltman [7] showed on photographs that cubes of gels became flat plates within half an hour at room temperature.

Schultz and Myers [185] measured the equilibrium shear modulus as a function of concentration of added borate. For borate concentrations of 2.5–10 mol/m^3, the modulus is approximately linear with the concentration of added borate. Above 10 mol/m^3 the modulus decreases. According to these authors, this is due to the electrostatic repulsion of the borate ions in solution, causing a retardation of further complexation.

3.4.2 More Recent Studies

Very recently Koike et al. [186] determined the viscoelastic behaviour of PVA/borate gels and concluded that it could be described by only one Maxwell-element (e.g. G' and G'' cross at the maximum in G'', the plateau value of G' at high frequencies is twice as high as the maximum in G'', and low frequency slopes on double logarithmic scales are 2 and 1 for G' and G'', respectively), its relaxation time depending on temperature and concentration. In Beltman's results (Fig. 49) G' and G'' did not cross at the maximum in G''; hence in his case the viscoelastic behaviour cannot be described by only one relaxation time. The relaxation times obtained by Koike et al. were approximately equal to the characteristic relaxation time of the slow mode calculated from dynamic light scattering. According to Koike et al. these results confirm the presence of dynamic coupling between concentration fluctuation and elastic stress. This is in agreement with the statement in the preceeding section that the crosslinks in PVA/borate are of a dynamic nature.

On the basis of the pH dependence of the complex formation between poly(vinyl alcohol) and borate (borax), Ochiai et al. [187] calculated the complex formation constants inclusive thermodynamic quantities for atactic poly(vinyl alcohol). Matsuzawa et al. [188] did the same for isotactic and atactic poly(vinyl alcohol). Data are given in Table 2. Although the results of both studies do not agree, the data by Matsuzawa et al. suggest that the reaction of borate ions with isotactically arranged adjacent hydroxyl groups is much easier than with two hydroxyl groups arranged syndiotactically. This is especially reflected by the differences in reaction entropy.

Recently the complex formation between poly(vinyl alcohol) and borate was studied with the aid of ^{11}Boron NMR. Sinton [191] concluded, in comparison with 2,4-pentanediol, to a didiol complex with a formation enthalpy of -36 kJ/mol. As model species Shibayama et al. [192] made use of 2-propanol and 2,4-pentanediol. It appeared that, first, a diol is needed for the formation of

Table 2. Thermodynamic parameters of the poly(vinyl alcohol)/borate didiol complex formation

Authors	PVA	$-\Delta G^\circ$ kJ/mol	$-\Delta H^\circ$ kJ/mol	$-\Delta S^\circ$ kJ/(mol·K)
Ochiai et al. [187]	a	9.1	52.3	0.15
Matsuzawa et al. [188]	a	9.5	38.5	0.091
Matsuzawa et al. [188]	i	11.1	51.4	0.135

a complex, second, that poly(vinyl alcohol) has a higher tendency to form complexes than the diol used (this is explained by the anchorage assistance of the hydroxyl groups of the poly(vinyl alcohol) chain, which stabilises additionally the complex product) and, third, that the commonly accepted 1:1 complex product with covalent bonds is not the most probable product. From ^{11}B NMR spectra it follows that the didiol complex is not present. The authors propose a model in which the Na^+ counterions form hydrogen bonds with hydroxyl groups of an adjacent poly(vinyl alcohol) molecule. However, in their model the functionality of the crosslinks formed in this way, would be 6 and that does not agree with later experiments [193], where it was concluded that the functionality is 4. Their ultimate conclusion of the composition of a crosslink is shown in Fig. 51. This means that the borate should be added in such a way that $[Na^+]/[B] = 1$ and that metaborate and perborate are better crosslinkers than borax. In the case of borax, extra sodium hydroxide has to be added. From their small angle neutron scattering studies, Shibayama et al. [193,194] concluded that the alkaline (borate) gels were stable over several months, and hence ageing apparently did not occur. This result is in strong contrast with hydrogels, where the scattering intensity at a constant q-value ($q = 2\pi \sin\theta/\lambda$) increased by a factor of more than 50 by ageing for two months at room temperature.

Cheng and Rodriguez [195] demonstrated that the addition of boric acid (H_3BO_3) to poly(vinyl alcohol) solutions has hardly any effect on the gelation properties. Addition of sodium hydroxide, by which $NaB(OH)_4$ is formed, results in gel formation. The maximum effect is observed for $[Na^+]/[B] = 1$; addition of more sodium hydroxide has no effect. Results are shown in Fig. 52, where $\log G'$ is plotted vs added sodium hydroxide: $\log G'$ rises linearly with added sodium hydroxide up to the point where the sodium/borate ratio is 1. From then on, the modulus is constant. This confirms the Shibayama model [193], where Na^+ is needed to form a crosslink. From the work of Kurokawa et al. [196] it became clear that the phase behaviour of the aqueous poly(vinyl alcohol)/borate system not only depends on concentration of polymer and borate, but also on the addition of alkali hydroxide and of indifferent electrolytes like sodium chloride.

From the temperature dependence of the storage modulus, Cheng and Rodrigues [195] conclude that the crosslinking enthalpy is 10.5 kJ/mol. With

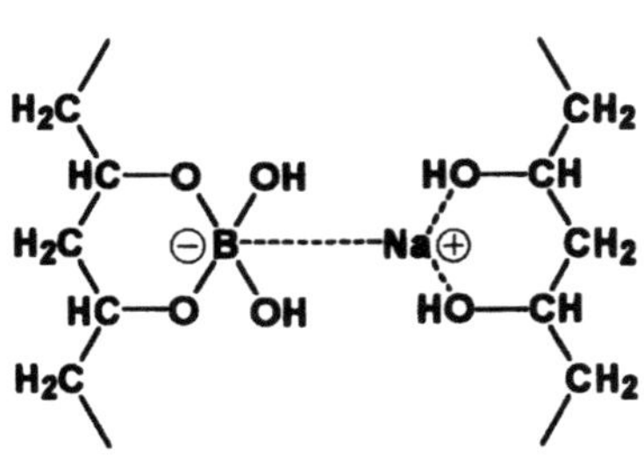

Fig. 51. Ionic bonding model of PVA-borax complex according to Shibayama et al. [193]

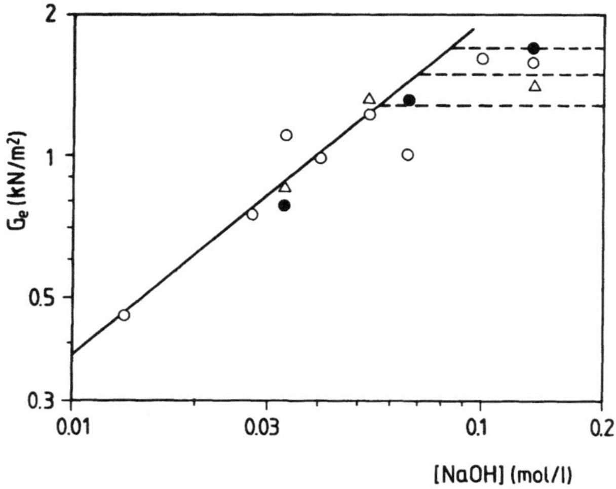

Fig. 52. Dependence of rubber storage modulus of PVA/borate gels on added sodium hydroxide; temperature 25 °C; PVA concentration 28.3 kg/m^3; initial borate concentrations in mol/m^3: ● 83.2; ○ 69.3; △ 55.9. Reproduced from data presented in [195].

increasing temperature an apparent increase is shown. However, this is probably a consequence of the dynamic nature of the crosslinks by which the rubber plateau is not extended to the frequencies used (see Beltman [7]): the "gel" becomes liquidlike. This could also be the explanation for the apparent increase of ΔH^o for sodium borate ratios smaller than 1 and probably not the consequence of extended crosslinks as the authors propose. The effect of other cations (Li^+, K^+ or Ba^{2+}) seems to be negligible in the formation of crosslinks [195] with respect to the effect of sodium ions.

According to Maerker and Sinton [197], linear viscoelastic behaviour of poly(vinyl alcohol)/borate gels is obtained at small deformation amplitudes only: in that case the complex viscosity is five times as high as the zero shear viscosity, whereas the complex viscosity measured at deformation amplitudes of 100% is equal to the zero shear viscosity.

Shibayama et al. [193] reported their measurements on the sol-gel transition of poly(vinyl alcohol)/borate complexes. They conclude that $\ln(c_{gel}\,Pb_o)$ plotted vs $1/T_m$ should yield a straight line (c_{gel} is the polymer concentration where the melting temperature is T_m; P is the degree of polymerisation and b_o is the concentration of added borate). From the fact that for constant P and b_o the plot $\ln c_{gel}$ vs $1/T_m$ is linear, the authors conclude that the functionality of the crosslinks is 4. The master curve they produced for poly(vinyl alcohols), varying in degree of polymerisation from 120 to 3230, polymer concentrations from 21 to 178 g/l, and added borate concentrations from 0.833×10^{-2} to 2.5×10^{-2} mol/l, is poor, because the authors did not take into consideration that borate is needed for the formation of crosslinks, and that instead $\ln(c_{gel}\,Pb)$ has to be plotted vs $1/T_m$. By calculating the amount of crosslinks with the

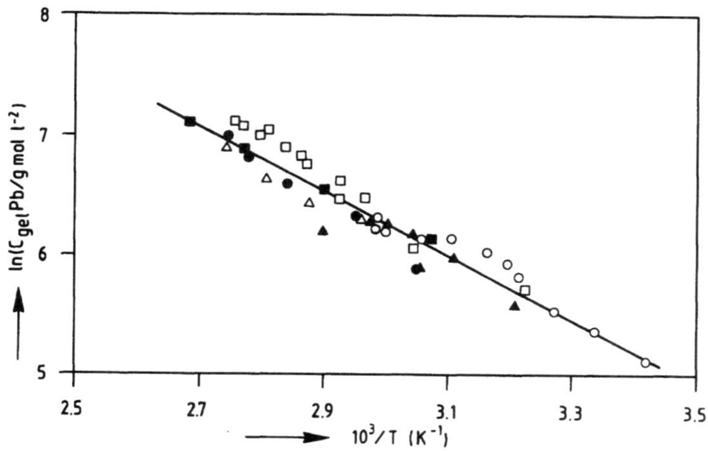

Fig. 53. $\ln(c_{gel}\,Pb)$ of PVA/borate gels of various compositions plotted vs reciprocal gel melting temperature, calculated from data presented by Shibayama et al. [193]

Symbol:	○	□	△	●	■	▲
mol wt (kg/mol):	8.8	13.2	78.8	113	142	13.2
added borate (mol/m^3):	25	25	8.33	8.33	8.33	16.7

authors' model [39–42, 127], it was possible to calculate the actual borate concentration b. Plots of $\ln(c_{gel}Pb)$ vs $1/T_m$ are improved in this way; however, calculations for PVA120 with a degree of polymerisation of 120 are worse (in most cases the calculated borate concentration is negative). If, instead of $P = 120$, use in made of $P = 200$, the results are in agreement with the other calculations. Results are shown in Fig. 53. The least squares fit yields

$$\ln(c_{gel}\,Pb) = 14.2 - \frac{2560}{T_m} \qquad (25)$$

from which it follows that $\Delta H = -21.3$ kJ/mol. This value is twice that calculated from measurements by Beltman [7], Schultz and Myers [185] and Cheng and Rodriguez [195]. This difference could be the result of the difference in network state: the high value was obtained at the gel point, whereas the low values were obtained far beyond the gel point, where the network formation is retarded by the presence of the crosslinks.

3.5 Gelation by Addition of Congo Red

3.5.1 Introduction

Much stiffer gels are obtained by addition of specific hydrogen bonding agents, which add crosslinks to the microcrystalline network of a "normal" gel, such as resorcinol or congo red [143]. The crosslinking mechanisms of these crosslinkers are entirely different. This is evident from the following experimental results:

- gelation of poly(vinyl alcohol) is counteracted by the presence of rhodanide ions or urea; in accordance with these phenomena the blue coloration caused by I_2/KI disappears;
- gelation of poly(vinyl alcohol) with the aid of congo red is thermoreversible, just as the much stronger blue coloration; both effects are hardly affected by rhodanide ions or urea;
- gelation of poly(vinyl alcohol) with resorcinol, which is very strong, is not counteracted by blue coloration with I_2/KI; this proves that no helical structures are formed during the gelation of poly(vinyl alcohol).

Experiments by Dittmar and Priest [198] show the side action of congo red only: if congo red in aqueous poly(vinyl alcohol)/congo red gels is chemically reduced in situ, no decrease of the modulus takes place, whereas addition of chemically reduced congo red to poly(vinyl alcohol) solutions does not cause gelation. Apparently congo red catalyses the gelation process, but it is not needed anymore if the gel is formed. Small ordered regions were observed in aqueous poly(vinyl alcohol)/congo red gels by Dittmar and Priest [198] and by Okada and Sakurada [199, 200]; these are absent in aqueous poly(vinyl alcohol)/borax gels.

3.5.2 Complexation in Aqueous Congo Red Solutions

Shibayama et al. [201, 202] studied the complexation of poly(vinyl alcohol) and congo red in aqueous solutions by means of viscometry, small-angle neutron scattering and small-angle X-ray scattering. The chemical structure of congo red is shown in Fig. 54. Its molecular weight is 696.68. According to the authors, congo red is a suitable candidate for a complexation reagent with poly(vinyl alcohol) because it has: a) a rather linear and plain molecular architecture; b) azo and amino groups capable of hydrogen bonding; c) high ionization in aqueous

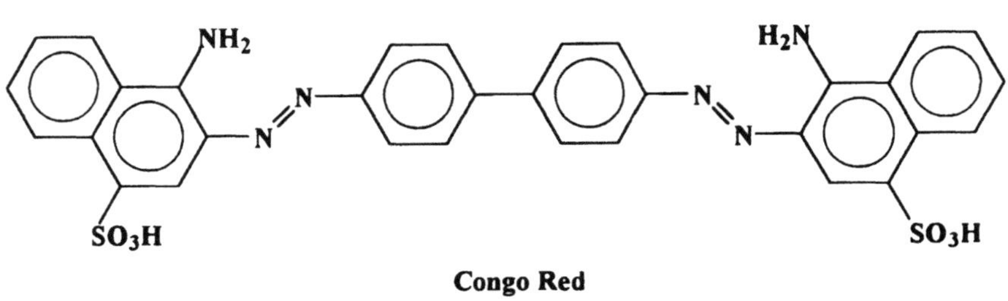

Fig. 54. Structure of congo red

solutions due to the presence of sulfate groups. Complexation might lead to weak and strong crosslinks, as shown in Fig. 55. A remarkable sol-gel transition behaviour is observed with the aid of the tilted-test-tube method. Results are shown in Fig. 56. At poly(vinyl alcohol) concentrations around 30 kg/m^3 there are sol-gel-sol-gel transitions with increasing congo red concentration between 1 and 30 kg/m^3. It is clear that two types of gel exist. At 50 °C, the gel region is separated into two domains and at 60 °C the gels are completely dissolved. From the polymer concentration dependence of the gelation temperature, the enthalpy of crosslink formation was calculated; it was concluded that two types are present: weak crosslinks with $\Delta H = -21$ kJ/mol and strong crosslinks with $\Delta H = -84$ kJ/mol. When the pH was raised by the addition of sodium hydroxide the anomalous sol-gel transitions were not observed; this is shown in Fig. 57.

From their SANS and SAXS studies, Shibayama et al. [202] propose that the anomalous behaviour of sol-gel transitions finds its origin in the electrorepulsive interactions and electrostatic screening effects of congo red, the

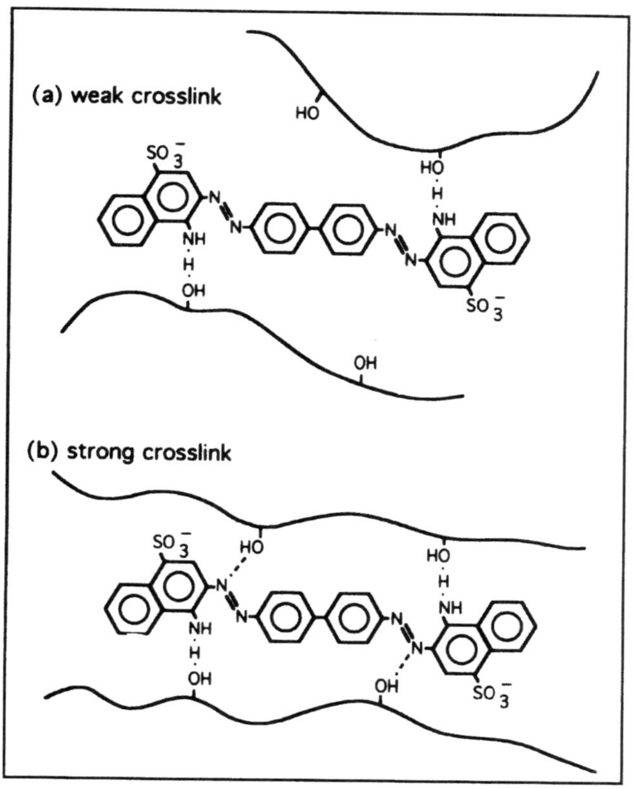

Fig. 55. Schematic representation of the weak and strong crosslink structures in congo red-poly(vinyl alcohol) complexes. Reproduced from Macromolecules [Ref. 201] by the courtesy of the authors and of The American Chemical Society

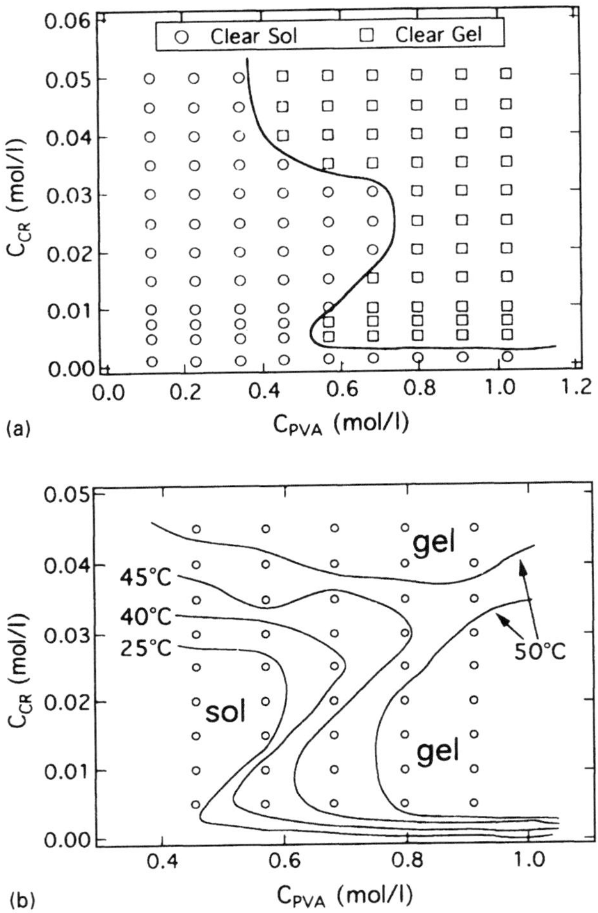

Fig. 56a, b. Phase diagrams showing the sol-gel transition of poly(vinyl alcohol)/congo red complexes in aqueous solutions aged for 120 h: **a** measured at 20 °C; **b** measured at various temperatures, as indicated. Reproduced from Macromolecules [Ref. 201] by the courtesy of the authors and of The American Chemical Society

predominant effect depending on the congo red concentration, giving rise to intrachain or interchain crosslinks. A schematic diagram is given in Fig. 58. In this picture the crosslink functionality is equal to 4. However, if the four hydrogen bonds tie four polymer molecules, then the crosslink functionality would be 8. Hence, it is reasonable to suppose that the average functionality might have a value between 4 and 8.

3.5.3 Viscoelastic Properties and Melting Enthalpies

Poly(vinyl alcohol)/congo red gels have quite different viscoelastic properties with respect to poly(vinyl alcohol)/borax gels. The poly(vinyl alcohol)/congo red

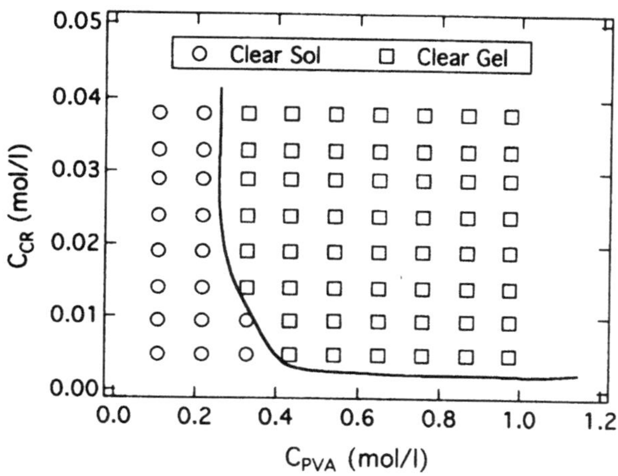

Fig. 57. Phase diagram showing the sol-gel transition of aqueous poly(vinyl alcohol)/congo red complexes at high pH (in comparison with Fig. 56a): [NaOH] = 0.095 mol/l. Reproduced from Macromolecules [Ref. 201] by the courtesy of the authors and of The American Chemical Society

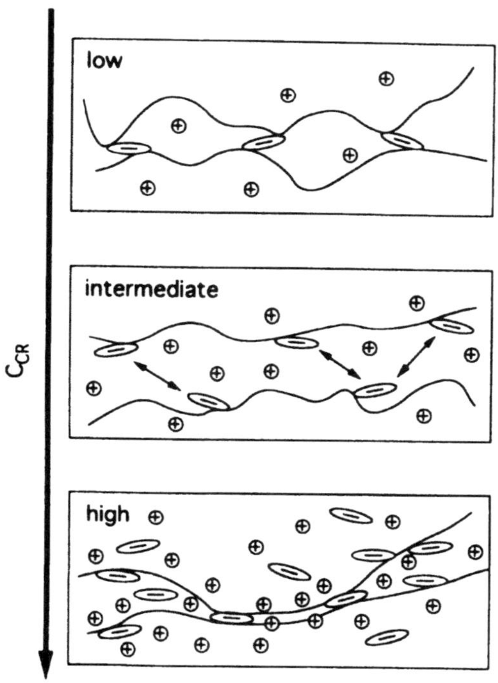

Fig. 58. Schematic representation of crosslink formation of poly(vinyl alcohol) (*solid lines*) with congo red ions (*ovals with minus signs*). The counterions are denoted by *circles with plus signs.* Reproduced from Macromolecules [Ref. 201] by the courtesy of the authors and of The American Chemical Society

gels have a long-drawn rubber plateau, which is shown by Beltman and Lyklema [7, 203] (see Fig. 59): G′ is almost constant over a frequency range of six decades; G″ is also almost constant, but much smaller ($\tan\delta = 0.023$). In order to determine whether the gels behave entropically elastically, Beltman and

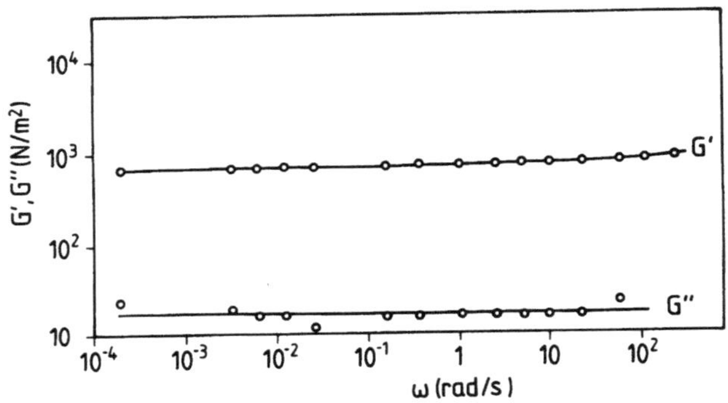

Fig. 59. Dynamic mechanical moduli of a PVA98.5/congo red gel after 200 h of ageing at 25 °C, plotted vs angular frequency; PVA concentration 4%; congo red concentration 2.4%; $\bar{M}_w$ PVA = 78 kg/mol. Reproduced from Discuss Faraday Soc [Ref. 203] by the courtesy of the authors and The Royal Society of Chemistry, Cambridge, UK

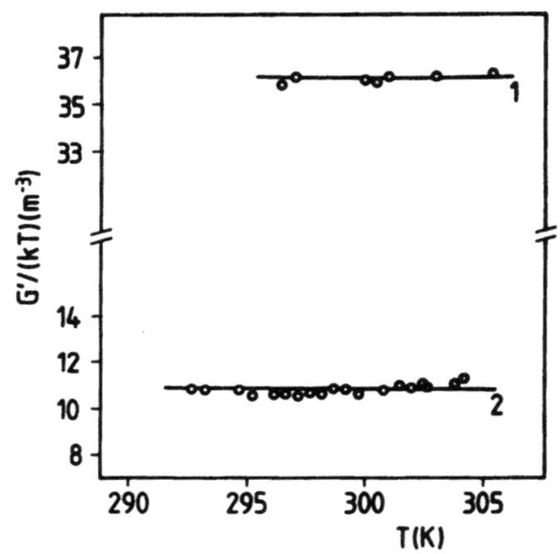

Fig. 60. Rubberlike behaviour of two PVA98.5/congo red gels; $\bar{M}_w$ PVA = 78 kg/mol; PVA concentration 4%; congo red concentration: *line 1* – 3.2%; *line 2* – 1.6%. Reproduced from Discuss Faraday Soc [Ref. 203] by the courtesy of the authors and The Royal Society of Chemistry, Cambridge, UK

Lyklema [7, 203] measured the temperature dependence of the equilibrium storage modulus and plotted the reduced storage modulus, i.e. G′/kT, vs temperature. They took into account that the structure of the gels changes with a temperature change: subsequent measurements at different temperatures were taken as a function of time and extrapolated to the moment of the temperature change, as was illustrated in Sect. 2 on poly(vinyl chloride). From Fig. 60 it appears that G′/kT has a constant value in the temperature range from 20 to 32 °C, in accordance with entropy elasticity. From the concentration and molecular weight dependences of the gel melting temperature, ΔH_m and ΔH_x of crosslink formation can be calculated [160], and also, subsequently, the number of polymer molecules, n, joined in a crosslinks, because $n = \Delta H_m/\Delta H_x + 1$. Experimental results from Beltman [7] are shown for poly(vinyl alcohol) gels with congo red as well as resorcinol. Results of calculations from ln c vs $1/T_m$

Table 3. Crosslinking enthalpies and number of polymer molecules joined in a crosslink for several poly(vinyl alcohol) gels

Gel	ΔH_m (kJ/mol)	ΔH_x (kJ/mol)	n
PVA/2% CR	− 81.8	− 30.3	3.7
PVA/6% Res	− 166	− 169	2.0

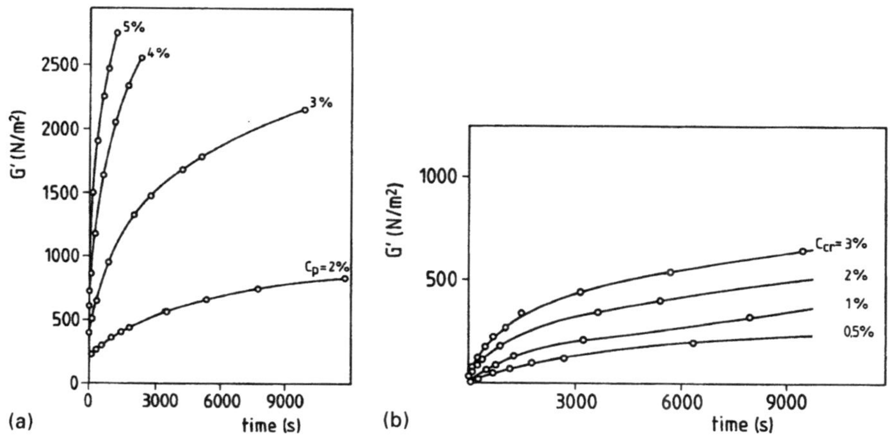

Fig. 61. Time dependence of the storage modulus of PVA98.5/congo red gels at 25°C for various PVA and congo red concentrations, as indicated; $\bar{M}_w$ PVA = 78 kg/mol; a) congo red concentration 5 wt%; b) PVA concentrations 5 wt%. Reproduced with permission of the author from 7.

and ln M vs $1/T_m$ plots are shown in Table 3. From these results, it appears, first, that the heat of formation of a crosslink in poly(vinyl alcohol)/resorcinol gels is twice as high as in poly(vinyl alcohol)/congo red gels and, second, that both are much higher than for poly(vinyl alcohol)/water gels (− 37 kJ/mol). The melting temperatures are also much higher. Apparently the crosslinks in the congo red gels are much more extended than in the aqueous gels. Their number in the congo red gels is also much larger, as the gels are much stiffer. The average functionalities of the crosslinks seem to be 7.5 and 4, for congo red and recorcinol, respectively.

3.5.4 Gelation Kinetics

Beltman and Lyklema [7, 203] have reported their investigations of the kinetics of the gelation of poly(vinyl alcohol)/congo red systems from measurements of the storage modulus as a function of time for various poly(vinyl alcohol) and congo red concentrations. Results of those measurements are shown in Fig. 61.

The authors assume that the gelation kinetics can be expressed as

$$\lim_{t \to 0} \frac{dG'}{dt} = k\,[PVA]^p\,[CR]^q \tag{26}$$

and conclude that $p = 4$ and $q = 1$. Their conclusion is that the rate of build-up of the network in the early stages of the network formation is dependent on $[PVA]^4$ and on [CR]. Two comments are necessary here, however. First, in general it is true that the best way to determine the rate equation is to measure the system properties before the network build-up is retarded by the network itself. Hence, taking the limit of dG'/dt for t tending to zero is correct. However, the determination of this limit is rather inaccurate, because the temperature quench is time dependent. Second, especially in the early stages of the network build-up, G' is not proportional to the number of crosslinks formed. Te Nijenhuis [127] calculated the number of crosslinks as a function of time and extrapolated it to the gel point, where $\bar{\gamma}_w = 2/(f-2)$. This gave $p = 3.8 \pm 0.2$, which means that the crosslink functionality would be equal to 7.6 ± 0.4 on average. This is in close agreement with the thermodynamic data obtained by Beltman, shown in Table 3: he concluded that, on average, 3.7 ± 0.3 polymer molecules are joined in a crosslink. Hence, the crosslink functionality would be 7.4 ± 0.6. From these results we might conclude that the average functionality of the crosslinks in poly(vinyl alcohol)/congo red gels is approximately 7.5 ± 0.5. This is in agreement with the suggestion made at the discussion of Fig. 58.

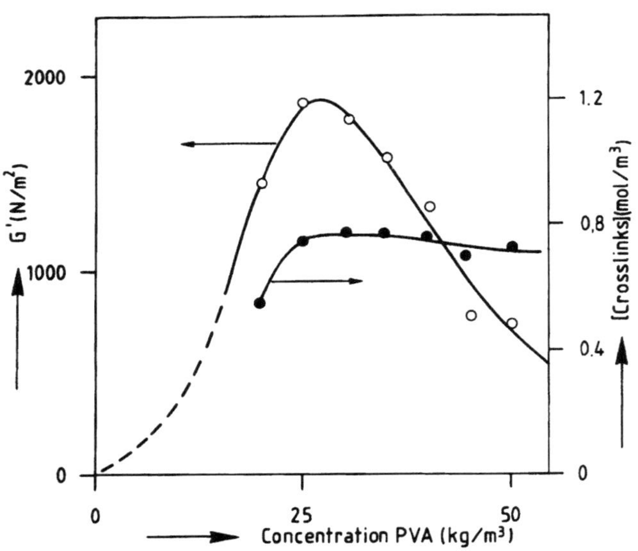

Fig. 62. Dependences of the storage modulus and concentration of crosslinks in PVA88/CR gels on the PVA concentration; congo red concentration 2%; $\bar{M}_w$ PVA = 78 kg/mol (calculations after data presented in [7])

In Sect. 3.4 it has already been mentioned that an increase in the amount of borate initially causes an increase in the storage modulus, but that eventually the result is a strong decrease. For poly(vinyl alcohol)/congo red gels of constant congo red concentration an increase of the poly(vinyl alcohol) concentration causes an initial rise of the storage modulus, but at a certain concentration the storage modulus strongly decreases again [7] (see Fig. 62). From the calculation of n_{cl}, it appears that the number of crosslinks rises initially, but this number becomes almost constant at a certain poly(vinyl alcohol) concentration. It seems that all congo red is used for the formation of crosslinks and that extra poly(vinyl alcohol) does not affect the number of crosslinks (the decrease of the modulus is the consequence of the decrease of the number of crosslinks per molecule and subsequently of the increase of the fraction of dangling ends and, probably, also the sol fraction). From a calculation it appears, however, that the number of crosslinks is $4.2 \times 10^{23}/m^3$ and the number of congo red molecules $3.5 \times 10^{25}/m^3$. For each crosslink 82 molecules of congo red are available. It is improbable that this excess is needed, unless it is the consequence of a protolysis equilibrium.

3.6 Conclusions

The gelation process in solutions of poly(vinyl alcohol) in water and other solvents is the result of crystallinity and hydrogen bond formation. The process is time and temperature dependent and it seems that a degree of saponification of the mother polymer of poly(vinyl alcohol) of at least 90% is needed for gel formation to occur. This means that the saponification process proceeds more or less randomly. This is known from acid catalysed hydrolysis, whereas in alkaline circumstances the saponification proceeds more or less via a zipper mechanism. As the gelation process of poly(vinyl alcohol) is determined by the syndiotacticity of the polymer chains, the stereoregularity of the mother polymer is of primary importance for the gelation behaviour of poly(vinyl alcohol) solutions. Although viscoelastic properties and small-angle neutron scattering of these gels have been reported, unfortunately a combination of these techniques has not been the subject of any study. Hence, it is not possible to calculate the functionality of crosslinks involved. For the gelation process in poly(vinyl chloride) solutions this proved to be a promising method to calculate the crosslink functionality.

Gels are also obtained by the addition of borate to aqueous poly(vinyl alcohol) solutions. The tetrafunctional crosslinks, consisting of didiol bonds with borate, are of a dynamic nature, with time constants of the order of 0.01 to 10 s, depending on temperature. The heat of crosslink formation varied from -10 to -20 kJ/(mol of crosslinks), which is much less than the values obtained in the gelation process of aqueous poly(vinyl alcohol) solutions, where it varied from -37 to -168 kJ/mol, depending on the syndiotacticity of the poly(vinyl alcohol) used.

The gelation in congo red solutions is the result of hydrogen bonding between the congo red molecules and hydroxyl groups of several molecules of poly(vinyl alcohol). From the kinetics of the gelation process of aqueous solutions of poly(vinyl alcohol) and congo red it appeared that the generally accepted crosslink functionality of 4 is not correct: an average functionality of 7.6 ± 0.4 seems to be more realistic.

4. Poly (methyl methacrylate)

4.1 Introduction

It is well-known that mixtures of solutions of isotactic and syndiotactic poly(methyl methacrylates) give rise to gel formation. However, the viscoelastic behaviour of poly(methyl methacrylate) gels has been the subject of only a few studies, most of them by Pyrlik et al. [205–208] and Rehage et al. [209–211]. Many publications deal with studies of the aggregation and gelation behaviour, especially of mixtures of isotactic and syndiotactic poly(methyl methacrylate) (i-PMMA and s-PMMA). Reviews on this subject have been written by Spěváček and Schneider [212] and by Spěváček [213]. The mechanism of the stereocomplex formation is not yet clear: often the largest effects were obtained at i/s ratios of 1/1 to 1/2. Measurements by NMR [214], calorimetry [215, 216] and GPC [217] are in favour of the ratio i/s = 1/1.5, whereas Schomaker et al. [218–220] and Brinkhuis [221] have shown that this ratio is 1/2 on a molecular scale. The ratio i/s = 1/1 seems to be an apparent one and it seems to be impossible at this moment to resolve the exact value of the ratio i/s (Spěváček, private communication).

Upon mixing relatively concentrated solutions of i- and s-PMMA (a few percentages), gel formation takes place almost instantaneously: the melting temperature is sharp and strongly dependent on the stereoregularity of s-PMMA in particular [222]. No gel formation occurs if the molecular weight is low (below 1000) [223, 224]. Gels, or at least aggregates, are also formed in mixtures of i-PMMA and a-PMMA [225]. Moreover, self aggregation takes place in solutions of s-PMMA [103, 212, 213, 226–237], of i-PMMA [103, 231–233] and even of a-PMMA [212, 213, 238, 239]. Research groups in Groningen [218–220, 227, 240–247] and Prague [103, 212–215, 226, 229–237, 239, 247–255] have carried out many studies dealing with the mechanism of these phenomena.

4.2 Parameters which Determine Aggregation and Gelation Behaviour

4.2.1 Ratio of i-PMMA and s-PMMA in the Mixture

In many publications, the maximum effect is reported to be i/s = 1/2, although also other ratios, 2/1 and 1/1, were found [205, 214, 215, 219, 220, 222, 238, 256–258]. From DSC and NMR measurements it appears that i/s = 1/1.5 to 1/2 is most probable [214, 215]. Imperfect stereoregularity and improper measurement techniques might be the origin of these discrepancies [212, 224, 253, 256].

4.2.1.1 Imperfect Stereoregularity

On the basis of viscometry and optical density in DMF and acetone solutions, Vorenkamp et al. [244] have shown that both components are always present in the ratio i/s = 1/2 on the level of monomeric units. In the complex itself the ratio i/s is fixed; for the complex formation a minimum regular sequence length is needed; relatively bad stereoregularity results in the presence of many too short regular sequences; a shift to smaller i/s ratios takes place if the stereoregularity of the s-PMMA used is insufficient [212, 214]: much of the polymer is not present in the stereocomplex in that case. Overall calculations of the i/s ratio yield wrong values, especially if the sequence length needed (dependent on the solvent) is relatively large. Hence, the overall mass ratio of both components in the complexed material cannot be a measure of the real stoichiometric ratio in the complex on the monomeric level [218]. In many cases, on using samples of low regularity an apparent stoichiometric ratio is found close to i/s = 1/1. According to Schomaker et al. [218–220], this is the consequence of the formation of contacts between two complementary chains at the very start of the aggregation process in a ratio i/s = 1/1. Subsequently, these complex nuclei can grow to i/s = 1/2 on the monomeric level. In samples of low stereoregularity only a few complexed sequences can be formed. If the concentration is higher than the minimum overlap concentration, then thermodynamic equilibrium will never be reached, due to the fast immobilisation of polymer chains. In addition, the work of Brinkhuis [221] has to be mentioned. This author studied Langmuir-Blodgett films of the stereocomplex and also concluded that on the monomeric scale the ratio i/s = 1/2.

With the aid of differential scanning calorimetry, Vorenkamp et al. [244] have shown that annealing results in a slow shift of the ratio i/s from 1/1 to 1/2. An example of this shift is shown in Fig. 63, where, for increasing annealing time, the heat of complex formation, ΔH_c, is shown as a function of x_s, i.e. the molar fraction of s-PMMA [218]. The maximum in the curve increases in the course of

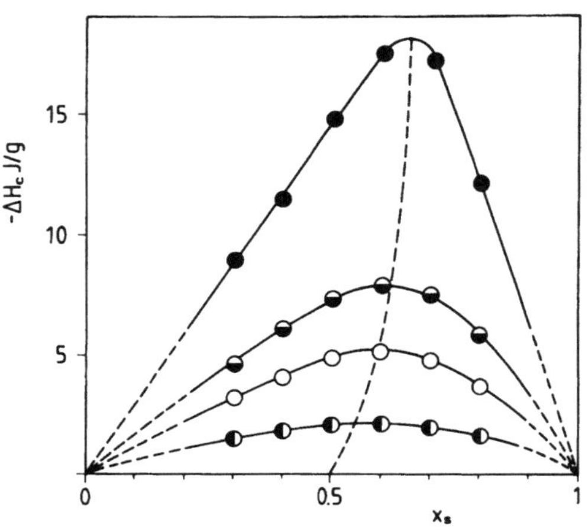

Fig. 63. Heat of complexation as a function of basemole fraction s-PMMA, x_s, evolved within: (◐) 100 s, (○) 200 s, (◒) 300 s, (●) 1 h after mixing 2.5 kg/m^3 DMF solutions of i- and s-PMMA at 30 °C. Reproduced, with permission of the author, from [218]

time and shifts from i/s = 1/1 to i/s = 1/2. If the circumstances for complex formation are favourable, i.e. if the concentration is higher than the minimum overlap concentration ($c > c^{\star}_{min}$) and if the critical sequence length is small (dependent on the solvent), immobilisation is attended with gelation of the system; in that case, the apparent i/s ratio is always 1/1 [205–211], because in the first instance the complex formation is determined by the chance that two chains meet.

The structure of the stereocomplex has been the subject of many studies [218, 244, 245, 247, 253, 256, 259–261]. Kusuyama et al. [262] suggested that the structure of the stereocomplex is closely related to the 10/1 double stranded helix of crystalline i-PMMA. Spěváček and Schneider [253] showed that their results of an NMR study could be explained with this model. On the basis of X-ray diffraction patterns and energy calculations, Bosscher et al. [247] came to the conclusion that the stereocomplex consists of a double-stranded helix in which an i-chain with a 30/4 helical conformation is surrounded by an s-chain with a 60/4 helical conformation. On the other hand, on the basis of fibre diffraction patterns of the stereocomplex, Schomaker et al. [218–220] suggest "as a starting point for further investigations" a two-state rotationally disordered 9/1 double helix, having an asymmetric unit consisting of 1 i-unit and 2 s-units.

4.2.1.2 Improper Measurement Techniques

The measurement techniques used are not always well correlated with the i/s ratio in the complex (e.g. osmometry, viscometry, turbidimetry, light scattering measurement); on the other hand, proton NMR seems to be able to distinguish properly between the associated and nonassociated fractions [212]. For this reason, high resolution proton NMR is a good tool to determine the stoichiometry of the stereocomplex [214]: it yields i/s = 1/1.5.

4.2.2 Minimum Sequence Length

For the stereocomplexation of PMMA, a minimum sequence length is required [209, 214, 223, 224], in general 8 to 10 monomeric units are needed, in some cases only 3 (CCl_4, CD_3CN e.g.) or 4 to 5 (toluene). A minimum sequence length also exists for self aggregation of s-PMMA, equal to 9 monomeric units (toluene, *o*-dichlorobenzene, butylacetate) [212, 235, 237].

4.2.3 Thermal Stability

Although the solvent plays an important role, in many cases the stereoregularity of the s-PMMA is the limiting factor; in general the gel melting temperature

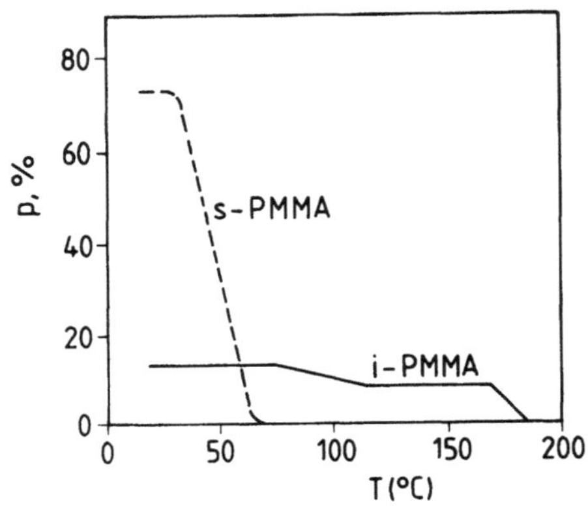

Fig. 64. Temperature dependence of the fraction of associated monomeric units for solutions of i-PMMA (——) and s-PMMA (- - - -) in *o*-dichlorobenzene. Reproduced from Adv Colloid Int Sci [Ref. 212] by the courtesy of the authors and of Elsevier Science-NL, Sara Burgerhartstraat 25, 1055 KV Amsterdam, The Netherlands

decreases with decreasing stereoregularity of the s-PMMA used [209, 215, 222, 256], because smaller and less regular sequences are present to a higher extent if the stereoregularity is worse. The upper limit is 120–130 °C. Most stable are gels in carbon tetrachloride and toluene; this is in agreement with the minimum sequence length, which is very low in these solvents [212]. At room temperature self aggregation occurs much more in s- than in i-PMMA, but in i-PMMA these aggregates are much more stable than in s-PMMA. This follows clearly from Fig. 64, where the temperature dependence of the fraction of associated monomeric units, p, is plotted vs temperature [103, 212, 230]. The melting temperatures of ordered structures formed in solutions of s-PMMA, of i-PMMA and of the stereocomplex (i/s = 1/2) are around 50, 160 and 120 °C, respectively. Most i-PMMA is aggregated in butylacetate at temperatures below 0 °C [231].

4.2.4 Molecular Weight Dependence

The effect of molecular weight has been a subject of several studies [219, 220, 223, 224, 258]; on the one hand a minimum molecular weight of M = 1000 (i.e. 10 monomeric units) has been reported [223, 224], on the other hand a minimum molecular weight of M = 6000 (i.e. 60 monomeric units); the value of 6000 is in agreement with the model of Bosscher et al. [247], in which a 60/4 helix of s-PMMA is proposed.

4.2.5 Effect of Solvent

Classification of solvents via minimum sequence length, q, was first done by Spěváček and Schneider [214]. Data by Challa et al. [242, 244], mainly based on viscometry, also show the existence of a solvent dependent minimum sequence length needed to form aggregates. For strongly complexating solvents, q = 3, whereas for weak complexating solvents q = 10; chloroform is a non-complexating solvent. This terminology was also used by Spěváček and Schneider [212], where it was shown that both classifications are identical, with the exception of toluene. According to these authors [212, 244] the solvent may affect the aggregation of PMMA in two ways: first, by a change of the conformational structure of stereoregular sequences via promotion or hindering of sterically complementary interactions and second, via hindering or promotion of the stereoassociation by specific interactions with certain functional groups (e.g. the ester groups), for it was shown by Spěváček and Schneider [212, 250] that the ester groups are involved in the complex formation. Later it was proved that both the α-CH_3 group of s-PMMA and the methyl ester group of i-PMMA play an important role [212, 243, 245, 246]. Moreover, specific polymer-solvent interactions leading to polymer-solvent complexes were shown from NMR and IR studies of s-PMMA self aggregation in mixed solvents [251, 252].

4.2.6 Kinetics of Aggregation

In general, the kinetics of aggregate formation are very fast when long syndiotactic sequences are present; subsequently the process slows down and the amount of aggregation increases with the aid of the smaller sequences; in carbon tetrachloride and acetonitrile, especially, the process is extremely fast. In other cases, this occurs only at the beginning of the ageing process. The rate increases strongly with decreasing temperature (the apparent overall activation enthalpy is around 200 kJ/mol, i.e. a factor of 13 in rate per 10 K temperature difference), somewhat dependent on the solvent [213]. From NMR measurements, it appears that the intramolecular rearrangements to crystal-like domains are very fast (primary aggregation), after which subsequent aggregation to larger structures takes place, as appears from the slow increase of the turbidity of solutions. This is shown in Fig. 65 [212, 232].

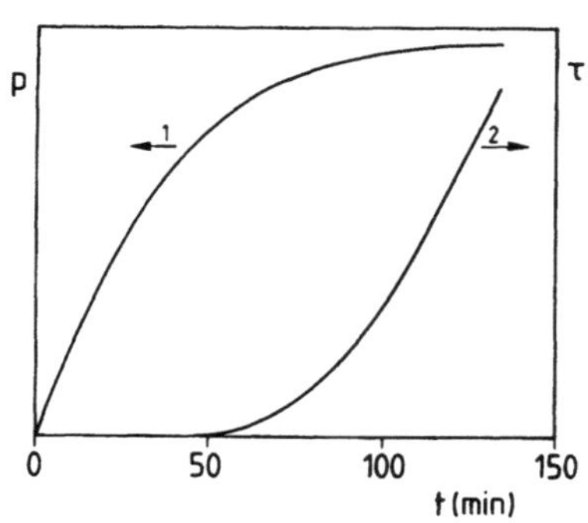

Fig. 65. Schematically shown temperature dependence of the fraction of associated monomeric units (p) and of turbidity (τ) for s-PMMA (s = 0.915) in butyl acetate; on molecular scale the process is very fast; subsequent aggregation needs more time. Reproduced from Adv Colloid Int Sci [Ref. 212] by the courtesy of the authors and of Elsevier Science-NL, Sara Burgerhartstraat 25, 1055 KV Amsterdam, The Netherlands

4.3 Viscoelastic Behaviour of PMMA Gels

4.3.1 Mixtures of i-PMMA and s-PMMA

Viscoelastic measurements on gels of mixtures of i-PMMA and s-PMMA were reported by Rehage et al. [205–208, 210]. They made use of i-PMMA ($\bar{M}_v = 230000$, 93% isotactic dyads) and s-PMMA ($\bar{M}_v = 50000$, 72% syndiotactic dyads) and made mixtures in a mass ratio of 4:5 in order to obtain an overall ratio i/s = 1/1. However, it has to be emphasized that, in general, this will not be the ratio of the regular sequences. Two solutions of 5.6% and 12.6% were made in *o*-xylene. From their published results interesting plots can be constructed. In Fig. 66 the storage modulus and log storage modulus of a 12.6% solution are plotted vs log ageing time for temperatures varying from 70 to 90 °C. The temperature appears to be an important parameter in the gelation process. This is also the case for the concentration, as becomes clear from Fig. 67 for 5.6% and 12.6% solutions. For prolonged ageing, the frequency dependences of the storage modulus and the loss modulus are weak [208]. This is shown in Fig. 68, where the storage modulus measured at various temperatures is plotted vs frequency in a window of five decades. It seems that the gels are behaving rubber elastically with a crosslinking density strongly dependent on temperature. Measurements on gels with different temperature histories are very interesting, as shown in Fig. 69, with ageing results (G′ vs $\log t_a$) for a 12.6% solution, which as usual was preheated at 145 °C, and also for an analogous

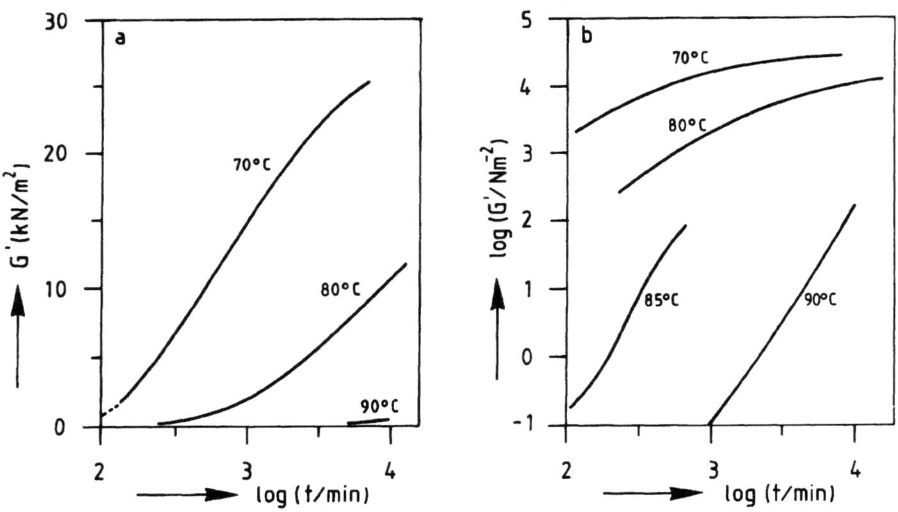

Fig. 66a, b. Storage modulus and log storage modulus of a 12.6% solution of PMMA (i/s = 1/1) in *o*-xylene plotted vs log ageing time, at various temperatures; ω = 1.57 rad/s. Data collected from figures shown in [205–208]

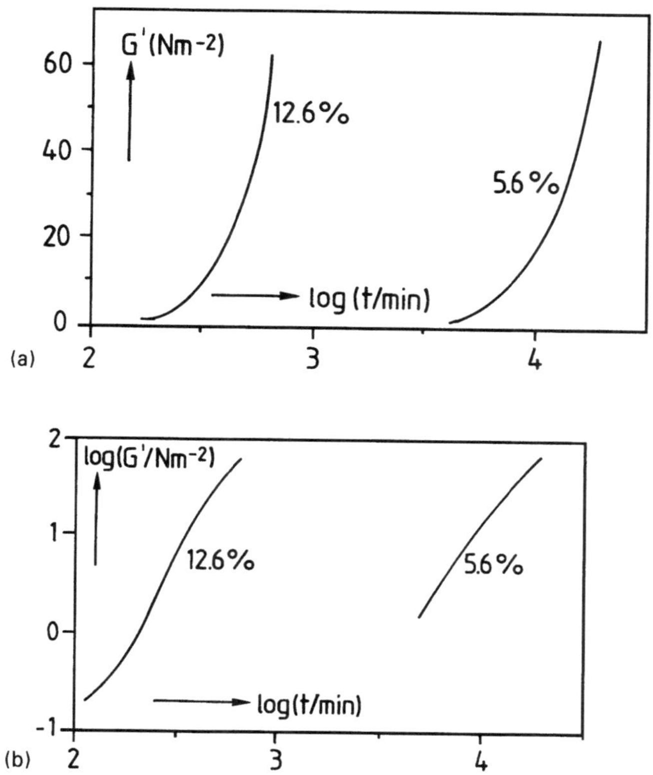

Fig. 67a, b. Storage modulus and log storage modulus of two solutions of PMMA (i/s = 1/1) in *o*-xylene (5.6% and 12.6%) plotted vs log ageing time; ageing temperature 80 °C; $\omega = 1.57$ rad/s. Data collected from figures in [205–208]

solution which was preheated at 120 °C before quenching to 80 °C. The solution that was preheated at 120 °C shows a higher tendency to form gels than the solution which was preheated at 145 °C. Although a temperature of 120 °C is sufficiently above the melting temperature of the gels, apparently, crystal or complex nuclei are still present, so that upon quenching to 80 °C a very fast growth results in gel formation. Apparently, this is not the case, or at least much less for solutions which were preheated at 145 °C. At this temperature nuclei are destroyed, so that subsequent ageing only proceeds after new nuclei have been formed. As 145 °C is the boiling point of *o*-xylene, it was not checked whether this temperature is high enough to erase completely the thermal prehistory. It might be done by preheating the solution under pressure. From Fig. 69b it appears that the slopes of the log G′ vs log t_a plots do not differ very much for both systems, but that the heights differ by a factor of almost 100. These measurements are a convincing and extremely sensitive method to check the erasure of the thermal prehistory.

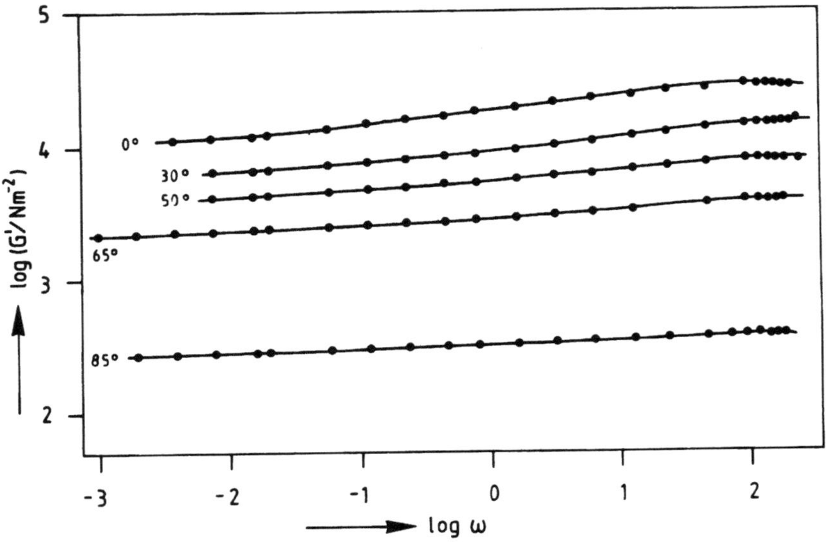

Fig. 68. Log storage modulus of a 5.6% solution of PMMA (i/s = 1/1) in *o*-xylene plotted vs log angular frequency for several ageing temperatures, varying from 0 to 85 °C. Reproduced from Colloid Polym Sci [Ref. 208] by the courtesy of Steinkopff Verlag Darmstadt, FRG

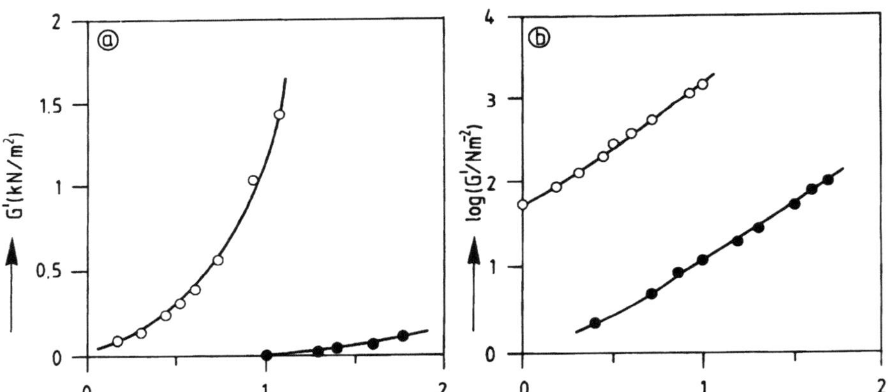

Fig. 69a, b. G′ and log G′ plotted vs log ageing time for a 12.6% solution of PMMA (i/s = 1/1) in *o*-xylene; ageing temperature 80 °C; dissolution temperatures: (○) 120 °C and (●) 145 °C. Constructed from data presented in [207]

In order to determine the dependence of the gel elasticity on the concentration of i/s = 1/1 PMMA mixtures without a variation in polymer concentration, Pyrlik and Rehage [208] made solutions in *o*-xylene of various mixtures of a-PMMA and i/s = 1/1 PMMA mixtures with an overall concentration of 5.6%. Results are shown in Fig. 70. The storage modulus decreases strongly with

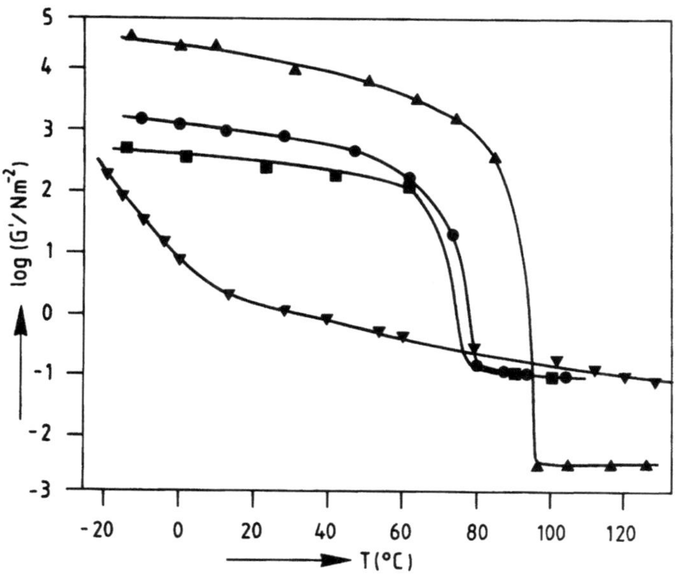

Fig. 70. Temperature dependence of log G' for 5.6% solutions of mixtures of a-PMMA and (i/s = 1/1)-PMMA in *o*-xylene; fractions of a-PMMA: (▲) 0%; (●) 60%; (■) 80%; (▼) 100% (constructed after results presented in [208])

decreasing concentration of the i/s mixture. Most striking is the tendency of the a-PMMA solution to form a gel, especially at temperatures below 0 °C. The i/s gels all melt around 80–100 °C.

A question arises in considering the fast gelation of mixtures of solutions of i-PMMA and s-PMMA. Gelation is the result of complex formation between isotactic and syndiotactic sequences of different polymer molecules. The complexes are helices of an isotactic chain surrounded by a syndiotactic chain and one could ask how these helical complexes are formed with the high rates needed for the fast gelation process. An elegant solution for this problem is given by Schomaker [218]. He suggests that the complexation could start with the formation of a kink of s-PMMA, which subsequently wraps around an isotactic chain. This complex nucleus can grow by means of a simple rotation, just like rotating two pieces of rubber band around each other between the thumb and the forefinger. This results in a left-handed and a right-handed double stranded helix next to each other. This kink nucleated complexation mechanism might be responsible for the fast gelation process.

4.3.2 Syndiotactic Poly(methyl methacrylate)

Self-association of s-PMMA has been mentioned before. It occurs in some solvents [103, 209, 211, 228, 232, 236, 237, 262] and the mechanism of this process resembles that of the formation of the stereocomplex in i/s-PMMA [228, 237, 247, 262]. On the other hand, according to Berghmans et al. [263] the gelation process is the result of the formation of single helices, followed by intermolecular association of these helices. Viscoelastic properties during the gelation process were reported by Berghmans et al. [228]. In their study the results of two gelation processes are shown: plain ageing at 63 °C and a process of the same solution with a special temperature history. In Fig. 71 both results are combined. It appears that intermediate cooling (at 58 and 23 °C respectively) followed by heating to the same temperature again (63 °C) has no influence on the value of the modulus: by plain ageing at 63 °C, the same value of the modulus is reached. The authors explain this phenomenon by a minimum sequence length, which decreases with decreasing temperature; the intermediate cooling process has no effect on the growth of the structures which are also formed at higher temperatures; only those smaller structures which are stable at the lower temperature are formed in excess: they disappear by increasing the temperature. Similar behaviour was reported by Te Nijenhuis [11, 16] for the system gelatin/water. In this study, he proposed an analogous temperature dependent mechanism; moreover, he concluded that the heat of activation of the gelation process is rather low, as intermediate cooling has no effect on the ageing process of the structures formed at higher temperature: the rate of the gelation

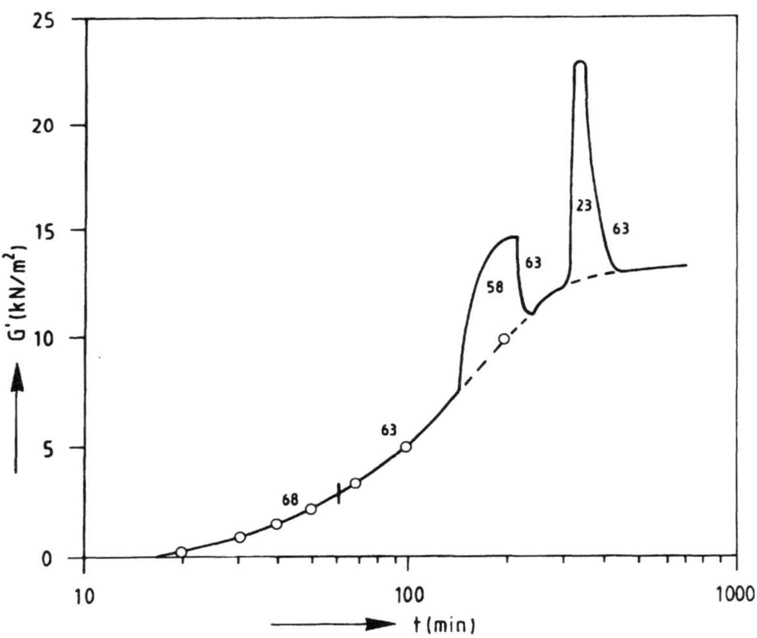

Fig. 71. Time dependence of the storage modulus of a 5% solution of s-PMMA in *o*-xylene for various processes; (— — ○ — —) plain ageing at 63 °C; *solid line*: special temperature history, as indicated: subsequently 68, 63, 58, 63, 23 and 63 °C. Constructed from data presented in [228]

process of certain structures is almost independent of temperature (see Section 10). The same conclusions with respect to the activation energy could be made in the case of the gelation process of s-PMMA.

Another view was recently proposed by Berghmans et al. [263] on the basis of their investigations on the system s-PMMA-toluene (DSC, NMR, IR, UV-vis, fluorescence, LS and rheology). The s-PMMA used contained around 90% syndiotactic triads. A 10% solution showed behaviour similar to that shown in Fig. 71 (apart from differences in the temperature regions and modulus values). At a temperature immediately below the gelation temperature (between 46 and 50 °C) a network was gradually, but slowly, formed. By intermediate cooling close to this temperature the process was only accelerated, whereas intermediate cooling to room temperature resulted in a very fast increase of the storage modulus. Above the gelation temperature (between 46 and 50 °C) the polymer is believed to be in the coiled state. By cooling to a temperature immediately below the gelation temperature a fast change in conformation takes place. This change in chain conformation is followed by a time dependent intermolecular association, which results in an increase of the storage modulus.

From infrared observations it was concluded that above the gelation temperature and below 80 °C the energetically most favourite TT conformation in the random coil polymer solution was present. The amount of this conformation increased suddenly at a temperature approximately 10 °C below the gelation temperature. This was attributed to formation of the more regular helix conformation. From a stationary fluorescence anisotropy study of pyrene labeled s-PMMA it became clear that fluorescence anisotropy was present only below the gelation temperature. Dilute solutions (0.1 wt%) that do not gellify showed no fluorescence anisotropy. This indicates that network formation and not conformational change leads to fluorescence anisotropy. From light scattering observations it was concluded that a second-order association mechanism is responsible for the formation of scattering elements at temperatures where the gelation proceeds slowly. At lower temperatures, where the gelation process is very fast, no conclusions could be drawn about the order of the intermolecular association.

In Fig. 72 the schematic representation of the two step gelation mechanism proposed by Berghmans et al. [263] is shown.

Above the gelation temperature the polymer is believed to be in the coiled state. At temperatures immediately below the gelation temperature, intramolecular helix formation, as observed by IR, is very fast. This is followed by a time dependent intermolecular association, which is the origin of the network formation and the gradual rise of the storage modulus. This slow step is a second-order transition. Upon cooling this network to room temperature, many of the remaining sequences will transform into helices without the formation of extra associates. Hence, extra crosslinks will not be formed during this process. The strong increase of the storage modulus is merely the result of the formation of stiff intramolecular "particles". These non-associated, stiff "particles" disappear ("melt") without hysteresis upon increasing the temperature,

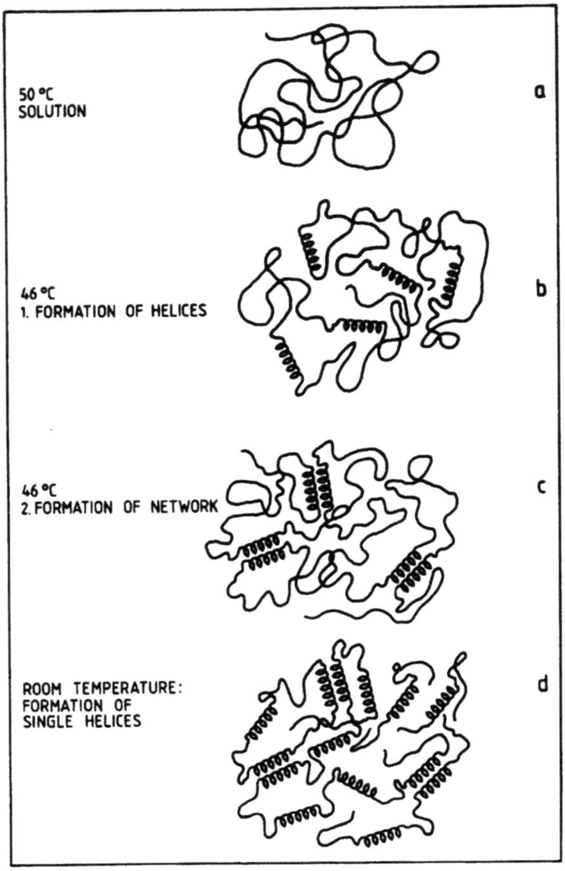

Fig. 72. Schematic representation of the gelation mechanism in moderately concentrated s-PMMA solutions. Reproduced from Macromolecules [Ref. 263] by the courtesy of the authors and of The American Chemical Society

whereas the agglomerated helices will melt at higher temperatures, with a certain degree of hysteresis.

Recently Fazel et al. [264, 265] studied the structure and viscoelastic behaviour of s-PMMA gels (s = 67%, i = 18%, h = 15%, $\bar{M}_w = 106$ kg/mol, $\bar{M}_n = 37$ kg/mol) in bromobenzene and *o*-xylene (in this respect it has to be emphasized that one has to be very careful in considering the gelation mechanism of poly(methyl methacrylates) with low stereoregularity: it is not an unrealistic possibility that the presence of 18% of isotactic dyads gives rise to complex formation with the syndiotactic sequences). Bromobenzene seems to be a good solvent and *o*-xylene a bad solvent: gels obtained in bromobenzene swelled after immersion in bromobenzene, whereas gels obtained in *o*-xylene did not swell after immersion in *o*-xylene. Gels were formed after dissolution at 155 °C (bromobenzene) or 149 °C (*o*-xylene) by quenching the solution to 0 °C; the gels were aged for a minimum of 1 h at this temperature. Compression stress relaxation experiments on these gels are shown in Figs. 73 and 74. In order to

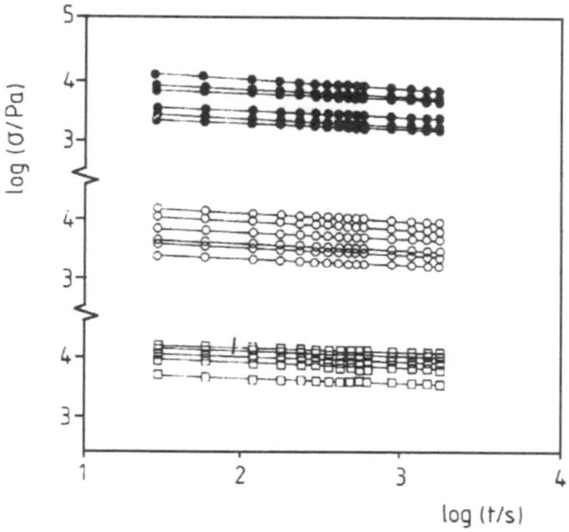

Fig. 73. Compression stress relaxation for PMMA/bromobenzene gels at a deformation $\lambda = 0.85$. *Upper curves*: as received PMMA gels; *middle curves*: "treated" PMMA gels; *lower curves*: "treated" PMMA gels swollen to equilibrium. Reproduced from J Phys II France [Ref. 265] by the courtesy of the authors and of Les Éditions de Physique, Les Ulis, France

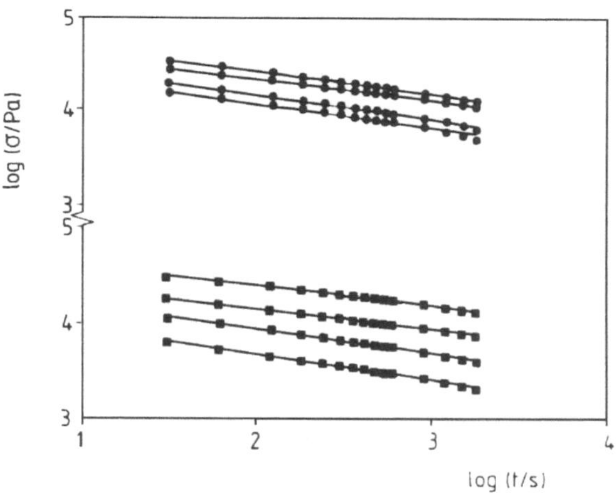

Fig. 74. Compression stress relaxation for PMMA/*o*-xylene gels at a deformation $\lambda = 0.85$. *Upper curves*: as received PMMA gels; *lower curves*: "treated" PMMA gels. Reproduced from J Phys II France [Ref. 265] by the courtesy of the authors and of Les Éditions de Physique, Les Ulis, France

determine the effect of structures that might have been build in during the polymerisation process, part of the polymer was dissolved in chloroform (a good solvent) and subsequently precipitated in methanol and dried. In this way all structures were destroyed. In Figs. 73 and 74 these polymers are indicated as "treated" PMMA. The rate of stress relaxation is lower in bromobenzene gels than in *o*-xylene gels: the slopes m are (for $t < 1000$ s):

in bromobenzene: $m = -0.13 \pm 0.01$ for s-PMMA
$m = -0.1 \pm 0.01$ for "treated" s-PMMA
$m = -0.08 \pm 0.005$ for "treated" s-PMMA in swollen gels
in *o*-xylene: $m = -0.23 \pm 0.02$ for all samples.

There is a clear dependence on the nature of the sample in the bromobenzene gels, but not in the *o*-xylene gels. Apparently small crystallites present in the "as received" PMMA are still present in the gels, but not in the "treated" PMMA

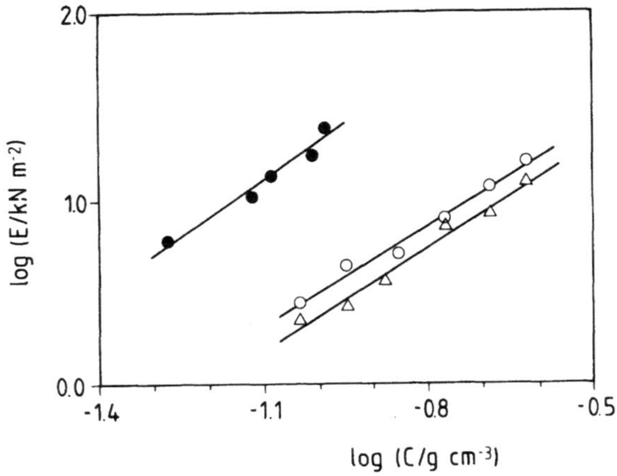

Fig. 75. Concentration dependence of the 120 s compression modulus, determined from compression stress relaxation experiments ($\lambda = 0.85$) for PMMA/bromobenzene gels; (△) as-received PMMA; (○) "treated" PMMA; (●) gels prepared from "treated" PMMA and then swollen to equilibrium. Reproduced from J Phys II France [Ref. 265] by the courtesy of the authors and of Les Éditions de Physique, Les Ulis, France

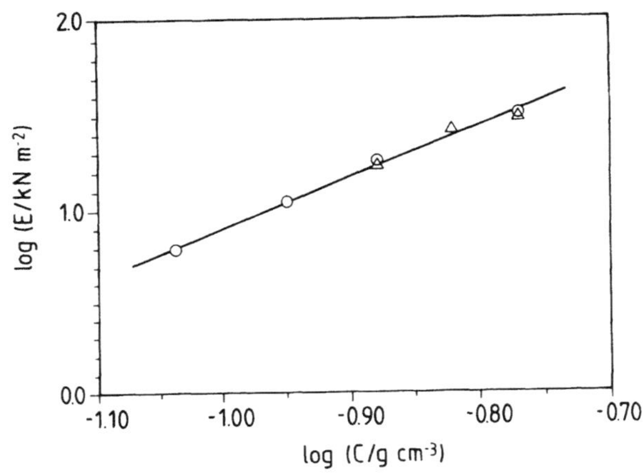

Fig. 76. Concentration dependence of the 120 s compression modulus, determined from compression stress relaxation experiments ($\lambda = 0.85$) for PMMA/*o*-xylene gels; (○) as-received PMMA; (△) "treated" PMMA. Reproduced from J Phys II France [Ref. 265] by the courtesy of the authors and of Les Éditions de Physique, Les Ulis, France

gels. These small crystallites are less stable under stress than the larger ones obtained during the gelation process. A close inspection of the relaxation curves betrays that in the *o*-xylene gels the relaxation rate increases after 1000 s: stress relaxation experiments up to 10^5 s show a considerable increase in relaxation rate with a slope $m = -0.46$ after 5000 s.

Concentration dependence of the 120 s compression modulus, measured from the stress relaxation experiments, was determined in the concentration range from 5 to 25 wt%; this is shown in Figs. 75 and 76. Results of linear regression are:

in bromobenzene: $E = 1.60 \cdot 10^5 c^{1.86 \pm 0.10}$ N/m^2 s-PMMA
$E = 2.08 \cdot 10^5 c^{1.86 \pm 0.05}$ N/m^2 "treated" s-PMMA
$E = 1.94 \cdot 10^6 c^{1.99 \pm 0.05}$ N/m^2 swollen "treated" s-PMMA
in *o*-xylene $E = 3.60 \cdot 10^6 c^{2.66 \pm 0.05}$ N/m^2 all samples.

From their small-angle neutron scattering experiments, the authors concluded that the s-PMMA gels were rigid gels somewhere in between gels of most synthetic polymers and biopolymer gels. Also the exponents of 1.86 and 1.99 are slightly lower than 2, i.e. the exponent obtained for gels of the enthalpic type with rod-like structures. The exponent of 2.66 obtained in the *o*-xylene gels is not yet clear in this respect (for further reading see [2], pp 196–199).

4.4 Conclusions

Many investigations of the complex formation in mixtures of i- and s-PMMA have dealt with the structure of the stereo-complexes. It was found that the i/s ratio is 1/2 on the molecular scale. However, it has been proposed that the complex formation starts with a 1/1 complex, which can change into a 1/2 stereocomplex in the course of time, if the kinetics are not hindered by gel formation. Moreover, in polymers with imperfect stereoregularity, many regular sequences are too short to be able to take part in the complex formation. In that case complex formation may result in gel formation. From these models one could conclude that the functionality of the crosslinks formed by stereocomplexation varies from 4 to 6.

The study of the viscoelastic behaviour of those gels have been the subject of only a few studies. A complication in the determination of the viscoelastic behaviour is the syneresis tendency of the gels formed. The gels seem to behave rubber elastically. Although the melting temperature of 12.6 wt% gels in *o*-xylene is approximately 120 °C, it appears that dissolution at 145 °C rejuvenates the gels much more. This becomes clear from subsequent quenching to low temperatures: the gelation process proceeds much faster for the solutions that were dissoluted at 120 °C than for the solutions that were dissoluted at 145 °C.

At high temperatures, gelation of s-PMMA solutions is the result of formation of small to large single helices (depending on the polymer stereoregularity), followed by intermolecular association. Hence, the crosslinks are multifunctional. By subsequent quenching to room temperature, single helices are formed which do not become intermolecularly associated, so that they do not take part in the network formation. Increase of the storage modulus is the result of stiffening of the polymer chains.

5 Atactic Polystyrene

5.1 Introduction

The gelation behaviour of solutions of atactic polystyrene has been the subject of relatively many studies. In the 1960s indications of the presence of aggregates in solutions of atactic polystyrene were obtained from light scattering experiments [266]. Guenet et al. [267, 268] it showed that the enhanced scattering at low angles disappeared on increasing the temperature and reappeared on lowering the temperature again. They also showed that solutions of atactic polystyrene that can form thermoreversible gels, also give rise to enhanced low angle scattering and vice versa. Although it was not always recognised, nowadays it is usually assumed that the mechanism of gel formation is a phase separation process.

Boyer et al. [269] explained the sharp, thermally reversible melting of polystyrene gels at T_{gel} as a local melting of segment-segment contacts, as was postulated by Lobanov and Frenkel [270] and by Boyer [271]. The gel temperature appeared to be dependent on the solubility parameters δ, a minimum in T_{gel} arising when $\delta_p \approx \delta_s$. The gelation behaviour was related to segment-segment interactions, which are hindered by good solvents and encouraged by worse solvents like carbon disulfide. Berghmans et al. [172, 272–274] proved that phase separation plays a dominant role in the gelation behaviour of solutions of amorphous polymers. A convincing proof of the structures of those gels was given by Keller et al. [275, 276]. In a short but interesting review on thermoreversible gelation of polymers by Hiltner [277], most attention was paid to the physical gelation behaviour of polystyrene solutions.

5.2 Experimental Data

In 1953, Ferry and Grandine [278] demonstrated that 5–10% solutions of atactic polystyrene in decalin showed phase separation and transformed into opaque gels upon cooling to 10 °C; at lower concentrations turbid, free flowing solutions developed.

Wellinghoff et al. [279] obtained clear gels in 1% solutions of atactic polystyrene (M = 670 kg/mol and 2000 kg/mol) in carbon disulfide by cooling the solutions to −94 °C. Both the gel formation temperature and the gel melting temperature increased with increasing polymer concentration: a 16% solution of polystyrene (M = 37 kg/mol) forms gels at −35 °C that melt at −20 °C, whereas for a 10% solution these temperatures are −78 and −40 °C, respectively. Molecular weight also has a strong effect upon the gel melting temperature: e.g. for 16% solutions T_m increased from −25 to +5 °C as the molecular weight increased from 4 to 670 kg/mol. This is not in disagreement with results reported by Clark et al. [280] on 9.5% solutions (molecular weight 900 kg/mol) in carbon disulfide, from which one can conclude that the gel formation temperature was about −15 °C. In contrast to gelation of poly(vinyl chloride) solutions, for example, the gelation of atactic polystyrene solutions is not caused by crystallization. This was revealed by Wellinghoff et al. [279], who reported that: a) high-sensitivity DSC measurements showed no endothermic or exothermic peaks; b) infrared spectra of atactic polystyrene solutions in carbon disulfide showed no significant changes upon cooling to −78 °C; c) FTIR spectra of atactic polystyrene glasses cast at −78 and 25 °C showed no detectable differences. These authors supposed that a fine dispersion of polymer rich regions developed upon quenching a solution to the unstable region. When the domains of relatively concentrated polymer solutions are in their glassy state, they are able to pin polymer chain ends. As the gels formed are clear, the size of the domains must be small (e.g. <100 nm). Moreover, the gel formation proceeds very rapidly, and hence one can conclude that it is caused by spinodal phase separation, by which a continuous network of a polymer poor phase is formed, the polymer rich, glassy domains being the multifunctional crosslinks. Melting of the gel is due to surpassing the glass temperature of the phase separated microdomains, freeing the polymer chains that were originally crosslinked by these "structures".

Clark et al. [280] reported dynamical mechanical measurements on 9.5% solutions of atactic polystyrene ($M = 900$ kg/mol) in carbon disulfide at temperatures varying from −10 to −22 °C. At the highest temperature the system behaved like a concentrated solution (G' and G'' strongly depending on the frequency) whereas, upon decreasing the temperature, the behaviour gradually transferred to rubberlike. (i.e. G' independent of the frequency) (see Fig. 77). The build-up of the network progressed so quickly that time effects were not observed over a period of minutes to many hours after the temperature change. With the aid of the Winter and Chambon method [25–30, 281], which was applied to physical gels by Te Nijenhuis and Winter [31], it might be concluded that the gelation temperature of this system is approximately −15 °C (see Fig. 78). For a gel with $G' = 600$ N/m^2 at −22 °C Clark et al. [280] decided on a crosslinking index of 2 to 3 trifunctional crosslinks per primary polymer molecule. However, the functionality of the crosslinks will be much higher than 3 and according to the Te Nijenhuis model [39–44] the crosslinking index would be $\gamma = 7.1$, 4.7, 2.95 and 2.36 for crosslink functionalities of 3, 4, 10 and ∞ respectively. On the other hand, from the equation $G_e = \nu kT$ it follows that the number of elastically effective network chains (EANCs) is equal to 1.73×10^{23}, whereas the number of polymer molecules is $cN_{Av}/M = 6.36 \times 10^{22}$. Hence, the average number of crosslinks per polymer molecule is 2.7. This means that on average every polymer molecule is part of at least 2.7 domains of high concentration.

Boyer et al. [269, 271] concluded that: a) the gelation temperature and the glass temperature are not identical; b) overlap concentration of polymer molecules is necessary; c) physical entanglements play an unimportant role at

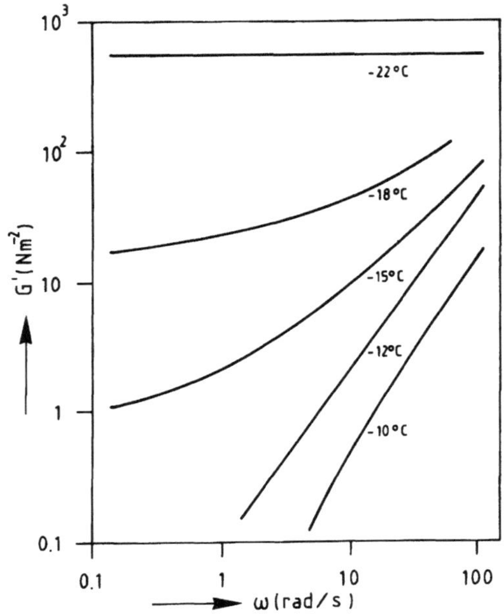

Fig. 77. The gradual formation of a rubberlike network (as depicted by the storage modulus) for a 9.5% solution of aPS ($\bar{M}_w = 900$ kg/mol) in CS_2 on lowering the temperature (constructed after data presented [280])

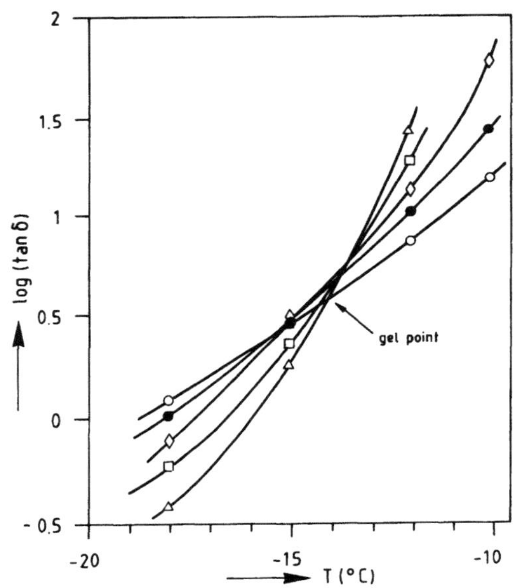

Fig. 78. Winter and Chambon plot of log tan δ vs temperature at various frequencies for the aPS/CS_2 system of Fig. 77; log(ω/rad · s^{-1}): (△) 0; (□) 0.33; (◇) 0.67; (●) 1; (○) 1.33

higher molecular weights and concentrations (formation of super gels); d) T_{gel} increases with decreasing solvent quality; e) gelation phenomena should be a characteristic feature of all atactic polymers; f) a critical gelation concentration exists.

A detailed, systematic study was carried out by Koltisko et al. [282] on the molecular weight, concentration and temperature dependences of the gelation behaviour of atactic polystyrene in carbon disulfide. They made use of almost monodisperse polystyrenes, with molecular weights varying from 2 to 900 kg/mol, concentrations from 12.5 to 200 kg/m^3 and temperatures from −100 to 0 °C. The storage modulus increased to much higher values than was reported by Clark et al. [280]. Koltisko et al. [282] measured the equilibrium shear modulus G_e and, in agreement with Clark et al. [280], they did not observe a time dependence.

The influence of both concentration and molecular weight on temperature dependence is shown in Fig. 79. Most striking is that a) at higher concentrations, the shear modulus reaches a temperature independent value and b) the low temperature shear moduli are equal as long as the molecular weight is higher than 35 kg/mol. Below this value the modulus strongly depends on molecular weight. The points indicated on the temperature axis, are gel temperatures as measured by Tan et al. [284]. It is curious that these gel temperatures are so strongly dependent on concentration. If the glass transition temperature of the polymer rich domains is determining the gel temperature, then T_{gel} should be independent of concentration. Measurements reported by Tan et al. are based on visual observations, and Keller et al. [275] wondered whether these measurements were made properly, in view of the small difference in refractive indices of atactic polystyrene and carbon disulfide. On the other hand, the determination method used by Koltisko et al. [282] is inaccurate for $G < 10\ N/m^2$, so that extrapolation to $G = 0$ is also unreliable. The differences in gelation temperatures of solutions of one concentration, but of different molecular weights, are very small, in particular for the higher molecular weights; lower molecular weight polystyrenes show somewhat lower gelation temperatures. This is in

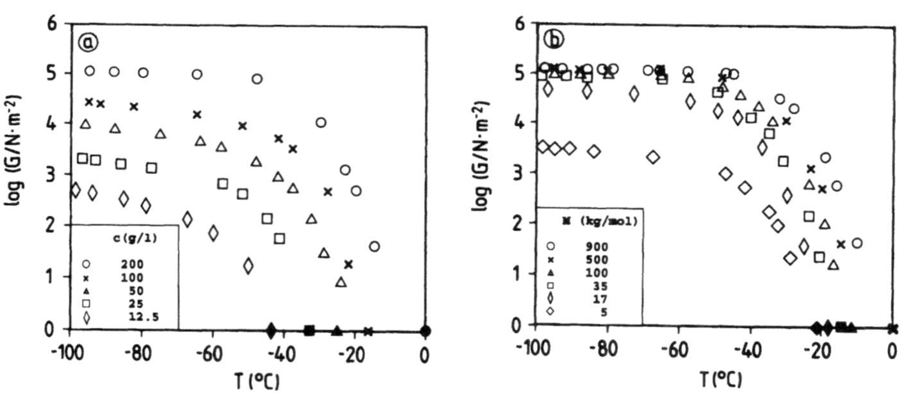

Fig. 79a, b. Effect of: **a** concentration (M = 500 kg/mol); **b** molecular weight (c = 200 kg/m^3) on the modulus-temperature behaviour of aPS/CS_2 gels; the *solid points* on the temperature axis indicate the gelation temperatures reported by Tan et al. [284]. Reproduced from Macromolecules [Ref. 282] by the courtesy of the authors and of The American Chemical Society

complete agreement with the width of the phase separation region, which is narrower for lower molecular weight polymers, so that the T_g-ϕ-curve crosses this region at lower temperatures. Koltisko et al. [282] defined a reduced temperature $T_r = T/T_{gel}$, where T_{gel} is the extrapolated gelation temperature (which of course is dependent on molecular weight, but in particular on concentration). In Fig. 80, log G is plotted vs log c (M = 500 kg/mol) for various values of T_r. It appears that straight lines are obtained, from which a quadratic concentration dependence of the shear modulus follows. Two shift factors were defined, the first for the concentration ρ/c, where ρ is the density of the pure polymer and the second one for the modulus $b = G^o/G$, where G^o is the plateau modulus of the highest molecular weight (i.e. 900 kg/mol). By using these shift factors, the authors constructed a master curve, where the effective modulus, G_R, defined as $G_R = Gb(\rho/c)^2$, is plotted vs the reduced temperature, T_r. This results in a standard curve for all molecular weights ($5 \leq M \leq 900$ kg/mol) and concentrations ($12.5 \leq c \leq 200$ kg/m^3), as shown in Fig. 81. It represents the modulus-temperature relationship for gels of atactic polystyrene (M = 900 kg/mol) without any solvent. This plateau modulus is 3×10^6 N/m^2, slightly higher than the pseudo equilibrium shear modulus of atactic polystyrene, but significantly lower than the modulus in the glassy state.

Recently Xie et al. [283] reported detailed studies on the gelation rate of atactic polystyrene/carbon disulfide solutions. They showed that a gelation time exists, depending on concentration and molecular weight, and they superimposed the concentration dependent gelation time vs temperature curves on to one master curve. Such a curve is shown in Fig. 82. It appears that the gelation

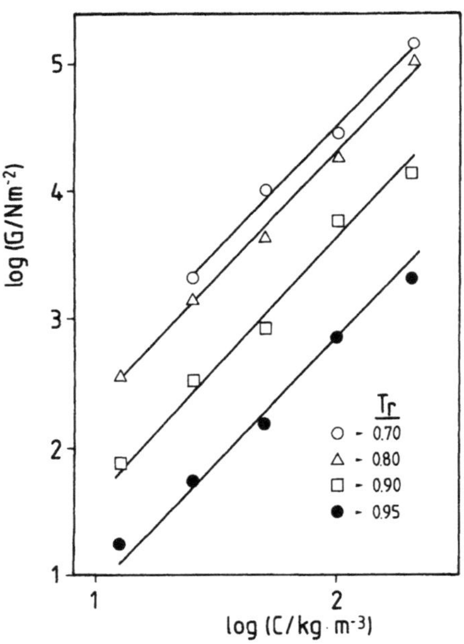

Fig. 80. Relationship between equilibrium shear modulus and concentration for aPS/CS_2 gels (M = 500 kg/mol) at four reduced temperatures. Reproduced from Macromolecules [Ref. 282] by the courtesy of the authors and of The American Chemical Society

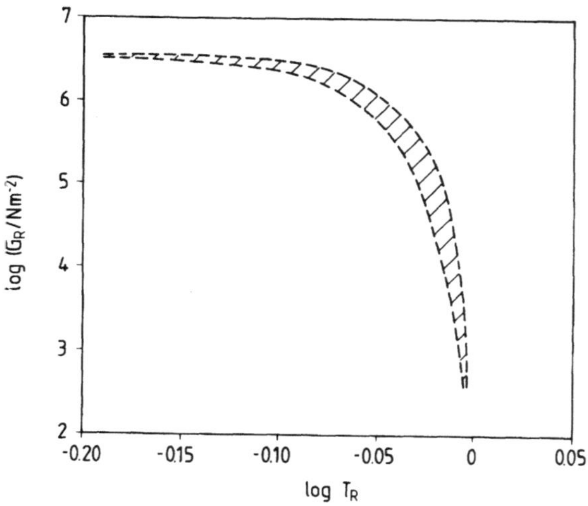

Fig. 81. Master curve ($\log G_R$ vs $\log T_R$) for aPS/CS_2 gels; M varying from 2 to 900 kg/mol; concentration varying from 12.5 to 200 kg/m^3. The *dashed region* covers all the experimental points. Reproduced from Macromolecules [Ref. 282] by the courtesy of the authors and of The American Chemical Society

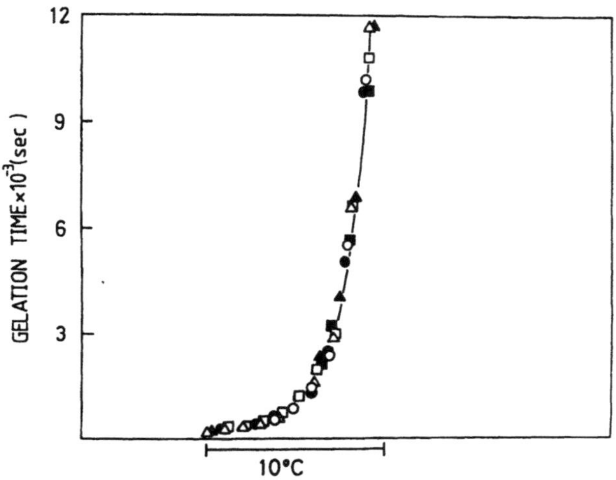

Fig. 82. Gelation time vs temperature curve for aPS ($\bar{M}_w = 280$ kg/mol)/CS_2 solutions, obtained by superimposing curves of different concentrations, varying from 150 to 400 kg/m^3. Reproduced from Polymer [Ref. 283] by the courtesy of the authors and Elsevier Science Ltd, The Boulevard, Langford Lane, Kidlington 0X5 1GB, UK

time is very short at low temperatures, but that it increases to very long times within a temperature increase of 10 °C. This means that at low temperatures the gelation process is so fast that no time effect is observed. They also found differences between T_{gel} and T_m; it appeared that $T_m - T_{gel} \approx 13$ °C, almost independent of concentration, as shown in Fig. 83. The authors conclude that the rate of the gelation process is determined by an Arrhenius type relationship, with an activation energy that strongly depends on the difference between the

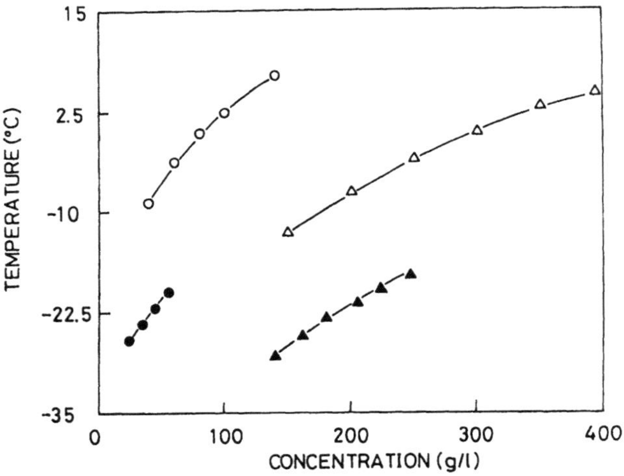

Fig. 83. Comparison of gel melting temperature (○, △) and critical gelation temperature (●, ▲) for two aPS/CS_2 systems: (○, ●) $\bar{M}_w = 2180$ kg/mol; (△, ▲) $\bar{M}_w = 280$ kg/mol. Reproduced from Polymer [Ref. 283] by the courtesy of the authors and Elsevier Science Ltd, The Boulevard, Langford Lane, Kidlington 0X5 1GB, UK

melting point of the gel and the temperature of the gelation process (at least 13 °C): the lower the gelation temperature, the lower the activation energy.

5.3 Mechanism of the Gelation Process

The gelation mechanism as a result of phase separation is schematically illustrated in Fig. 84 for solutions of monodisperse polymer. Upon cooling a diluted solution, as presented, for example, by point A, phase separation will take place as soon as the temperature crosses the binodal at point B. The rate of the phase separation process depends on the presence of nuclei as long as the spinodal curve (not illustrated in Fig. 84) is not reached. When the point C is reached, two phases will develop: C_1, the polymer poor phase and C_2, the polymer rich phase. In principle, in the long run two clearly distinct phases will be created: the solvent rich phase is clear, whereas, in general, the polymer rich phase will be turbid due to the presence of solvent rich droplets which, even after months, are not phase separated due to the high viscosity of the polymer rich phase.

If the formation of two distinct layers is somehow opposed, then a dispersion of polymer rich microdroplets develops in a polymer poor liquid phase. The possibility that parts of polymer molecules present in the dilute phase form part of the droplets and even form connections between the droplets, is not excluded. In that case, droplets are not able to coalesce, because a network will be formed consisting of polymer molecules which are connected by the polymer rich droplets, behaving as polyfunctional crosslinks. However, it is possible that polymer molecules free themselves, particularly if the viscosity in the droplets is

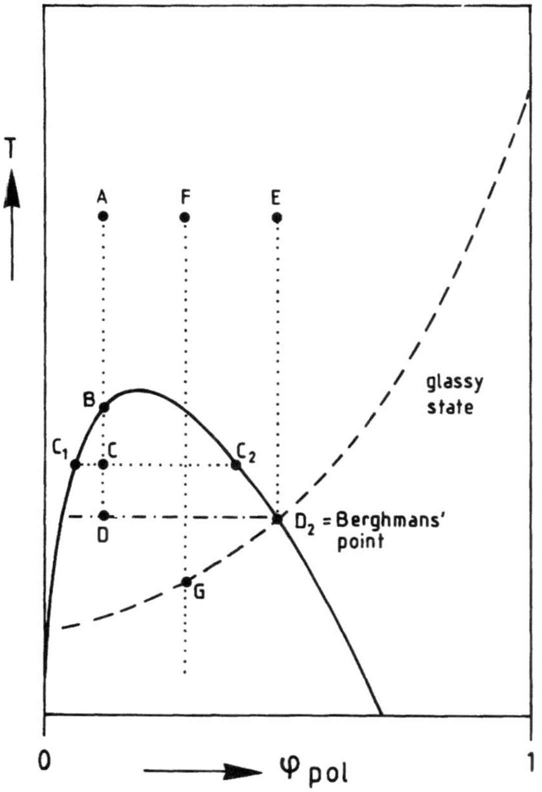

Fig. 84. Phase diagram, shown schematically, of solutions of monodisperse polymer for which the glass transition curve (- - - -) crosses the binodal boundary curve

not too high. With decreasing temperature the viscosity will increase because the polymer concentration in the polymer rich phase will increase. If the temperature becomes equal to the glass transition temperature, then the viscosity will become so high that polymer molecules are really fixed, so that a real gel network will be developed with infinite relaxation times. In Fig. 84 the concentration dependence of T_g is indicated by the broken line. The $T_g - \phi$ curve crosses the binodal curve in D_2. Upon cooling a solution from A to D, polymer rich droplets in their glassy state are obtained (point D_2). It has to be emphasized that the line DD_2 is horizontal only for solutions of mono-disperse polymer. For polydisperse polymer the line starting from D slightly deviates to higher temperatures (see e.g. [285]). Upon further cooling no further phase separation will occur in the polymer rich phase, due to the immobility in the system; on the other hand, in the solvent rich phase further phase separation will take place. However, the amount of polymer rich phase which will be added to the phase D_2 is extremely small and will not affect the glass transition temperature of the polymer rich phase. The consequence is that upon cooling a polymer solution of $\phi < \phi_E$ to a temperature T_D a system will always be obtained with a glass transition temperature equal to T_D. Further cooling will hardly affect the value of T_g. However, if the cooling process proceeds very fast to $T < T_D$, then a polymer rich phase will be developed with a lower T_g. According to Keller et al. [276], in that case the glass transition temperature will increase in the course of time until $T_g = T_D$ as phase separation certainly will take place with a very low speed. This is observed in the system atactic polystyrene (M = 3 kg/mol) /cyclohexanol (the molecular weight of this polymer is very low, so that phase separation processes even in the glassy phase will take place relatively quickly). Upon quenching to a temperature below the $T_g - \phi$ curve, no phase separation will take place: the mass is homogeneous. Upon increasing the temperature a glass transition temperature equal to T_G will be observed.

The experimental diagram at Fig. 85 for the system aPS/ *trans*-decalin [286] supports the model shown in Fig. 84.

5.4 Gel Structure

The structure of atactic polystyrene gels was studied by Keller et al. [276, 287, 288]. The results will be demonstrated on the basis of Fig. 86, where the upper region of the atactic polystyrene ($\bar{M}_w$ = 2750 kg/mol)/cyclohexanol phase diagram is shown.

Upon cooling a dilute solution C_1 to a temperature below the binodal curve, but above the spinodal curve, phase separation will take place via formation of nuclei, so that a dispersion of polymer rich droplets develops in a sea of polymer poor solution. If the droplets (domains) are not separated too much, thus if they are small and numerous enough, then the possibility arises, particularly if the

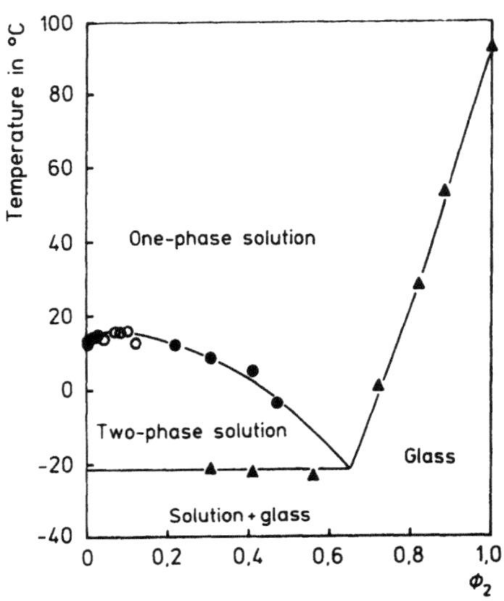

Fig. 85. Phase diagram for the system aPS/*trans*-decalin; M_n, M_w and M_z: 144, 164 and 184 kg/mol, respectively; (○) optically obtained temperatures; (●) demixing temperature (onset DSC peak); ϕ_2 = polymer weight fraction; (▲) glass transition temperature (DSC). Reproduced from Makromol Chem [Ref. 286] by the courtesy of the authors and of Hüthig & Wepf Verlag Publishers, Zug, Switzerland

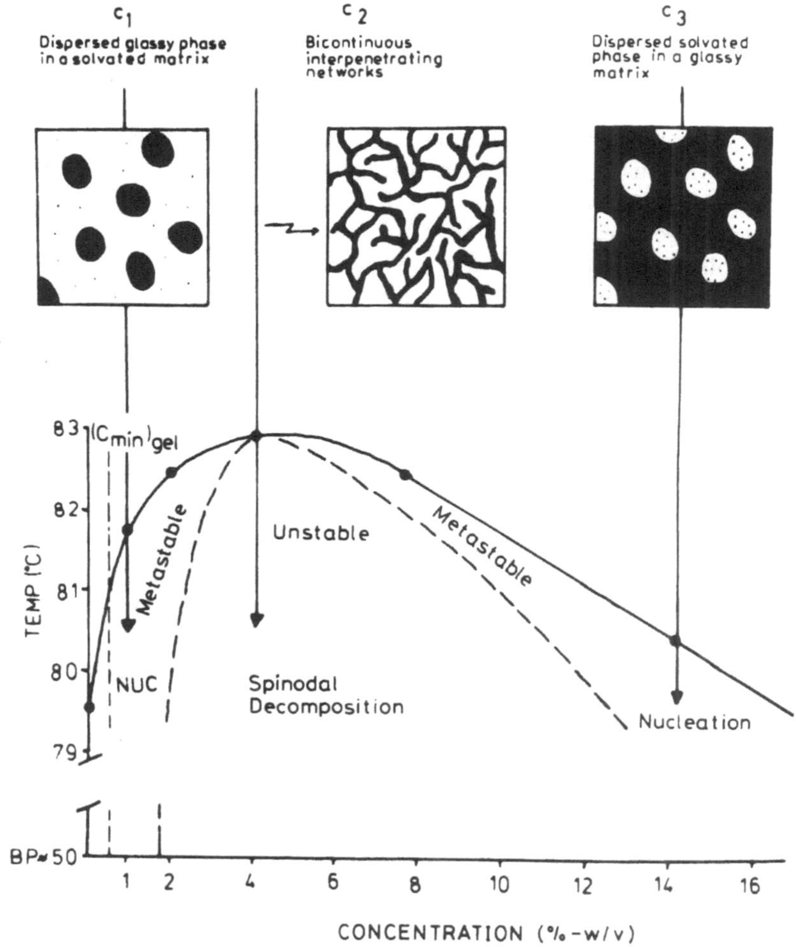

Fig. 86. Detail of Upper Critical Temperature phase diagram of aPS ($\bar{M}_w = 2750$ kg/mol) in cyclohexanol, including expected variation in (gel) morphology with concentration. Reproduced from Makromol Chem [Ref. 276] by the courtesy of the authors and of Hüthig & Wepf Verlag Publishers, Zug, Switzerland

molecular weight is high, that polymer molecules act as connective fibres between the small domains. For that purpose it is necessary that the average distance between the domains is not much higher than the average end-to-end distance of the polymer molecules. If the phase separation takes place above the glass transition temperature of the domains, then polymer molecules might be loosened, particularly if they are under stress. This will be a real possibility if the distance between the domains is relatively large. If the phase separation takes place at the glass transition temperature of the domains, then parts of the polymer molecules are fixed in the domains, so that they may be elongated and generate a tensile stress in the polymer molecules themselves. The consequence is that the gel is not purely rubberlike anymore, because an energy term contributes to the elasticity. Of course, a minimum concentration exists below which no gel will be formed because in that case the number of domains is too small to be bridged by polymer molecules. This c_{min} depends on the molecular weight of the polymer used, as has been reported by Keller et al. [276] for atactic polystyrene/cyclohexanol systems (see Fig. 87). For M > 500 kg/mol a linear relationship exists between c_{min} and $\bar{M}_w^{-0.5}$, hence, between c_{min} and $\langle r_o^2 \rangle^{-0.5}$. For smaller molecular weights this does not hold.

Analogous results are obtained in the reverse situation, i.e. solvent rich droplets in a polymer rich matrix, by quenching a C_3 solution into the metastable region. At intermediate concentrations, e.g. C_2, a cocontinuous interpenetrating network arises as a consequence of the very fast spinodal decomposition. If the temperature is higher than the glass transition temperature of the polymer rich phase, prolonged standing yields two separated layers. However, if the temperature is lower than the glass transition temperature, a network of polymer strands remains. Upon freeze drying a porous structure of network strands is obtained. If the molecular weight is high enough, then the gel is formed by a continuously phase connected network, whereas in the case of low molecular weight the network is formed by adhesive contact of hairy particles adhesively phase connected. From the results presented in Fig. 79, it can be calculated, with the aid of $G_e = \nu kT$, that for the highest molecular weight polymer (i.e. 900 kg/mol) the polymer molecules are part of at least 400 domains of high concentration. This is a rather high value and one has to conclude that the highly concentrated domains, which presumably are in their glassy state or present as a continuous glassy matrix, contribute considerably to the value of the modulus. For this reason it is unlikely that the shift procedure mentioned

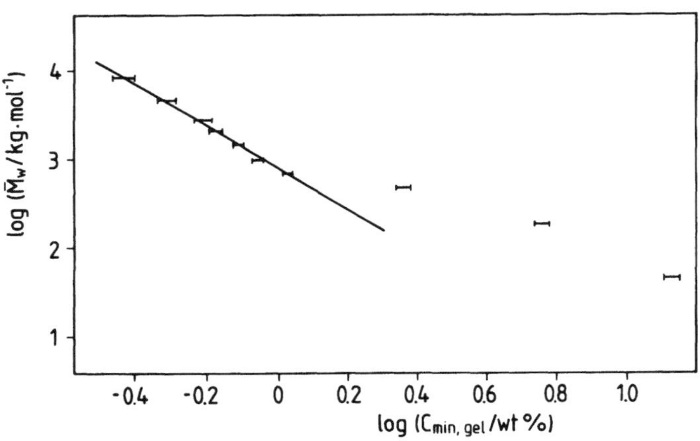

Fig. 87. log $\bar{M}_w$ plotted vs log $c_{min,gel}$; the *solid line* for the highest molecular weights corresponds to the condition $c_{min,gel} \propto \bar{M}_w^{-0.5}$. Reproduced from Makromol Chem [Ref. 276] by the courtesy of the authors and of Hüthig & Wepf Verlag Publishers, Zug, Switzerland

before has any physical background. The three kinds of gel networks are shown schematically in Fig. 88. The different morphologies of the gels were demonstrated with electron micrographs [275, 276] that show the mentioned morphologies very clearly. The broad lines of this mechanism are supported by the L-L-demixing region (Fig. 89) and the beautiful SEM pictures (Fig. 90) published by Arnauts et al. [286].

Another view is offered by Guenet et al. [267, 289–293]. These authors suggested that the gelation probably proceeds via the creation of three-dimensional structures, arising from the formation of a polymer-solvent intercalate. This would explain why the phenomenon is often enhanced in good solvents, in contrast with studies by Wellinghoff et al. [279, 280]. From DSC measurements, Guenet et al. [267, 291] were able to demonstrate quite clearly the formation of exotherms and melting endotherms. This was also shown by Tan et al. [284] for a 16% solution of atactic polystyrene in toluene: a transition was detected at about −80 °C, which corresponds to the gelation temperature of the same solution as measured by the tilted test tube method. For the exotherm associated with this transition, a heat of gelation of the order of 1 to 2 J/g was calculated. According to Guenet et al. these DSC measurements showed conclusively that the gelation is due to a first order transition which implies the

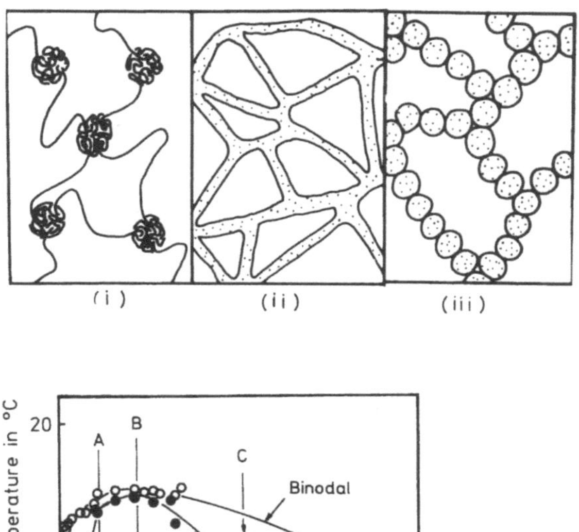

Fig. 88i–iii. Schematic illustration of the three classes of gels: **i** molecularly connected; **ii** phase connected (continuous); **iii** phase connected (adhesive). Reproduced from Makromol Chem [Ref. 276] by the courtesy of the authors and of Hüthig & Wepf Verlag Publishers, Zug, Switzerland

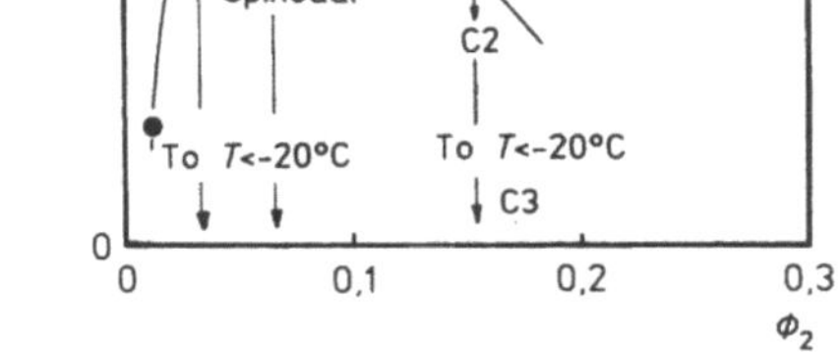

Fig. 89. Binodal (○) and spinodal (●) for the system aPS/*trans*-decalin, mentioned in Fig. 85. Reproduced from Makromol Chem [Ref. 286] by the courtesy of the authors and of Hüthig & Wepf Verlag Publishers, Zug, Switzerland

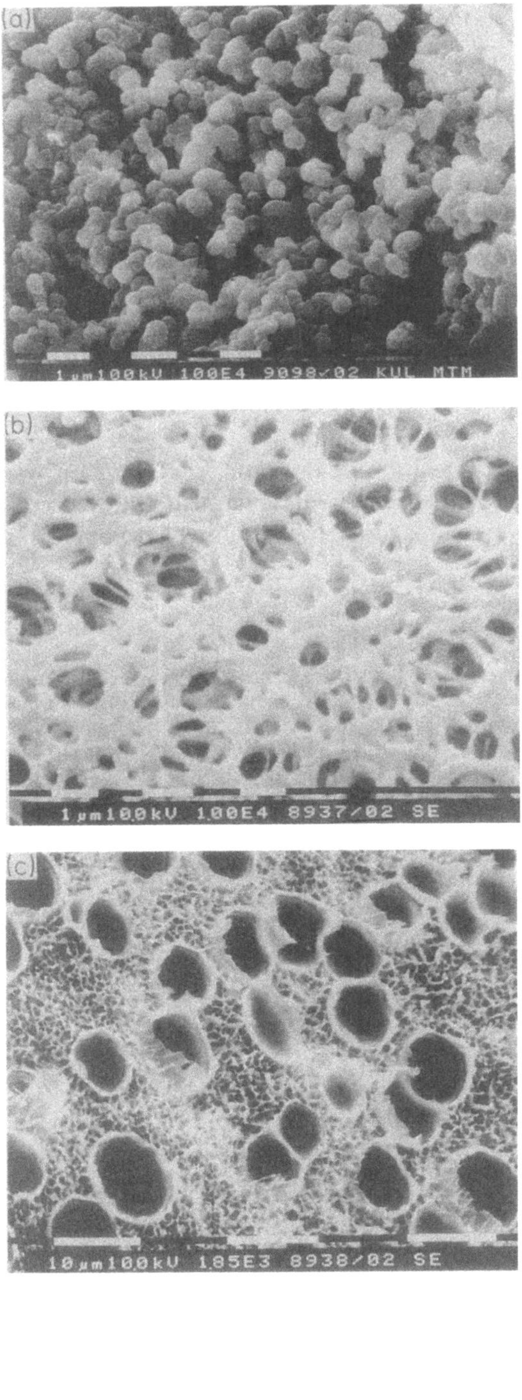

Fig. 90a–c. Scanning electron micrographs of dried aPS/*trans*-decalin samples with different overall concentrations: **a** $\phi_2 = 0.044$; **b** $\phi_2 = 0.067$; **c** $\phi_2 = 0.159$. Reproduced from Makromol Chem [Ref. 286] by the courtesy of the authors and of Hüthig & Wepf Verlag Publishers, Zug, Switzerland

existence of some three-dimensional structures in which the solvent molecules play an important role (intercalates).

Recently Lehsaini et al. [294, 295] studied the branching effect on the physical gelation behaviour of polystyrene solutions. They synthesized linear polystyrenes and polystyrenes with three or four long chain branches as well. The authors concluded that the sol-gel transition of atactic polystyrene solutions shifted to higher temperatures at a given concentration and molecular weight when passing from linear to star polymers, and that the concentration dependence is slightly higher than found for linear polystyrene carbon disulfide gels. It shows that the chemical crosslinks in branched star polymers play an important part in the gelation behaviour so that less physical crosslinks are required to reach the gel point.

5.5 Conclusions

Gels are obtained in solutions of atactic polystyrene as a result of liquid-liquid demixing. The gels are more or less permanent if the phase separated highly concentrated polystyrene domains are in their glassy state. Measurement of the viscoelastic properties of gels of atactic polystyrene has been the subject of only a few studies. A systematic approach would be very helpful, in order to elucidate the mechanism of the gelation process. The work of Keller et al. and of Berghmans et al. show that the mechanism is complicated and strongly dependent on the solvent used. This is the result of various kinds of interactions between the concentration dependent glass transition temperature and the phase separation region.

6 Polyacrylonitrile

6.1 Introduction

Polyacrylonitrile exhibits gelation from various solutions. This was definitely a disadvantage for the traditional fibre solution spinning process. However, in recent years, solvents have been sought for gel spinning of polyacrylonitrile. For that reason the gelation behaviour of PAN solutions has been the subject of many studies. Although some crystallinity is present in PAN, from earlier studies on PAN-DMF (i.e. *N,N*-dimethylformamide) and PAN-DMA (i.e. *N,N*-dimethylacetamide) by Bisschops [296–298], Ziabicky et al. [299, 300], Paul [301] and others, it is not always clear whether the junction zones in PAN gels are formed by crystallization. Bisschops [296–298] mentioned the possibility of the existence of nuclei or small crystallites as crosslinking loci in PAN-DMF gels. On the other hand, on the basis of their light scattering and X-ray diffraction experiments on PAN-DMF gels, Ziabicky et al. [299, 300] suggested that the strong dipole-dipole forces between the nitrile groups are responsible for gelation: they could not detect any differences between PAN-DMF solutions and PAN-DMF gels.

In more recent work [302–305] using propylene carbonate (PC) as a solvent, X-ray patterns showed the presence of crystallites in PAN-PC gels. These gels are particularly interesting because it has been suggested that the junctions are not the normal crystallites of pure polymer, but crystallites of a polymer-solvent complex.

6.2 Crystallinity

More recently the study of the crystal morphology of PAN, as powder as well as in gels, has been the subject of many investigations. The crystal structure of PAN is controversial with many scientists finding a hexagonal and some an orthorhombic lattice (see [7–14] in [302]. Bashir is of the opinion that it depends on the type of sample: the hexagonal lattice is the normal one for dry PAN, whereas, in the presence of solvents such as PC, large orthorhombic lattices might be obtained.

Bashir et al. [302–305] assumed the existence of a hexagonal packing of rodlike segments (molecular cylinders) with a diameter of 0.6 nm, with the nitrile groups protruding at all angles when viewed down the chain axes due to the mutual intramolecular repulsion of the nitrile groups. According to these authors, the crystal structure of PAN is influenced by the type of sample studied (reactor powder, oriented fibres, single crystals grown from dilute solutions, gels and swollen films). Moreover, incorporation of (residual) solvent might also be a reason for the differences. Especially in PAN crystals grown from propylene carbonate solutions, the unit cells reported are larger than those of the reactor powder. From the results reported by Klement and Geil [306], Bashir et al. [302–305] suggested that propylene carbonate might be incorporated in between the molecular PAN cylinders because of the strong dipole-dipole interactions between the C=O groups of this solvent and the nitrile groups of PAN. The unit-cell parameters a and b of pure PAN are 1.04 and 0.60 nm [307] (which is in agreement with $a/b = \sqrt{3}$ for a hexagonal lattice), whereas in the presence of propylene carbonate they are 2.118 and 1.16 nm, respectively. The "length" of propylene carbonate (i.e. the distance between the carbonyl oxygen and the opposite C–C bond in the ring) is 0.332 nm. These authors postulated that the size of the unit cell in the presence of propylene carbonate allows the propylene carbonate molecule to be incorporated into the lattice (see Fig. 91). In a recent paper Bashir et al. [308] showed that water might also be incorporated into the lattice.

From their NMR and X-ray diffraction investigations on dry PAN powder Bashir et al. [305] concluded that their results are consistent with an intermediate type of order, lying somewhere in between truly crystalline and truly amorphous. For that reason, a paracrystalline or laterally ordered structure has also been suggested as a possibility for the crystal structure, because it has been recognized that in general PAN has only two-dimensional order: only equatorial reflections are present in oriented fibres (two-dimensional lateral order instead of true crystalline order [305]). Hence, it is concluded that there is no long periodicity along the chain axes. This is in agreement with the almost atactic character of the polymer molecules. According to Bashir [302], a few authors claimed a three-dimensional crystalline order, as they observed equatorial as well as non-equatorial reflections in drawn fibres and in single crystals (see [1, 12, 13, 15–19] in [302]). In a recent paper, Bashir [307] discusses the various unit cell parameters reported in the literature: the two-dimensional hexagonal model is the normal one for dry PAN, and he suggested that the

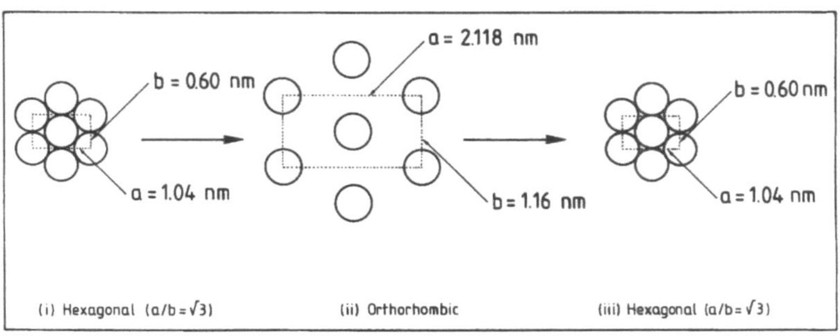

Fig. 91. Model illustrating the change in molecular packing in going from: **i** reactor powder (hexagonal) to **ii** a gel with solvent (orthorhombic) and back to **iii** powder after extraction of the solvent from the gel (hexagonal). The *circles* are the cross-sections of cylinders representing the PAN molecules (a modified version of [305], taking into account results presented in [307]

three-dimensional orthorhombic cells, proposed as the "true" crystal structure of PAN by some authors, are probably artefacts due to the presence of solvent and the formation of polymer-solvent complexes.

6.3 Crystallinity in Gels

Although solutions of PAN in DMF can give rise to gel formation, it was very difficult to prove that the junction regions in these gels are crystallites. The same applies to PAN-DMSO (i.e. dimethyl sulfoxide) and PAN-DMA gels. Gelation from solutions in DMA, DMF and DMSO is very slow and is also affected by the presence of water. The solvents are hygroscopic and solutions can gel due to slow absorption of water. X-ray diffraction and DSC measurements did not reveal the existence of any crystallinity. Beckmann and Zenke [309] were unable to detect any crystallinity from X-ray diffraction studies: the PAN-DMF gels only showed an amorphous halo, which was also present in the pure solvent. Nevertheless, their careful DSC measurements showed a broad endothermic peak, from which they calculated an overall melting enthalpy of 101 J/(mol of monomeric units). From his model calculations Beevers [310] decided on a melting enthalpy of 4.5 to 4.8 kJ/(mol of monomeric units). From a combination of the measured and theoretical values of the melting enthalpy, Beckmann and Zenke [309] came to the conclusion that not more than 2% of the PAN is present as crystallites in the gels. Therefore, one can imagine that X-ray diffraction patterns do not show any crystallinity. This had been recognized a few decades ago by Ziabicky et al. [299, 300]. According to Beckmann and Zenke [309] the junction regions in PAN-DMF gels might be a) composed of stable nuclei, where further growth is hindered in the chain direction due to stereo-irregularity of the PAN molecules or b) the result of nucleation processes due to phase separation in the binodal region of liquid-liquid demixing, although no phase separation was observed upon cooling PAN-DMF solutions.

From their ^{13}C NMR studies on PAN gels in mixtures of DMF and water, Lindstedt and Flodin [311] concluded that gelation of PAN solutions was due to crystallization of stereoregular sequences of monomeric units. According to Flodin [312], Lagerkvist showed in his doctoral thesis that the length of these stereoregular (syndiotactic) sequences amounts approximately to 4 to 6 monomeric units. The existence of crystallites with a strongly disturbed orthorhombic structure was suggested by Finke and Zachmann [313] on the basis of their neutron scattering experiments during gelation of 30% PAN-DMF gels.

It might be concluded that it is difficult to prove that crystallinity is the basis for the gelation process of PAN solutions. This is the consequence of the atactic character of conventional PAN. Chiang [314] mentioned that infrared spectra, X-ray diffraction patterns and densities of various samples may be similar, but that they nevertheless may differ in gelation behaviour: small differences in the molecular structure of the samples may have a considerable influence on their gelation tendencies. In reactor powder, it is rather easy to observe crystallinity.

However, once dissolved in solvents like DMA, DMF and DMSO, it seems to be impossible to regain the original crystallinity. This is not an uncommon phenomenon: the same problems arise in regaining the crystallinity of poly(vinyl chloride), once dissolved in a solvent or plasticizer. Nevertheless, crosslinking loci seem to be the result of a nucleation process, resulting in intermolecular crystallites with long polymer chains in between, so that a rubbery network is formed. Due to the generally atactic character of PAN, growth of crystallites in the chain backbone direction is not possible. For that reason equatorial reflections in X-ray diffraction will be observed, whereas non-equatorial reflections are difficult to observe. An exception seems to be the use of propylene carbonate as a solvent. It was found that PAN has a low solubility in this solvent at room temperature but a good solubility at elevated temperatures. This was already recognized by Holland et al. [315], who observed that PAN precipitates and crystallises in the form of single crystals upon cooling such a hot solution. Also ethylene carbonate and *N,N*-dimethylformamide-acrylonitrile mixtures are exceptions: single crystals might also be grown from PAN dissolved in these solvents [316, 317]. From their DSC, X-ray diffraction and NMR studies, Bashir et al. [302–304] concluded that crystallinity is also present in PAN-propylene carbonate gels. However, the crosslinking loci are not very strong, as it was observed that the polymer gradually dissolves during swelling experiments in the solvent used. It has to be emphasized that PAN-PC gels are unusual, because it is suggested that solvated crystallites (or crystallo-solvates) rather than crystallites of the pure polymer are the junctions. The basis for this is the very different X-ray pattern in the gel, which reverts to the normal hexagonal pattern when the gel is extracted. Lately Bashir et al. [318] concluded that a polymer peak is also hidden in the diffuse halo due to the solvent in X-ray diffraction. This was shown by measurements on concentrated solutions (films) of PAN in ethylene carbonate, propylene carbonate and γ-butyrolactone, where the polymer peaks become more important than the solvent peaks when the polymer content increases above 60 wt%.

It is worth mentioning that the gelation of PAN-PC is extremely rapid (seconds rather than hours for PAN-DMF gels). As a result, the rapid thermoreversible gelation of PAN-PC solutions can be exploited for gel spinning and plasticized melt spinning [319].

In a recent paper, Ko et al. [320] showed by means of calorimetric and thermomechanical measurements that two endotherms are present in PAN-propylene carbonate gels when the ageing temperature is lower than 70 °C: the peak temperature of the lower melting endotherm increases linearly with ageing temperature, while that of the higher melting endotherm does not change with the ageing temperature (but it increases with polymer concentration). As the higher melting temperature corresponds to the gel melting temperature, it was concluded that the corresponding crystallites are the cause of the gelation process. This is in agreement with the increase in the gel melting temperature with increasing polymer concentration. From literature studies they concluded that the crystals that melt at the higher temperature have a lamellar structure.

On the other hand, the crystals that melt at the lower temperature are more bundlelike: their melting temperature and amount increase during an ageing process. It was shown that melting of these crystallites also causes a small decrease in the gel rigidity.

6.4 *Viscoelastic Behaviour of Gels*

Although the gelation behaviour of PAN solutions has been the subject of many studies, viscoelastic properties of PAN-DMF and similar gels have been investigated by only a few authors [296–298, 301, 321]. The viscoelastic behaviour of the nowadays so important PAN-PC gels has not yet been studied and hence this may be a fruitful area of study for making original contributions. An exception is the work by Ko et al. [320], who made a hesitating start and concluded that the gelation rate (defined as dG′/dt) of a 4% solution in propylene carbonate increases with ageing temperature (or decreases with increasing supercooling). Bisschops [296–298] studied the viscoelastic properties during the gelation process of PAN ($M_v = 40\,000$ g/mol) in *N,N*- dimethylformamide. The viscosity of a 20% solution showed normal behaviour in the temperature range down to 10 °C: by plotting log η vs 1/T a straight line was obtained (activation energy = 24.4 kJ/mol). Below 10 °C the viscosity increased above the normal value; moreover, at constant temperature an increase with time was observed, whereas extrapolation of the measured values to time = 0 (i.e. the time where the temperature was decreased from above 10 °C to the new temperature) the above mentioned straight line was reached (see Fig. 92).

Creep curves in shear were measured at different stages of the gelation process. At temperatures above 0 °C the rate of gelation was too small for a reasonable measurement. Initially, the creep curves showed liquidlike behaviour, whereas after some time, depending on temperature, rubberlike behaviour

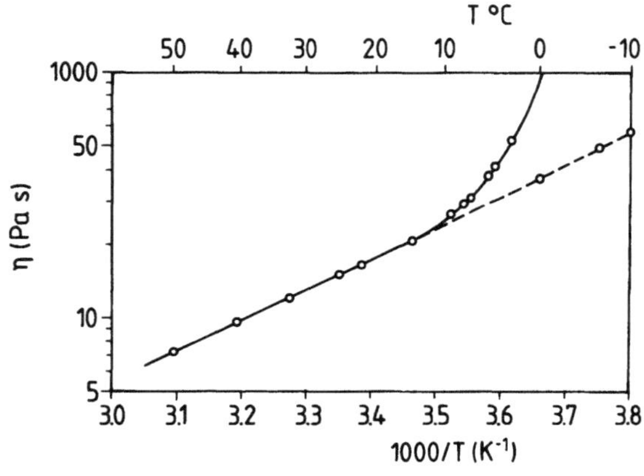

Fig. 92. Viscosity logarithmically plotted vs reciprocal temperature. The points on the *full curve* have been obtained by slow cooling. The points on the *dotted line* below 10 °C are extrapolated values at zero time after having cooled the solutions abruptly to the corresponding temperature. Reproduced from J Polym Sci [Ref. 296] by the courtesy of John Wiley & Sons, Inc

was obtained with complete recovery (see Fig. 93). The shear moduli, calculated from the horizontal regions in the creep curves (called E_∞ in Bisschops' papers), are plotted vs time in Fig. 94, from which it can be concluded that lower temperatures give rise to faster gelation processes and that an equilibrium value is not obtained in a reasonable time. Gelation did not occur at high temperatures e.g. at 25 °C.

However, when a solution was gelled at −9.5 °C for 100 h and subsequently heated to 25 °C the gel melted immediately, whereas after some time gelation occurred. This is shown in Fig. 95.

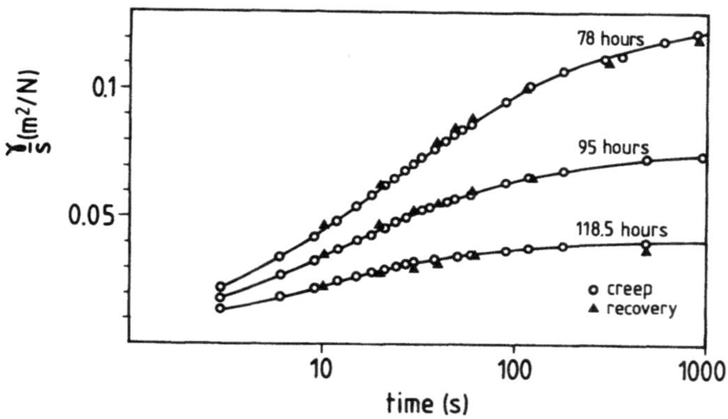

Fig. 93. Reduced creep and recovery curves (ratio of angle of shear and shearing stress vs time) of 20% PAN-DMF rubberlike gels at 0.4 °C, after various times of ageing. Reproduced from J Polym Sci [Ref. 296] by the courtesy of John Wiley & Sons, Inc

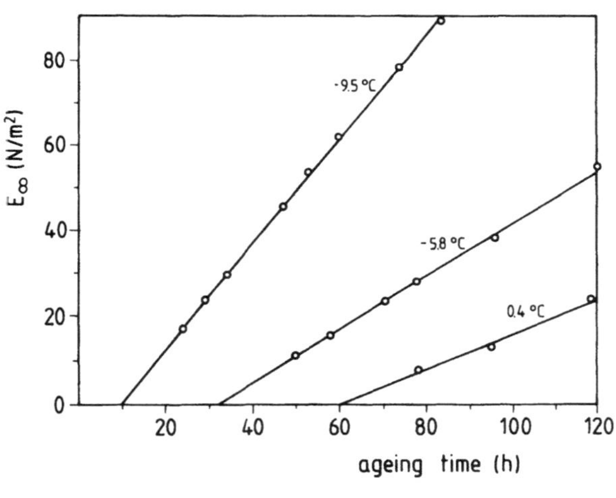

Fig. 94. Equilibrium shear modulus of 20% PAN-DMF gels plotted vs ageing time during ageing at various temperatures. Reproduced from J Polym Sci [Ref. 296] by the courtesy of John Wiley & Sons, Inc

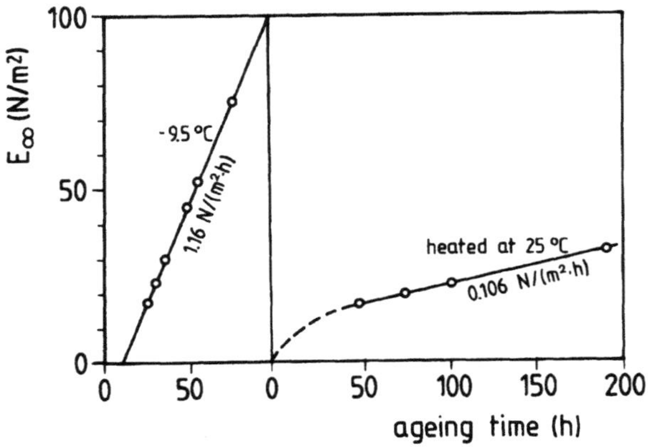

Fig. 95. Equilibrium shear modulus at 25 °C (*right*) after preceeding gelation at −9.5 °C (*left*), plotted vs ageing time. Reproduced from J Polym Sci [Ref. 296] by the courtesy of John Wiley & Sons, Inc

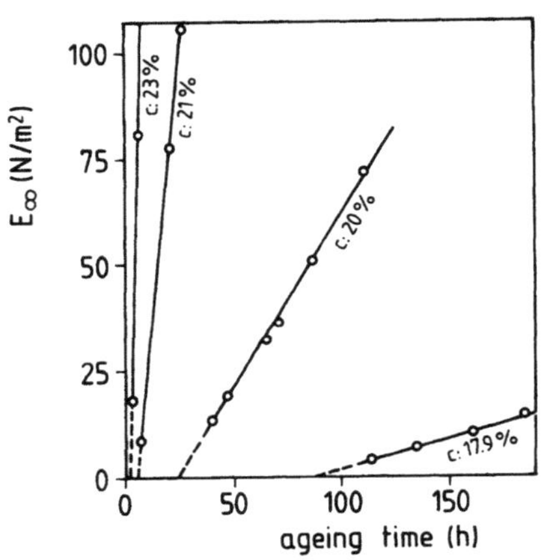

Fig. 96. Equilibrium shear modulus vs ageing time of PAN-DMF gels of various concentrations around 20%, measured at −10 °C. Reproduced from J Polym Sci [Ref. 298] by the courtesy of John Wiley & Sons, Inc

Apparently, crystallization nuclei remain present after melting the gel. Those nuclei will grow at 25 °C despite the fact that it seems impossible to form crystallites during plain maturation at 25 °C. The rate of gelation increases by a factor of 150 when the concentration is varied from 18 to 23% (see Fig. 96) and by a factor of 3 for a 20% solution when the molecular weight is varied from 100 to 150 kg/mol. Bisschops showed that the nature of the solvent also affects the gelation behaviour: when pure DMF is used as a solvent, the gelation time of a 20% PAN solution at −10 °C is 20 h, whereas by addition of 2% water to DMF gelation occurs immediately at −10 °C. Beckmann and Zenke [309] measured the critical gel concentration as a function of temperature for various weight average molecular weights and plotted log c* vs the reciprocal gel temperature (see Fig. 97). From the slope of the straight lines they calculated the

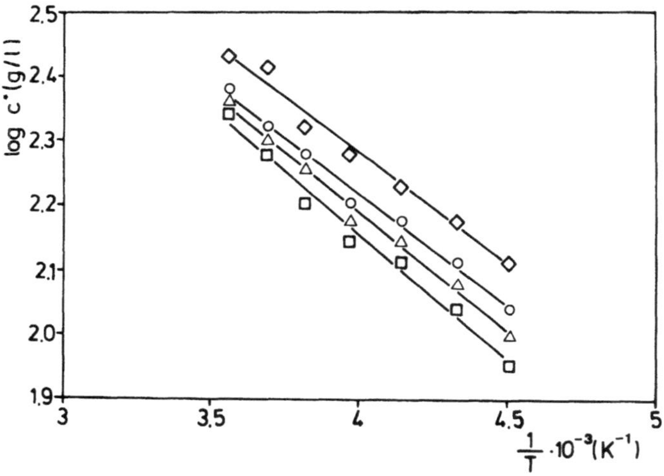

Fig. 97. Critical gel concentration of solutions in DMF of various PAN samples, differing in molecular weight, plotted vs reciprocal temperature. $\bar{M}_w$ in kg/mol: (◇) 192, (○) 286, (△) 360, (□) 462. Reproduced from Colloid Polym Sci [Ref. 309] by the courtesy of the authors and of Steinkopff Verlag Darmstadt, FRG

average heat of formation of a junction zone, assuming pairwise association, according to the method of Tanaka [322]: $\Delta H = -6 \pm 1.5$ kJ/(mol of junctions) (increasing somewhat with increasing temperature). It has been mentioned before that the melting enthalpy of PAN was calculated by Beevers [310]: $\Delta H_m = 4.5$ to 4.8 kJ/mol. For this reason Beckmann and Zenke [309] concluded that an ordered junction has a low level of thermodynamical stability.

As the gelation phenomenon of PAN solutions is a nuisance for fibre solution spinners, work has been done on the gelation behaviour of acrylonitrile-vinyl acetate copolymers. A systematic study has been performed by Paul [301]. The addition of a small amount of comonomer generally does not change the hexagonal lattice of PAN: it is incorporated in the lattice as defects. In the dry solid state the X-ray diffractograms are hardly altered compared with the homopolymer. According to Frushour [323], up to about 12% of vinyl acetate could be incorporated into the hexagonal lattice: the polymer becomes amorphous only when the comonomer level rises to approximately 25%. However, addition of a strange comonomer can have a tremendous effect on the gelation tendency of PAN, depending on the nature of the comonomer, due not only to the defects in the crystal structure but also to the change in interaction with the solvent. Figure 98 shows that the gel melting point of a 27% solution in *N,N*-dimethylacetamide decreases from 128 to approximately 60 °C, while the vinyl acetate content increases to only 9%. From Fig. 5 in [301], where the gel melting point of a copolymer with 7.7% vinyl acetate in various solvents is plotted vs the polymer weight fraction, a plot of ln c vs 1/T is constructed (see Fig. 99) from which the gel melting enthalpy was calculated with the aid

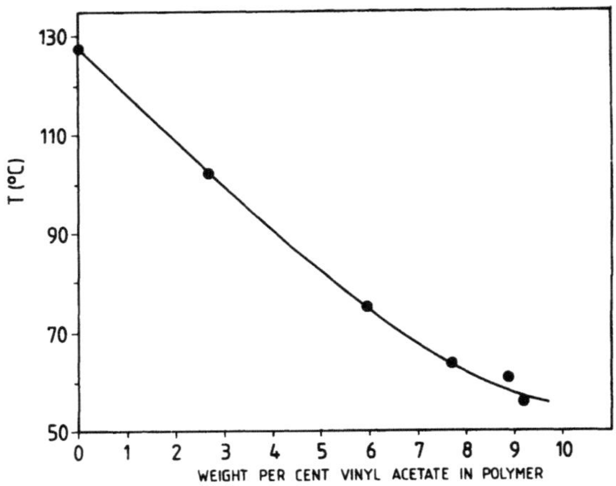

Fig. 98. Dependence of the gel melting point on the vinyl acetate content in 27 wt% gels of acrylonitrile-vinyl acetate copolymers in *N,N*-dimethylacetamide. Reproduced from J Appl Polym Sci [Ref. 301] by the courtesy of John Wiley & Sons, Inc

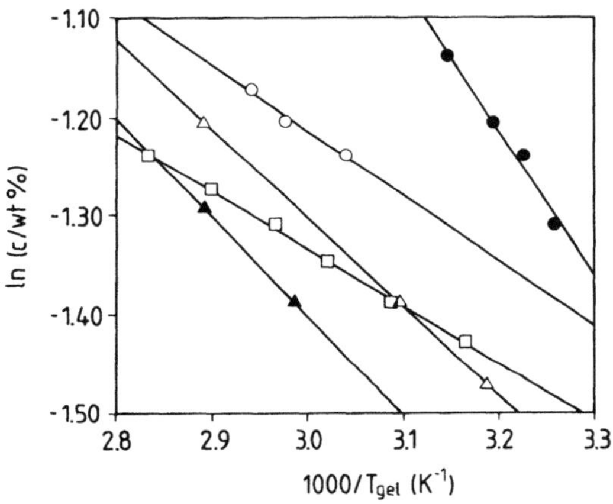

Fig. 99. Logarithm of the critical gel concentration plotted vs reciprocal gel melting temperature of acrylonitrile-vinyl acetate (7.7 wt%) copolymers dissolved in various solvents: (△) ethylene carbonate, (▲) γ-butyrolactone, (○) *N,N*-dimethylformamide, (●) dimethyl sulfoxide, (□) *N,N*-dimethylacetamide (constructed after data presented in [301])

Table 4. Gel melting enthalpy and gel melting temperature of acrylonitrile-vinyl acetate (7.7 wt%) copolymers in various solvents

Solvent	$\Delta H_{m,gel}$ (kJ/mol)	$T_{m,gel}$ (°C)
N,N-dimethylacetamide	4.8	80
N,N-dimethylformamide	5.5	58
ethylene carbonate	7.5	40
γ-butyrolactone	8.3	62
dimethyl sulfoxide	12.1	30

of linear regression. Results are given in Table 4 together with the gel melting temperatures. There is a slight tendency for a decrease in melting temperature with increasing melting enthalpy.

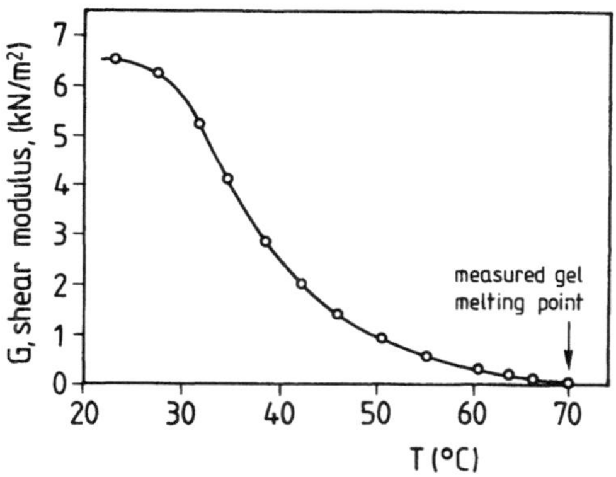

Fig. 100. Shear modulus of a 27.7 wt% acrylonitrile-vinyl acetate (7.7 wt%) copolymer-DMAc gel as a function of temperature. Reproduced from J Appl Polym Sci [Ref. 301] by the courtesy of John Wiley & Sons, Inc

It appears that this PAN-vinyl acetate copolymer has a strong gelation tendency in dimethyl sulfoxide. This is a solvent in which PAN has a strong gelation tendency, although still much less than PAN-PC solutions.

The shear modulus of a 27.7 wt% solution of the same copolymer in *N*,*N*-dimethylacetamide was measured as a function of temperature (see Fig. 100). The gelation measurements were performed starting at room temperature and then increasing the temperature in steps. The gel was held at the new temperature for 1 h at least before measurements were started. Unfortunately, it was not mentioned whether or not an equilibrium state was reached in that time. From the measurements by Bisschops [296–298] on solutions of pure PAN it is clear that an equilibrium value was not reached within 100 h, but it is possible that in gels of the copolymers, with the smaller tendency for gel formation, an equilibrium state is reached earlier. Above 30 °C the decrease of the shear modulus is fast.

6.5 Conclusions

The gelation behaviour of PAN solutions is the result of crystallinity of parts of the generally atactic polyacrylonitrile. In addition, dipole-dipole interactions seem to play a role. Moreover, the solvent molecules may also be incorporated in the crystals present in gels, especially when propylene carbonate is used as a solvent. The result is the presence of crosslinks with a high functionality. Although the gelation of PAN solutions is a disadvantage in the solution spinning process, use is made of these gelation properties in the gel spinning process.

Measurement of the viscoelastic properties of PAN gels has been an underdeveloped area. This is understandable, because non-gelling solvents have been sought for the solution spinning process in the past. However, now that gelation spinning could become an important process, it would be worthwhile to investigate viscoelastic behaviour during the gelation of PAN solutions more systematically.

7 Polyethylene

7.1 Introduction

Although much research has been done concerning the gel spinning behaviour of high molecular weight polyethylenes, most reports deal with the properties of the resulting strong fibers (see e.g. [324–326]). Only a few studies have been performed concerning the gelation behaviour of polyethylene solutions, and most of them did not take into consideration the viscoelastic properties of those gels.

Usually polyethylene precipitates in crystalline lamellae upon cooling of dilute solutions. However, in non-dilute solutions crystallisation may take place without precipitation of crystals, but with the formation of a gel in which tiny crystals act as crosslinks.

It is well known that the crystallisation behaviour of the various kinds of polyethylene, i.e. high density polyethylene (HDPE), low density polyethylene (LDPE) and linear low density polyethylene (LLDPE), is quite different. Hence, it is understandable that the polymers also differ in gelation behaviour. For that reason distinction will be made in the description of these polyethylenes.

7.2 Phase Behaviour

Properties of polyethylene gels (or of crystalline gels in general) depend on the way the gels are formed: via crystallisation in a good solvent or via liquid-liquid-demixing (L-L-demixing) in a bad solvent. In good solvents (e.g. decaline and *o*-xylene), used in the production of strong fibres, the melting curve and crystallisation temperatures (T_m and T_{cr}) as a function of the volume fraction of dissolved polymer are far above a possible two-phase region. On the other hand, in bad solvents (e.g. diphenyl ether), used for the production of porous membranes, these temperatures are in the vicinity of the two-phase region, i.e. just above this region, or the $T_m(\phi)$ and the $T_{cr}(\phi)$ curves cross the two-phase region (see Fig. 101), ϕ being the volume fraction of polymer in the system [327].

A more detailed picture of the interference between the crystallisation curve and the two-phase region is shown in Fig. 102, where the crystallisation curve crosses the two-phase region [327, 328]. The dotted parts of the binodal lines show the metastable part of the L-L-demixing region, when crystallisation does not occur for some reason. If a solution is cooled from point A (volume fraction of polymer ϕ_1), then phase separation will just start at point A_1. Upon continuing cooling L-L-demixing will take place (at point A_2 into phases ϕ_{1a} and ϕ_{1b}). An invariant situation is reached at temperature T_i, where a three-phase equilibrium exists and where phase separation changes into crystallisation. This results in a solid polymer phase and a very dilute liquid polymer solution with polymer

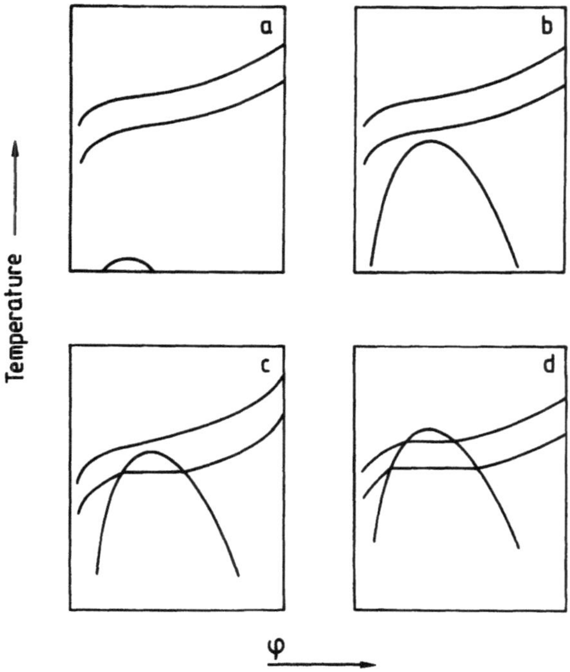

Fig. 101a–d. Schematic representation of phase diagrams with interference between melting (upper curves) and crystallisation curves (lower curves) and the two-phase region. Reproduced, with permission of the author, from [327]

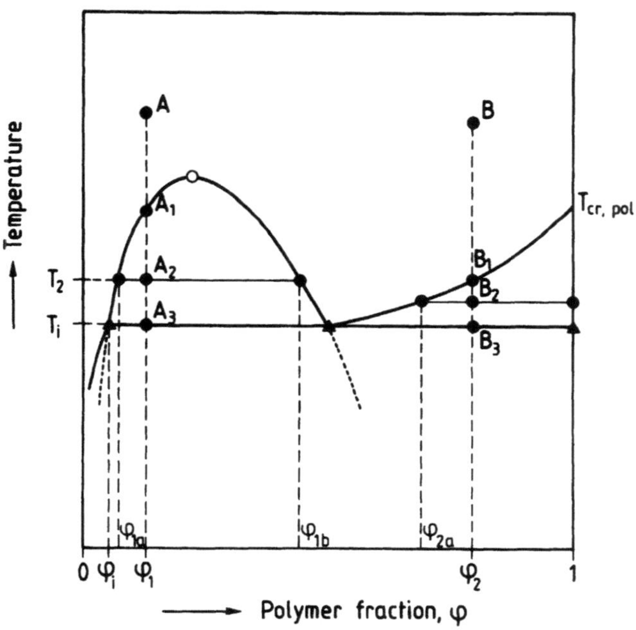

Fig. 102. Phase diagram for a monodisperse polymer/solvent system, with interference between liquid-liquid-demixing and liquid-solid-demixing. Reproduced from Makromol Chem [Ref. 328] by the courtesy of the authors and of Hüthig & Wepf Verlag Publishers, Zug, Switzerland

fraction ϕ_i. It has to be emphasized that this is the case when real thermodynamic equilibrium is reached. In general, however, densely packed lamellar crystals will contain a considerable amount of solvent in the interlamellar zones. Thermodynamic equilibrium on the microscopic scale might be reached, although on the macroscopic scale the system does not necessarily have to be phase separated into two distinct phases, because a cocontinuous network might also be formed. When the whole system is separated into ϕ_i and $\phi = 1$, then further cooling results in phase separation, following the lower left part of the crystallisation curve.

Upon cooling a highly concentrated polymer solution from point B (polymer volume fraction ϕ_2), crystallisation just starts when, at point B_1, the crystallisation curve is reached. Upon further cooling crystallisation continues. When point B_2 is reached, the system will be separated into pure polymer crystals ($\phi = 1$) and a polymer solution, ϕ_{2a}. When at point B_2 the temperature T_i is reached, the system will be separated into two phases $\phi = 1$ and the dilute solution ϕ_i; crystallisation continues until the whole system has been separated in pure polymer crystals and dilute solution.

7.3 High Density Polyethylene

A few decades ago, Pennings et al. [329] mentioned the gelation of dilute solutions of high molecular weight linear polyethylenes. They discovered that stirring a solution during cooling was important in the formation of gels. Smith and Lemstra [330] measured weak but sharp reflections in the X-ray diffraction patterns of those polyethylene gels. These were identified with 110 and 200 reflections of the orthorhombic crystal structure of PE. According to Pennings [331], the viscosity of PE solutions increased prior to the gelation up to a maximum value, whereupon a decrease was observed. This was interpreted as buildup and subsequent breakdown of the gel network. This phenomenon was confirmed by Narh et al. [332].

Coombes and Keller [333] observed that solutions of high molecular weight polyethylene often formed macroscopic gels after they had been stirred for a certain period before cooling, while without stirring prior to cooling gelation did not take place. A systematic study of this phenomenon was performed by Barham et al. [334] and by Narh et al. [332]. A 0.6% polyethylene ($\bar{M}_w \geq 1.5 \cdot 10^6$) solution in xylene was sheared in Couette geometry. After arresting the cylinder, the solution was cooled to room temperature and gelation occurred, while without stirring no gelation took place. It turned out that stiffer gels were formed when the previous stirring temperature was lower. Even after 4 h of rest following the stirring process at 134 °C, gels were still formed after cooling. A minimum stirring time was needed for gelation at room temperature; this time was shorter when the stirring temperature was lower. A maximum temperature exists above which stirring does not result in gelation at room temperature. This temperature increases with concentration (see Fig. 103). The authors concluded that, during stirring, elongated structures are formed which behave as nuclei for the gelation (i.e. crystallisation) process at room temperature. Those structures remained present for a long time after stirring, even more than 4 h after stirring at 120 °C (see Fig. 104).

The structure of the crosslinks in a polyolefin gel and the nature and mechanism of the thermoreversible sol gel transition are very complex. Smith and Lemstra [330] studied the thermal properties of a 2% PE-decaline gel. DSC-thermograms showed two endothermic peaks around 90 °C (see Fig. 105). The higher temperature peak grows, becomes sharper, and shifts to higher temperatures during maturation at the cost of the lower temperature peak (87 °C). Similar results were obtained by Narh et al. [332]. However, these authors also observed a broad endotherm from 110 to 150 °C, which corresponded to the gel melting temperature range (see Fig. 106). This endotherm was not present when the system was kept at 150 °C for 2 h, prior to the gelation process.

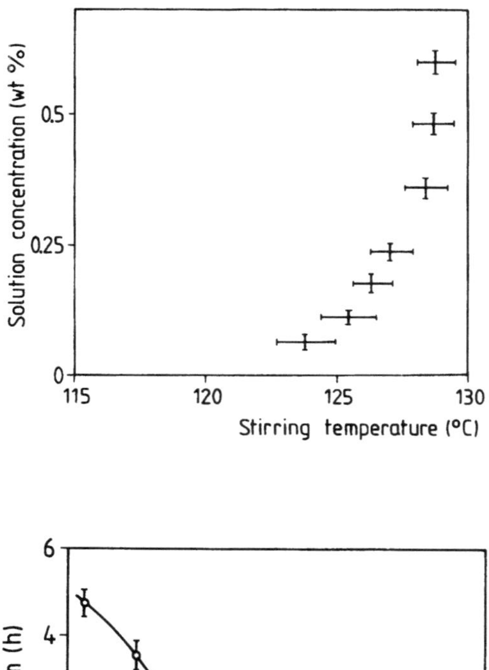

Fig. 103. Concentration of PE ($\bar{M}_w = 1.5 \cdot 10^6$) solutions in xylene vs maximum stirring temperature from which solutions will gel upon cooling; the *horizontal error bars* indicate the uncertainty in determining the final state of the cooled product and the *vertical bars* the uncertainty in determining the actual solution concentration. Reproduced from Colloid Polym Sci [Ref. 334] by the courtesy of the authors and of Steinkopff Verlag Darmstadt, FRG

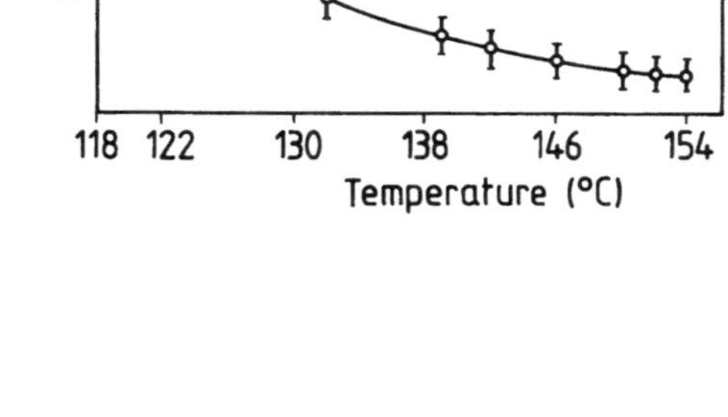

Fig. 104. Time required to destroy the memory of the stirring effect on a high molecular weight polyethylene ($\bar{M}_w = 1.5 \cdot 10^6$) solution in decalin (0.6%) vs stirring temperature. Reproduced from Macromolecules [Ref. 332] by the courtesy of the authors and of The American Chemical Society

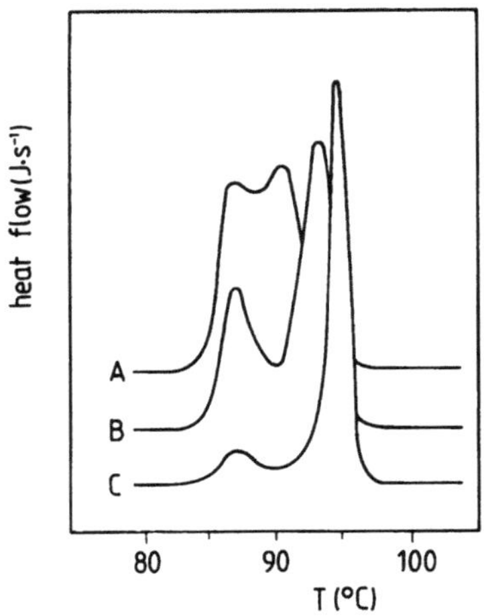

Fig. 105. DSC endotherms of a 2% HDPE-decalin gel. Heating rates: A: 40 °C/min; B: 20 °C/min and C: 5 °C/min. Reproduced from J Mater Sci [Ref. 330] by the courtesy of the authors and of Chapman & Hall

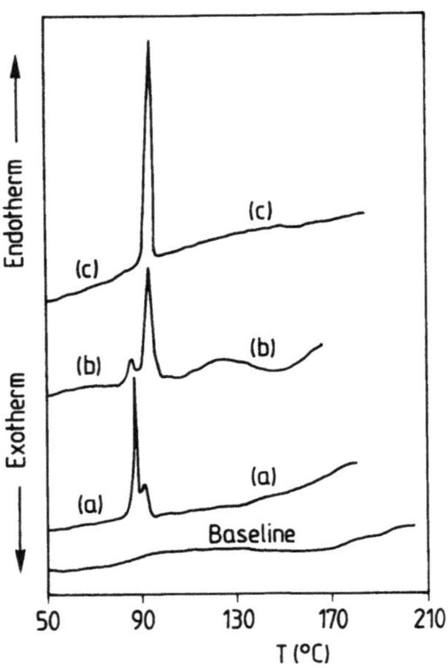

Fig. 106. DSC thermograms for both PE single crystals and gels in decalin; (a) single-crystal suspension; (b) gel in decalin; (c) same as (b) but kept for 2 h at 150 °C before cooling. Heating rate 5 °C/min. Sensitivity range 0.4 mJ/s. Reproduced from Macromolecules [Ref. 332] by the courtesy of the authors and of The American Chemical Society

The sharp peaks around 90 °C were attributed to melting of chain folded crystals.

7.4 Linear Low Density Polyethylene

On the basis of their thermodynamics and structure studies (SAXS and Raman LAM), Mandelkern et al. [335–339] concluded that the crystallites found in gels are essentially identical to those formed from the dilute solutions, as long as the molecular weights of the polymers are the same. They concluded that the generally accepted concept that a fringed micellar crystalline structure is necessary for gelation to occur by a crystallisation mechanism is not correct. Structural properties and thermodynamics of the copolymer crystallites formed in dilute solution and those formed in gels are identical over the complete range in co-unit composition. This holds for well defined lamellae and for very small crystallites as well. Hence, the fringed micelle concept is fairly limited.

By comparing the results of Mandelkern et al. [335–339] with those of Narh et al. [332], the conclusion can be reached that the crosslinking sites in the gel network consist of fringed micelles of relatively small, extended chain crystals. They melt at higher temperatures than the chain folded crystals that are also present in the gels and probably also contribute to the crosslink sites.

Concerning the role of stirring a system prior to gelation, Mandelkern et al. [335–339] concluded from their study of gelation of a wide molecular weight range of linear polyethylene fractions (M = 10–6000 kg/mol) that there is no necessity for any type of stirring or pre-stirring: gels were also obtained without any (pre-) stirring. According to these authors, stirring is necessary only if the concentration is not correctly chosen.

The critical concentration c* needed for gelation was also studied as a function of molecular weight, for linear polyethylenes as well as for ethyl branched polyethylenes (hydrogenated polybutadienes). It was concluded that the concentration dependence of the linear polymers was smaller than that of the branched polyethylenes, especially at the lower gelation temperatures. The expected molecular weight dependence of overlap concentration $c^* = KM^{-a}$, with $a \approx 0.5$, was not confirmed. It was concluded that, for the homopolymers and the copolymers, molecular overlap is not a sufficient condition for gelation, as was deduced by Tan et al. [284] from their study of the gelation of atactic polystyrene in carbon disulfide (the result of L-L phase separation), where it turned out that $a = 0.5$. Gelation of solutions of crystalline polymers apparently also depends on the way crystallites are formed: whether or not connected by non-crystallised polymer chains. In that sense, large, well developed crystallites are less effective as crosslinks than the smaller fringed micelles.

Thermodynamic properties and characteristic dimensions of the crystallites in linear PE gels and those formed in dilute solution are essentially identical. Data for copolymers extrapolate smoothly to that for homopolymers: in Fig. 107 the influence of the content of small branches on the (gel) melting temperatures of 9.5% (w/v) gels in toluene (hydrogenated polybutadiene gels, ethylene-vinyl acetate gels and hydrogenated polybutadiene solution crystals) is demonstrated. The melting temperature decreases smoothly with increasing branches content and extrapolates well to that of linear polyethylene. The molecular weights of the polybutadienes are comparable. By chance, the results of the ethylene-vinyl acetate copolymers are identical to those of the other polymers.

The analysis of the melting enthalpy of these polymers is shown in Fig. 108, where the degree of crystallinity is plotted vs the branches content. As could be

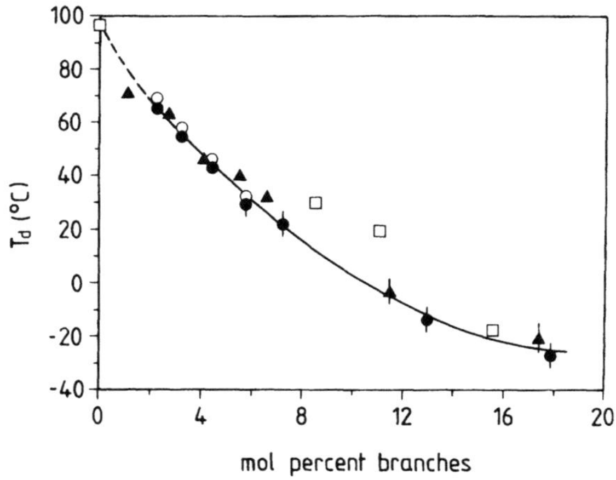

Fig. 107. Dissolution or melting temperature of 9% solution of gels and of dilute solution crystals vs mole percent branches: hydrogenated polybutadiene gels (●); ethylene-vinyl acetate gels (▲); hydrogenated polybutadiene solution crystals (○); chlorinated polyethylene gels of [340] (□). Reproduced from Macromolecules [Ref. 338] by the courtesy of the authors and of The American Chemical Society

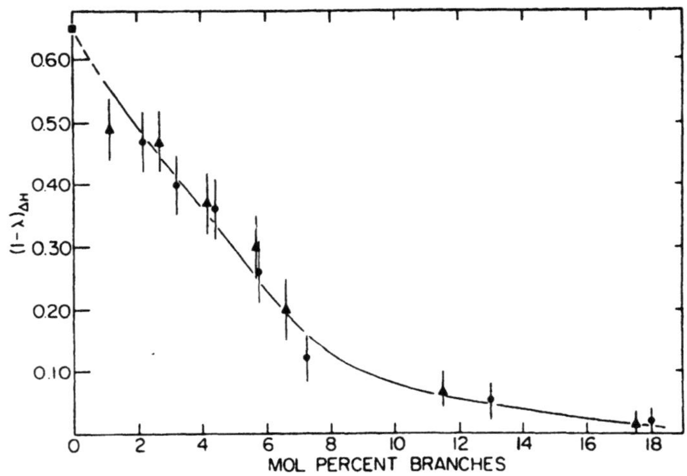

Fig. 108. Degree of crystallinity calculated from the enthalpy of fusion of 9% PE gels vs mole percent of branches: hydrogenated polybutadiene (●); ethylene-vinyl acetate (▲). Reproduced from Macromolecules [Ref. 338] by the courtesy of the authors and of The American Chemical Society

expected, the crystallinity decreases with increasing branches content and at 8% of branches the crystallinity has practically disappeared, but is still present. Polyethylenes with more than 9 mol% of co-units are not crystalline at room temperature after rapid crystallisation from the melt, but it has been reported that about 10% crystallinity is developed in 9 mol% copolymer after solution crystallisation. Moreover, gel formation is in general possible, because not much crystallinity is needed for gel formation: well defined crystals are not needed for gelation to occur.

Okabe et al. [341] observed the presence of spherulites in LLDPE gels by polarizing microscopy. They also studied the influence of the kind of comonomers in LLDPE gels: it appeared that the gel temperature increases by approximately 8 °C in the series butene-1, 4-methyl pentene-1 and octene-1. Hence, the gel temperature of LLDPE depends not only on concentration and molecular weight but also on the kind of comonomer.

7.5 Low Density Polyethylene

In the early 1980s, Matsuda et al. [342] observed that solutions of low density polyethylene with long branches in a mixture of cyclohexane and carbon disulfide transformed into a gel upon cooling. Later, it was found that thermoreversible gels were also formed from solutions in decalin, tetralin, benzene, toluene and *o*-xylene [343]. From X-ray diffraction patterns it was concluded that the gels were crystalline. Experimental results showed that the gel formation temperature depended on concentration and on the molecular weight of the polymer. They also found that the gelation temperature increased with concentration and molecular weight of the polymer: $1/T_{gel}$ plotted vs $\ln c$ yielded straight lines. With the aid of the method described by Takahashi et al. [344], it was concluded that $\rho = 3$ to 4 (i.e the number of polymer molecules present in a crosslink site) whereas $\zeta \approx 9$ (i.e. the number of monomeric units along the chain in a crystalline crosslink site). This appears to be nearly independent of molecular weight in a series of fractionated samples [20, 345]. Hence the crosslink functionality amounts to 6 to 8.

In polyethylenes with a special copolymeric nature, in that branches of various lengths are present, gel formation takes place. According to Takahashi et al. [19], the melting temperature $T_{m,gel}$ of such a gel is lower when the number of long chain branches is higher. They also showed that the melting enthalpy of the gels, ΔH_m, decreased with increasing number of long chain branches, as is clear from Fig. 109. The melting enthalpies, ΔH_m, calculated from the slopes of the various lines are 252, 52 and 32 kJ/(mol of crosslinks) for 0, 4.07 and 7.18 long branch points on average per polymer molecule, respectively. From their experiments the authors conclude that the average functionalities of the

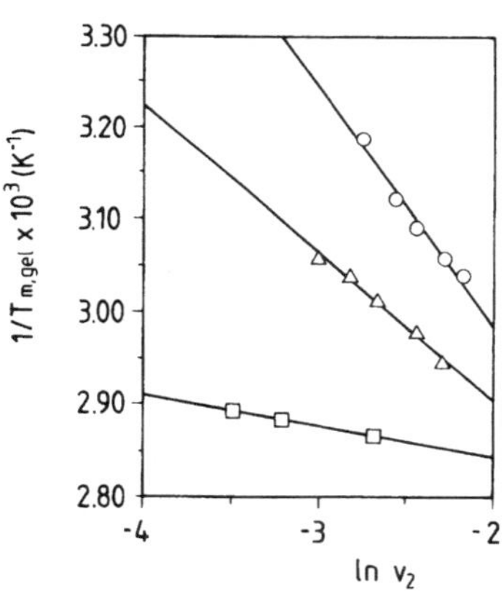

Fig. 109. Reciprocal gel-melting temperature vs $\log v_2$ (v_2 = polymer volume fraction) for low density polyethylenes of various long branches content: (○), (△) and (□) 7.18, 4.07, and 0 branch points of long branches per polymer molecule, respectively. Reproduced from Polym J [Ref. 19] by the courtesy of the authors and The Society of Polymer Science, Japan

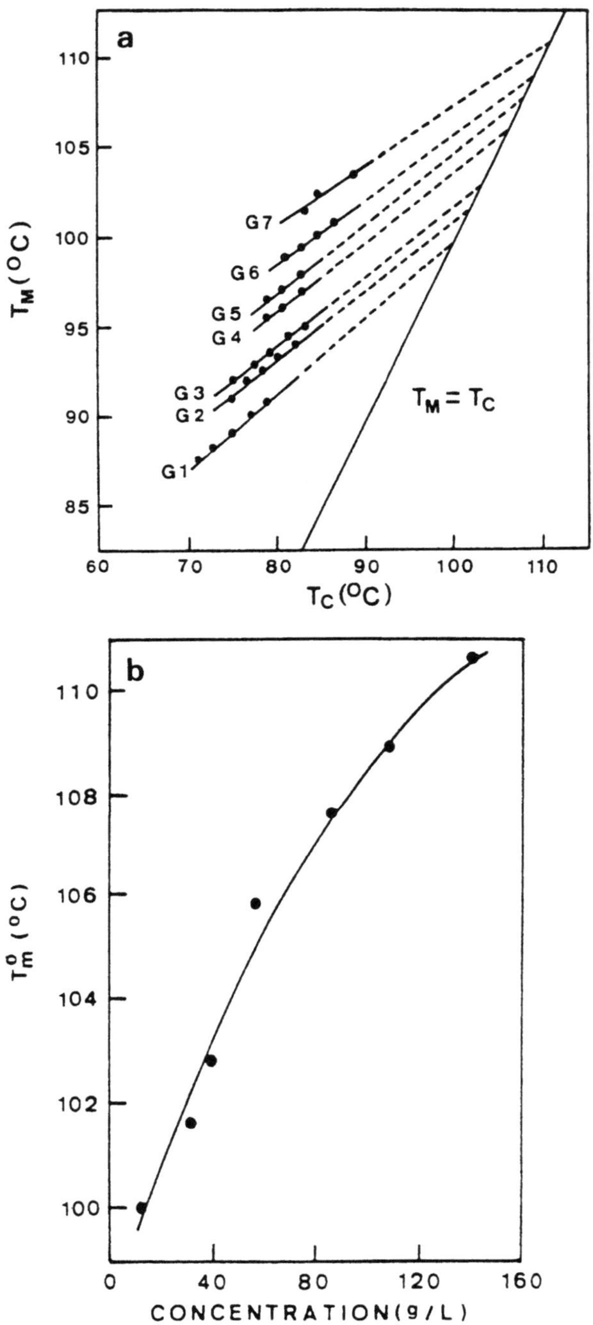

Fig. 110. **a** Hoffmann-Week plots of UHMWPE-decalin gels. **b** Equilibrium melting temperature vs UHMWPE concentration. Reproduced from Polym Eng Sci [Ref. 346] by the courtesy of the authors and of The Society of Plastics Engineers, Brookfield, UK

crystalline crosslinks varied from 2.2 up to 14. Another approach is to measure the equilibrium shear modulus G_e of those gels, complemented, for example, by the number of crystallites (crosslinks), by SAXS investigations. In that way, the functionality of the multifunctional crystalline crosslinks might be calculated just as for poly(vinyl chloride) gels [40, 127].

Jo et al. [346] measured the sol-gel transition and gelation kinetics of UHMWPE (M = 4000 kg/mol) in decalin. Results for T_{gel} are almost the same from DSC and TMA. Hoffmann-Week plots (T_m as a function of T_c) for various concentrations yielded a T_m^o vs concentration plot (see Fig. 110).

7.6 Chlorinated Polyethylenes

Tan et al. [340] chlorinated polyethylene in solution and in suspension. It appeared that the gelation tendency decreased with increasing chlorine content. For the solution chlorinated polyethylenes this decrease was much more dramatic than for the suspension chlorinated polyethylenes. Apparently, in suspension chlorination only the amorphous parts of the polyethylene are chlorinated. This has little influence on the crystallinity of the polymer. However, in solution chlorination of the polymer will be randomly attacked by chlorine, and thus the amorphous parts and the crystalline regions are chlorinated in the same measure. Naturally, this causes a decrease in crystallinity. Hence, the tendency to form gels decreases tremendously. As a result, both the gelation temperature and the heat of gelation decrease with increasing chlorine content. It must be emphasized that the tendency to form gels is present, even in samples where no crystallinity could be detected by differential scanning calorimetry. The conclusion is again such that not much crystallinity is needed for gelation to occur and, moreover, that ill-defined crystallites are able to act as crosslinks. These chlorination effects resemble closely the chlorination of poly(vinyl chloride) in solution and in suspension, as was reported by Dorrestijn et al. [123].

7.7 Viscoelastic Properties

The number of studies on the viscoelastic properties of polyethylene gels is minute. In a detailed study, Li et al. [347] investigated for a large range of polyethylenes, differing in molecular weight, molecular weight distribution and composition, the compression modulus as functions of the concentration and molecular weight. The concentration dependence of the compression modulus appeared to be quadratic up to 2.3 wt% for polyethylenes with broad and narrow molecular weight distributions (see Fig. 111). For gels just above their critical gelation concentration, the molecular weight dependence of compression modulus is shown in Fig. 112. It turns out that the compression modulus scales with the number average molecular weight, independent of the molecular weight distribution. It is unrealistic, however, to draw straight lines through the points:

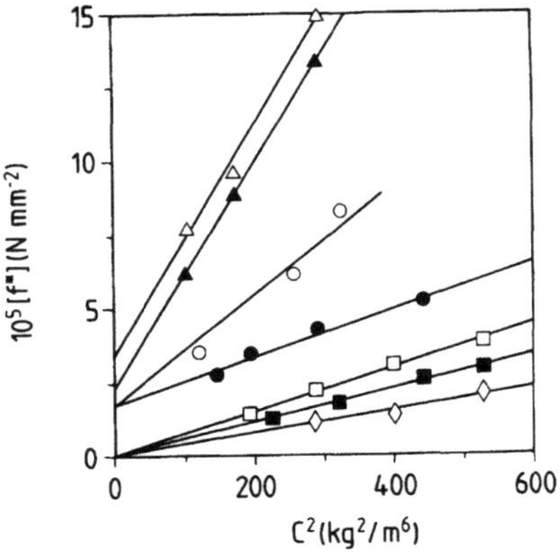

Fig. 111. Compression moduli of various polyethylene-decalin gels vs square of the polymer concentration; $\bar{M}_n$ and $\bar{M}_w$ (kg/mol): (◇) 5.80 and 10.1; (■) 22.5 and 53.5; (□) 11.0 and 150; (●) 91.0 and 500; (○) ? and 1000; (▲) 250 and 2000; (△) 800 and 8000, respectively. Reproduced from Macromolecules [Ref. 347] by the courtesy of the authors and of The American Chemical Society

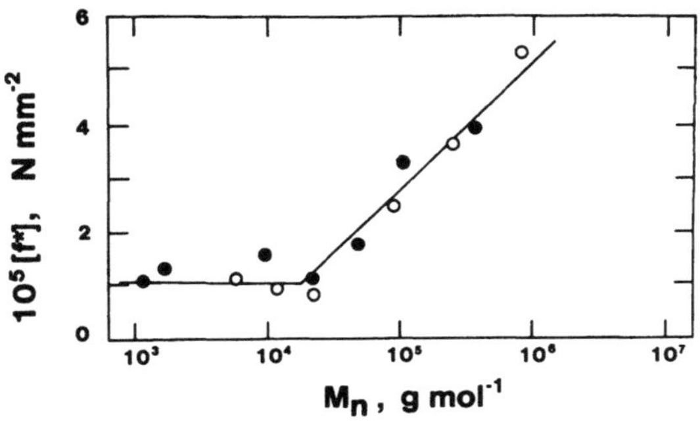

Fig. 112. Compression modulus of polyethylene-decalin gels, extrapolated to c*, vs number average molecular weight; *open circles*: whole polymers; *solid circles*: fractionated (nearly monodisperse) polymers. Reproduced from Macromolecules [Ref. 347] by the courtesy of the authors and of The American Chemical Society

a curved line would be more appropriate. Calculations of $\bar{M}_c$, the molecular weight between crystalline crosslinks, yielded too small values. One of the reasons is that the authors did not take into consideration that the gels not only consist of real network but that a sol fraction will also be present, especially for low values of the compression modulus. Moreover, $\bar{M}_c$ was calculated with a crosslink functionality $f = 4$, which certainly will not be the real functionality in these crystalline gels. Approximate calculations with Te Nijenhuis' model [39–42] show that the sol fraction amounts to about 0.8–0.9, depending on the chosen functionality of the crystalline crosslinks. The gels turned out to be very fragile and, although their concentration was very low, high compression moduli were measured, up to $E \approx 10^5$ N/m^2.

Chung and Zachariades [348–350] have determined the viscoelastic behaviour of ultra-high molecular weight polyethylene (M varying from 2000 to 8000 kg/mol) in paraffin oil (2–5 wt%) by rheological and rheo-optical techniques. The authors observed spherulites and lamellar single crystals when the gels were prepared under quiescent conditions, whereas shish-kebab crystals were observed under stirring conditions and a mixture of single and shish-kebab crystals under uncontrolled conditions. Their measurements of the dynamic moduli also reveal that these gels are very fragile and cannot withstand small shearing forces: the storage modulus of a certain gel decreases during continuing frequency sweeps. In this way it is difficult to obtain equilibrium values. The authors attribute the decrease of the storage moduli to reduced rigidity of the crystals and simultaneous loss of trapped entanglements under the action of the hydrodynamic forces. On the other hand, the authors showed [350] from X-ray experiments that the two crystal peaks (at 2θ angles of about 21.5° and 24°) were sharper after dynamic measurements, whereas DSC analysis revealed that the melting enthalpy of the gels had increased from 8.42 to 15 J/g, and the melting temperature remained at about 116 °C. This could be the consequence of recrystallisation, by which large crystals grow at the cost of smaller ones.

Recently a rather thorough study of the phase behaviour and morphology of HDPE in good and bad solvents in relation to viscoelastic behaviour was carried out by Aerts et al. [327, 328]. These authors made use of various HDPE samples. Data are shown in Table 5.

In Figs. 113 and 114 the dynamic moduli of the three polyethylenes dissolved in the good solvent decalin are shown as a function of angular frequency. From these figures it becomes clear that the dynamic moduli increase with increasing molecular weight.

The storage moduli are not frequency independent in the frequency range shown, neither for HDPE1 and HDPE2 gels nor for UHMWPE gels, although the slope for the latter is smaller than for the former gels. Moreover, the loss moduli of the HDPE1 and HDPE2 gels show minimum values at low frequencies, which means that at still lower frequencies a relaxation mechanism must be present. This is not clearly perceptible in the UHMWPE samples. These differences in viscoelastic behaviour are due to differences in morphology of the gels. The molecular weights of HDPE1 and 2 are relatively low, so that during nucleation and crystallisation the diffusion rate of the polymer molecules is high enough to enable a rapid crystallisation of relatively few nuclei, which will become relatively large crystals without much molecular coupling between the crystals. The crystals consist of lamellar material concentrated in spherulite-like structures. Connectivity between the crystals is not molecular but is due to Van der Waals attraction, which might be rather strong as a result of the large surface areas of the crystals. This will result in an adhesively phase connected particle network. On the other hand, the molecular weight of UHMWPE is much larger, so that the rate of diffusion of the polymer molecules is too small to

Table 5. Molecular weights (kg/mol) of the high density polyethylenes, as used by Aerts et al. [327, 328]

Sample	$\bar{M}_n$	$\bar{M}_w$	$\bar{M}_v$
HDPE1	12	80	
HDPE2			240
UHMWPE	400		2000

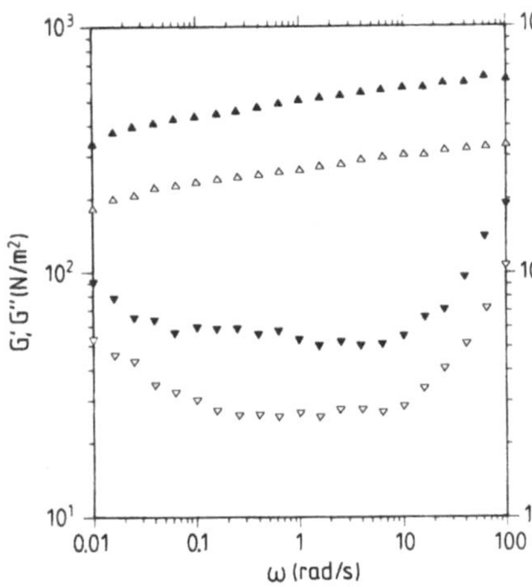

Fig. 113. Dynamic moduli of HDPE1/decalin and HDPE2/decalin gels vs angular frequency (volume fraction of polymer $\phi = 0.05$) deformation amplitude = 0.5%); (△) and (▲) storage moduli; (▽) and (▼) loss moduli; *open symbols* HDPE1 gel; *filled symbols* HDPE2 gel. Reproduced, with permission of the author, from [327]

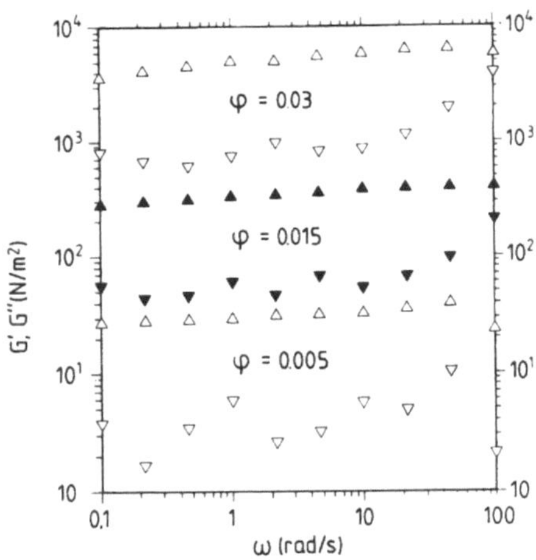

Fig. 114. Dynamic moduli of UHMWPE/decalin gels of various volume fractions of polymer (ϕ = 3%, 1.5% and 0.5%) vs angular frequency; (deformation amplitude = 0.5%); (△) and (▲) storage moduli; (▽) and (▼) loss moduli. Reproduced, with permission of the author, from [327]

enable a rapid crystallisation. The result is the formation of such a large number of growing nuclei that parts of the same polymer molecule are present in different tiny lamellar crystallites. Hence, a network will be formed with crystalline crosslinks that are molecularly connected. A schematic picture of the two different networks is shown in Fig. 115 (for atactic polystyrene the same kind of structures were also proposed by Keller et al. [276], as shown in Sect. 5).

A further distinction between both types of network is the difference in ability to withstand deformation forces: the gels obtained from HDPE1 and 2 cannot withstand deformation amplitudes higher than 0.4% without being disturbed, whereas this amplitude is approximately 2% for the UHMWPE networks.

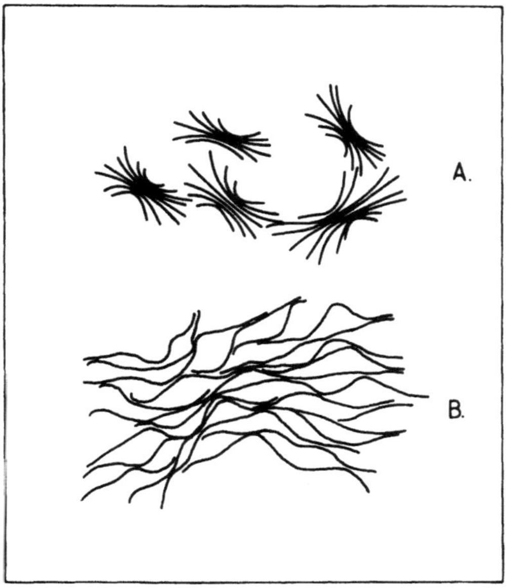

Fig. 115A, B. Schematic representation of polyethylene gels in a good solvent: **A** low molecular weight gel; **B** high molecular weight gel. Reproduced, with permission of the author, from [327]

The transition from the particle networks to the molecularly connected networks is due to the increase in molecular weight. Another way to change the type of network formed in the low molecular weight HDPE is to disturb its crystallinity. This was shown by Aerts et al. [328] by making use of polyethylenes that were copolymerised with vinyl acetate (EVA polymers) to various degrees (according to [328] by E. Paredis, 1991, Doctoral Thesis, Leuven Belgium): EVA1 to EVA3 with 5–28 wt% of vinyl acetate (i.e 1.7–11.2 mol%). The amounts of crystallinity obtained by dynamic cooling from the melts read 45, 22 and 13%, respectively. This is in close agreement with the results shown in Fig. 116. The decrease in crystallinity will affect the character of the gels obtained. This becomes clear from Fig. 115 for 10% solutions of EVA in *o*-xylene. From a comparison of the results shown in Figs. 113, 114 and 116, one might conclude that networks formed vary from a particle network (EVA1) to a network consisting of molecularly connected tiny crystalline crosslinks (EVA3).

For the system HDPE1-diphenyl ether (DPhE, a bad solvent) the temperature-concentration diagram is shown in Fig. 117: the cloud point, crystallisation and melting curves. They are in agreement with the general picture shown in Fig. 102, apart from the horizontal invariant that has changed in a line with a slight upward curvature with decreasing concentration. This is the consequence of the polydispersity of the polymer: the three phase equilibrium is only temperature independent if the polymer is monodisperse [327, 328]. This behaviour was also reported by Vandeweerdt et al. [285] for the intersection between the glass transition curve with a L-L-demixing region. It appears that for HDPE1

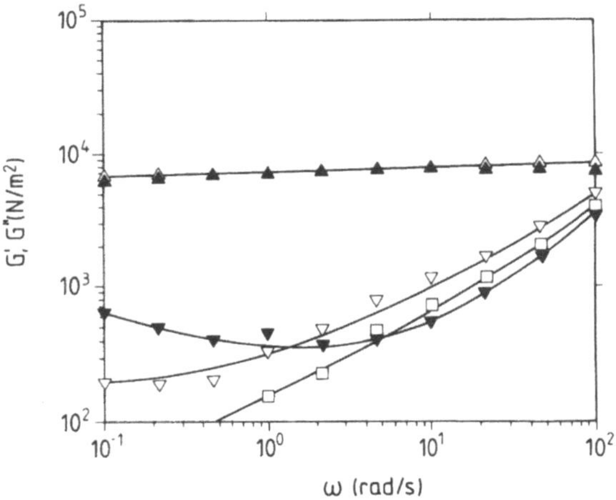

Fig. 116. Dynamic moduli of gels of ethylene-vinyl acetate copolymers ($\phi = 10\%$) in decalin vs angular frequency; deformation amplitude = 0.5%. The moduli of the EVA1 and EVA2 gels were scaled to the storage modulus of the EVA3 gel; (▲) and (△) storage moduli, (▼), (▽) and (□) loss moduli of EVA1, EVA2 and EVA3, respectively. Reproduced, with permission of the author, from [327]

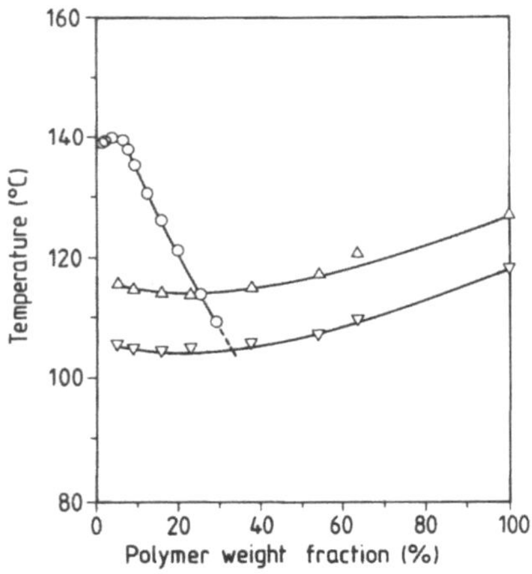

Fig. 117. Temperature-concentration diagram for the system HDPE1-diphenyl ether, as derived from DSC and optical cloud point measurements; (▽) onset of crystallisation; (△) end of melting endotherm; (○) optical cloud point. Reproduced from Makromol Chem [Ref. 328] by the courtesy of the authors and of Hüthig & Wepf Verlag Publishers, Zug, Switzerland

solutions with $\phi < 0.30$ upon cooling L-L-demixing will take place; continued cooling results in crystallisation when the temperature of 105 °C is reached. At $\phi = 0.30$ and T = 105 °C the crystallisation curve and the L-L-demixing coincide and crystallisation will take place upon cooling. For systems with $\phi > 0.30$ crystallisation will take place.

Upon cooling of dilute solutions of HDPE1 in DPhE, first L-L- demixing will take place and upon further cooling L-S-demixing. Solutions with a polymer concentration of less than 5% do not form coherent structures, irrespective of the cooling history. Gels are only formed if the polymer concentration is higher than 5%. The gelation process is accompanied by syneresis, to the inside as well as to the outside of the gel.

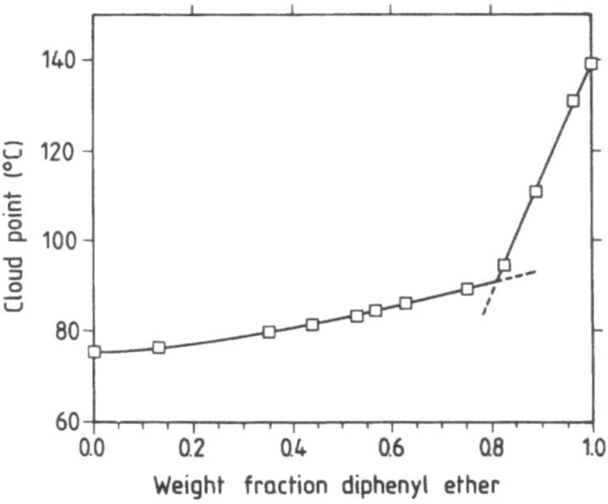

Fig. 118. Dependence of the cloud point of HDPE1 solutions ($\phi = 5\%$) in mixtures of decalin and diphenyl ether vs the composition of the solvent mixture. Cooling rate 3.5 °C/min. Reproduced, with permission of the author, from [327]

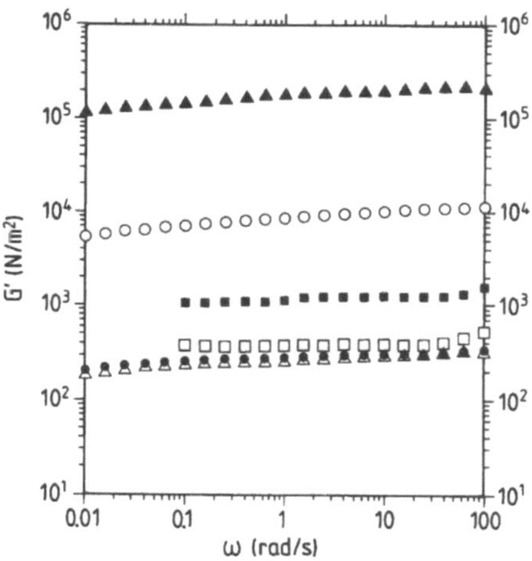

Fig. 119. Frequency dependence of the storage moduli of HDPE1 gels ($\phi = 5\%$) in mixtures of decalin and diphenyl ether; deformation amplititude = 0.5%; (△) decalin; (●), (□), (■) and (○) volume ratio Dec/DPhE 6/4, 4/6, 3/7 and 2/8, respectively; (▲) diphenyl ether. Reproduced, with permission of the author, from [327]

Rheological studies have been concentrated on gels from mixtures of a good and a bad solvent (decalin and diphenyl ether, respectively). For HDPE1 solutions ($\phi = 0.05$) the cloud point is shown in Fig. 118 as a function of the composition of the solvent, where two lines can be distinguished. For $\phi = 0.05$ the low temperature line for $0 \leq \phi_{DPhE} \leq 0.8$ corresponds to the crystallisation curve of HDPE1 in decalin to which DPhE is added. The high temperature curve for $\phi_{DPhE} \geq 0.8$ corresponds to the L-L-demixing curve obtained by the addition of decalin to DPhE. The frequency dependence of the storage modulus of the HDPE1 solutions in various solvent mixtures is shown in Fig. 119. At high volume fractions of decalin this modulus is hardly dependent on solvent composition. Only from $\phi_{DPhE} = 0.7$ is a strong increase of the storage modulus visible. In Fig. 120 both dynamic moduli, as measured at $\omega = 1$ rad/s, are shown as a function of the solvent composition. Again around $\phi_{DPhE} = 0.7$ an increase

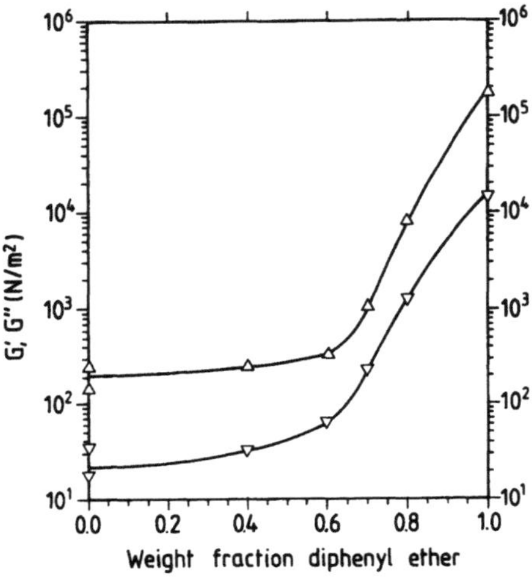

Fig. 120. Dependence of the storage moduli of the gels mentioned in Fig. 119 vs composition of solvent; angular frequency $\omega = 1$ rad/s; deformation amplitude 0.5%; (△) storage modulus; (▽) loss modulus. Reproduced, with permission of the author, from [327]

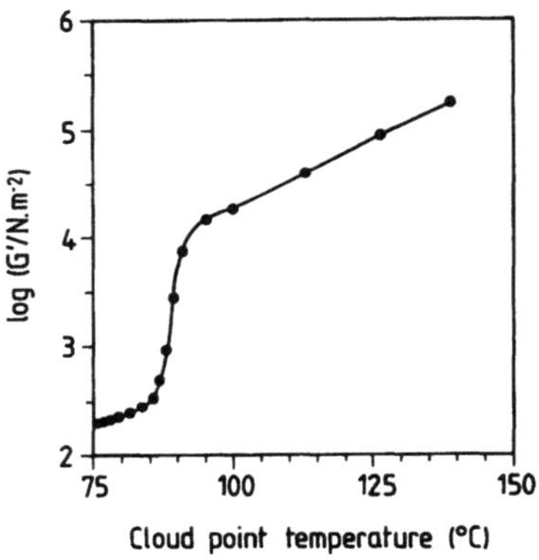

Fig. 121. Combination of results presented in Figs. 118 and 120, resulting in a storage modulus vs cloud piont plot

in the dynamic moduli is visible. A combination of both results is shown in Fig. 121, where the storage modulus, as measured at $\omega = 1$ rad/s, is shown as a function of the cloudpoint temperature; a clear distinction between both regions is visible: for $T_{cloud} < 85\,°C$ the moduli are relatively low in the DPhE added decalin solution and for $T_{cloud} > 90\,°C$ the moduli are high in the decalin added DPhE solution. There is a transition region of only 5 °C. Similar results were obtained for HDPE2 and UHMWPE [328].

7.8 Conclusions

The gelation behaviour of polyethylene solutions is the result of a combination of L-L-demixing from a bad solvent, followed by L-S demixing (i.e. crystallisation) or by L-S-demixing from a good solvent. The first method is used for the preparation of porous membranes, the second for the production of strong fibres. Polyethylene networks may be formed in several ways: first by Van der Waals attraction of crystallites, which results in a very fragile network of adhesively phase connected particles, and second, if the molecular weight is high enough, by the presence of crystalline crosslinks that are molecularly connected. This variation in network structure closely resembles the results obtained in atactic polystyrene gels (see Sect. 5).

Data of viscoelastic properties of polyethylene gels have been scarcely reported. Compression moduli were measured as a function of concentration and molecular weight. Interesting work was done by the Leuven (Belgium) group, to discriminate between dynamic mechanical properties of gels of HDPEs of various molecular weights, obtained from bad and good solvents (or mixtures).

The conclusion is that, for polyethylene gels, the measurement of viscoelasticity is underexploited.

8 Block Copolymers

8.1 Introduction

When triblock copolymers of the type ABA are dissolved in a selective solvent, which is a solvent for polymer B and a precipitant for polymer A, the A-blocks have a tendency to associate into swollen domains. In dilute solutions, micellar structures are formed, while in concentrated solutions more or less rigid organized structures can be formed (cylinders or lamellae) depending on the nature of copolymer and solvent [351]. If the micelles are molecularly dispersed in solution, they are called monomolecular micelles; if, on the other hand, they are associated, they are called polymolecular micelles or aggregates. In many cases, the monomolecular micelles are spherical and nearly monodisperse in size [351]. The micelles are composed of a relatively compact core of the "insoluble" blocks and a highly swollen shell of the "soluble" blocks [352]. The A-blocks of one molecule can be incorporated in the same or in different domains. However, in ABA block copolymers it is best that the A-blocks are incorporated in different domains, because in that case the entropy of the B-block is not lowered too much, as would be the case when forming a loop, by backfolding to the same domain [353]. Thus a three-dimensional network will be formed: the A-block domains behave as multifunctional crosslinks and the flexible B-blocks as network chains. If the concentration is not high enough, only unimers (i.e. single molecules of the block copolymer in solution: see Fig. 122) are present and micelles are not formed.

Nevertheless, in that case there are also interactions between the A-blocks. In that case a random network will be formed. A schematic view of these

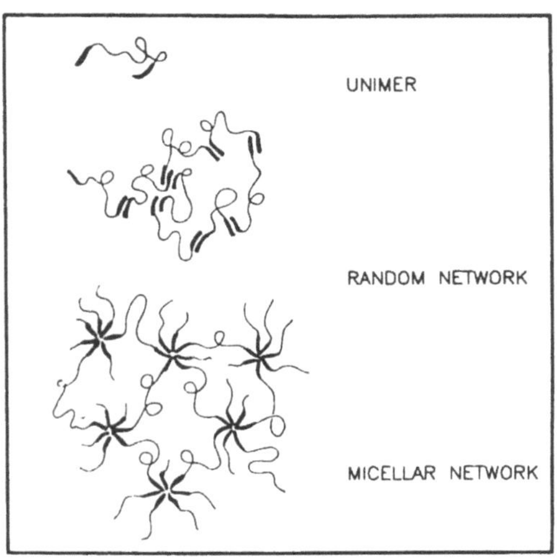

Fig. 122. Schematic representation of different phases in triblock copolymer solutions; unimers and two structures leading to networks: random network and micellar network. Reproduced from Macromolecules [Ref. 354] by the courtesy of the authors and of The American Chemical Society

possibilities of network formation is shown in Fig. 122 [354]. Beautiful pictures, resulting from computer simulation of solutions of block copolymers (diblocks, triblocks and multiblocks) in solvents varying from good solvents for both polymers to precipitants for both polymers, have been published by Glotzer et al. [355]. A short but interesting review of polymer-polymer phase behaviour is presented in a paper by Bates [356].

The permanence of networks of tri- or multi-block copolymers depends on the dynamics of the equilibrium between the aggregates or micelles and unimers. If the dynamics are fast, then the continuous change of the temporary network is fast: the system behaves like an elastic liquid. If the dynamics are slow, then a pseudo rubber plateau may arise, the longest relaxation time depending on the dynamics of the equilibrium. If the temperature becomes low enough for the swollen domains to be in their glassy state, the dynamics are practically absent, and the network is a permanent network with infinite relaxation times as long as the temperature is low enough. The dynamics of the aggregation equilibrium of di- and tri-blocks has been the subject of a thorough study by Spitteler and Mijnlieff [357, 358]. They investigated equilibrium and dynamics of *n*-tetradecane solutions of the diblock copolymers AB, consisting of polystyrene (A) and polyisoprene (B). *n*-Tetradecane is a solvent for polyisoprene and a non-solvent for polystyrene. One of their main conclusions was that the time constant of the dynamics of the equilibrium process is dramatically dependent on temperature and on the molecular weight of the styrene block. From Figs. 123 and 124, where SI-x-y is defined as a polystyrene block with molecular weight x kg/mol and a polyisoprene block with molecular weight y kg/mol, it follows that the time constant at 15 °C increases by a factor of 10^5 on increasing the molecular weight of the styrene block from 9.4 to 15.2 kg/mol and decreases by a factor of approximately 250 by increasing the temperature to 25 °C. Hence, it is logical that the viscoelastic behaviour of solutions of ABA block copolymers

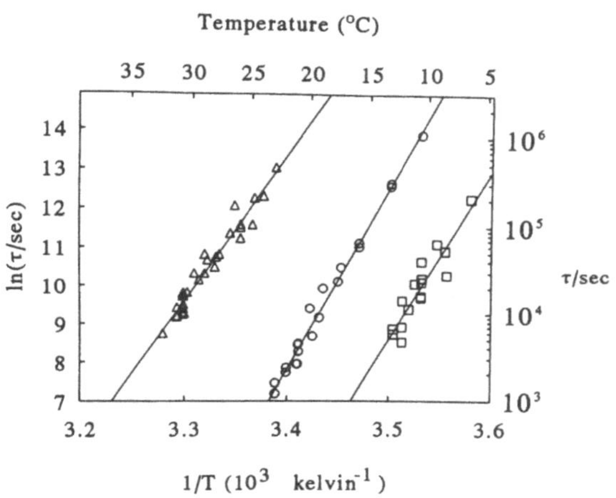

Fig. 123. Temperature dependence of time constant of the equilibrium process in *n*-tetradecane solutions of diblock SI copolymers; SI-15-90 (△); SI-12-80 (○); SI-9-80 (□). Reproduced, with permission of the author, from [357]

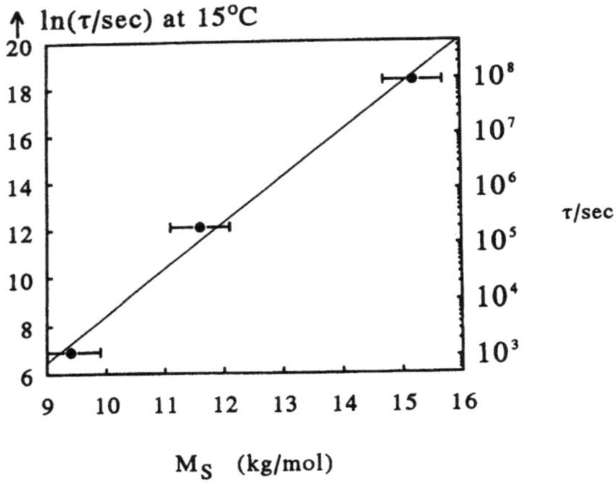

Fig. 124. The time constant of the equilibrium process in *n*-tetradecane solutions of diblock copolymers vs the molecular weight of the polystyrene block of the three samples SI-9-80, SI-12-80 and SI-15-90. Reproduced, with permission of the author, from [357]

in selective solvents strongly depends on the molecular weight of the A-blocks and on temperature.

8.2 Styrene-Isoprene-Styrene Triblock Copolymers

Rheological measurements on SIS triblock copolymer solutions were reported by Spitteler [357], Mijnlieff et al. [358, 360, 361] and Visscher [359]. An example of the dynamic moduli of a 1% solution of the styrene-isoprene-styrene triblock copolymer SIS-12-320-12 in *n*-tetradecane is shown in Fig. 125. At temperatures of 20 °C and higher, the system behaves like an elastic liquid (low frequency slopes of 2 and 1 for G′ and G″, respectively) but with a tendency for a rubber-like plateau at higher frequencies. At 6.2 °C a permanent rubber plateau seems to exist with a slope of less than 0.01. The loss modulus increases with decreasing frequency, however, so a maximum will be present at lower frequencies, which will be attended by a decrease of the storage modulus resulting from a relaxation process due to the dynamics of the styrene blocks in the dense styrene domains. Thus, a real permanent network is not present, although the time constant (the longest relaxation time) is quite high. The network changes into a permanent network at temperatures where the swollen polystyrene aggregates are in their glassy state, with infinite time constants or longest relaxation times. As the time constants depend on the structure of the aggregates formed, in a first approximation their temperature dependence is determined by the molecular weight of the styrene blocks only, provided the structure of the aggregates does not depend on concentration. This is in agreement with the time constants obtained from viscoelastic experiments shown in Fig. 126: the curves for SIS-12-320-12 almost coincide, irrespective of the concentration, with an exception for the lowest concentration. It is clear that the molecular weight of the styrene block is the most important parameter; an

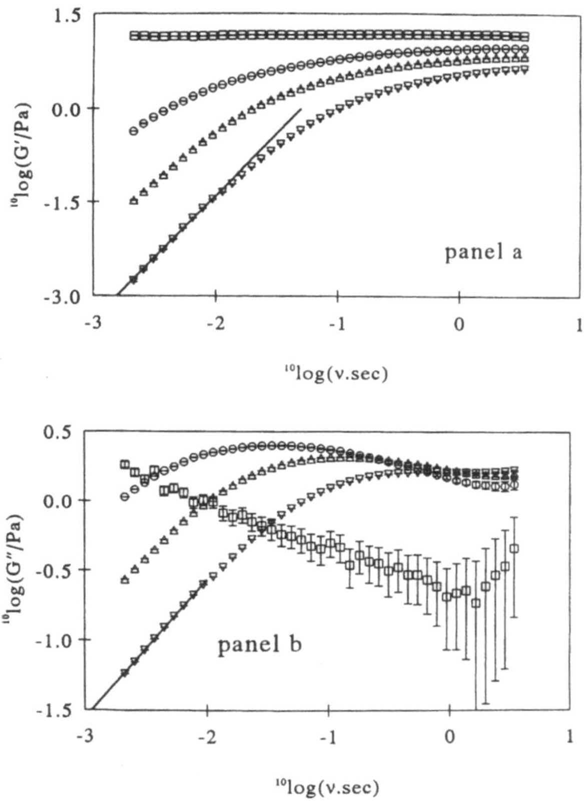

Fig. 125a, b. Dynamic moduli of 1 wt% SIS-12-320-12 solutions in *n*-tetradecane at different temperatures: (□) 6.2 °C; (○) 20.0 °C; (△) 22.5 °C; (▽) 25.0 °C: **a** storage modulus; **b** loss modulus. Reproduced, with permission of the author, from [357]

increase in the molecular weight of the isoprene block appears to be of minor importance. In this respect it is worth mentioning that Nguyen-Misra and Mattice [362] very recently developed a model for micellisation in and gelation of ABA-triblock copolymer solutions. On the basis of calculations on an on-lattice Monte Carlo model, they concluded that the micellisation and gelation properties in selective solvents (athermal for the middle B-block and a precipitant for the end A-block) are primarily dependent on the size and solvent insolubility of the A-blocks and only weakly dependent on the middle block sizes. This is in agreement with the results obtained by Spitteler and Mijnlieff [357, 358].

The time constants for $M_s = 9$ and $M_s = 12$ kg/mol differ by a factor of 90 approximately. Interesting is the difference in time constants between the SI-diblocks and SIS-triblocks with the same value of the molecular weight of the styrene blocks: the time constants of the SI-solutions are lower roughly by

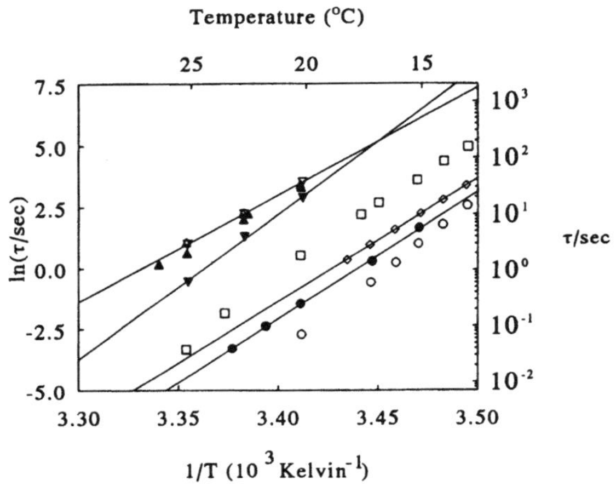

Fig. 126. Temperature dependence of the time constants of dilute *n*-tetradecane solutions of a series of SIS copolymers with almost the same polyisoprene masses, but with different polystyrene masses. *Triangles*: SIS-12-320-12; (▼) 0.847 wt%; (▲) 1.00 wt%; (△) 1.19 wt%; (▽) 1.52 wt%; *Squares*: (□) SIS-10-330-10; 1.54 wt%; *Diamonds*: (◇) SIS-9-330-9, 1.50 wt%; *Circles*: SIS-9-250-9; (○) 2.51 wt% (●); 3.32 wt%. Reproduced, with permission of the author, from [357]

a factor of 200 compared with the SIS-solutions (or gels). It is worth mentioning that the energies of activation of both processes are approximately equal, notwithstanding the big differences in time constants of the diblock and triblock solutions.

Also interesting are the results obtained on mixtures of solutions of two SIS copolymers with different polystyrene masses and equal polyisoprene masses. One would expect that the loss peak would broaden due to the strong dependence of the relaxation times on the polystyrene mass (see Fig. 124). However, the resulting loss peak is of the same form as those of the unmixed polymer solutions: the G″ values of the mixture are an intermediate between the G″ values of either polymer solution separately [357]. This unexpected result was attributed to the decrease in the time constants of polymers with relatively large S-blocks in the mixed domains and an increase in the time constants of the relatively small S-blocks. The resulting relaxation times seem to depend on the domain size of the S-blocks (rather than the individual S-sizes), which in turn will depend on the (average) S-block mass.

8.3 *Styrene-Butadiene-Styrene Triblock Copolymers*

Determination of the flow behaviour and the structure of SBS solutions have been the subject of many investigations (see e.g. [363–367]). However, the dynamic mechanical behaviour of solutions has been little reported. Pico and Williams [368] describe the gelation behaviour of highly concentrated SBS solutions in dipentene on the basis of the dynamic viscosities η'' ($=G'/\omega$) and η' ($=G''/\omega$). Although these authors observed the presence of rubber elasticity, from inspection of their results one can conclude that, in the rubber elastic region, $\eta'' < \eta'$. Watanabe et al. [369] studied the dynamic mechanical viscoelastic behaviour of Kraton 1101, i.e. SBS-16-64-16 with $\bar{M}_n = 95$ kg/mol (styrene content = 33%) in *n*-tetradecane. In Fig. 127 the dynamic moduli G′ and G″ are plotted vs angular frequency for various concentrations: 12, 20, 30 and 40%. The dependence of the dynamic moduli on the strain amplitude was examined and appeared to be negligible in the applied amplitude range. This becomes clear from Fig. 127: doubling the amplitude of the cup oscillation did not affect the results of the measurement of the dynamic moduli. Furthermore, the storage moduli have reached their plateau values which are almost independent of frequency, whereas the corresponding values of the loss modulus are lower by a factor of 10, approximately. The phenomenon of the increase of the loss modulus with decreasing frequency betrays that a relaxation process is present at still lower frequencies, attended by a decrease in the storage modulus. This in agreement with results obtained by Spitteler and Mijnlieff [357, 358] for the SIS triblock copolymer in *n*-tetradecane. One could expect such behaviour, because the phase separated polystyrene domains are, at 23 °C, far above their glass transition temperature, and thus a temporary network is present. Only at temperatures below the glass transition temperature of the swollen polystyrene domains is the network a permanent network with infinite relaxation times. The concentration dependence of the storage modulus is almost quadratic ($G_e = 28.2c^{2.06}$) as shown in Fig. 128. The authors describe the system as

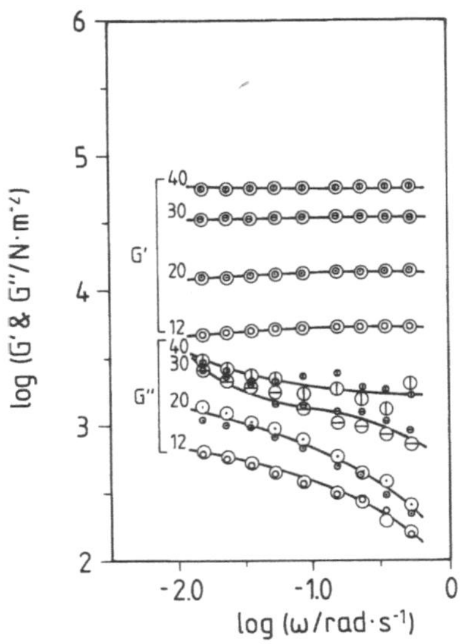

Fig. 127. Dynamic moduli at 23 °C of *n*-tetradecane solutions of SBS Kraton 1101 vs angular frequency for various concentrations, as indicated in wt%; the dependence on strain amplitude was checked by measurement of two amplitudes of cup oscillation: *small symbols* 1.1 degrees, *large symbols* 2 degrees. Reproduced from J Rheol [Ref. 369] by the courtesy of The American Institute of Physics, Woodbury, NY

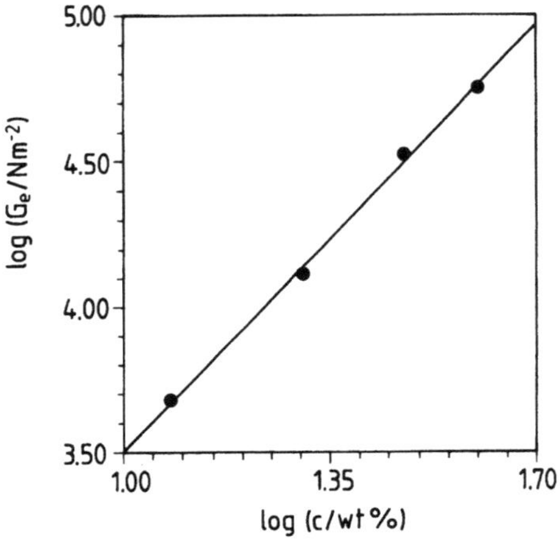

Fig. 128. Concentration dependence of the equilibrium shear modulus of *n*-tetradecane solutions of SBS Kraton 1101 at 23 °C (constructed from data presented in [369])

a swollen state, where

$$G_e = (1 + 2.5\,\phi_s + 14.1\phi_s^2)\,\frac{c_B RT}{\bar{M}_e} \tag{27}$$

where φ_s = the volume fraction of the polystyrene domains in the system; c_B = the concentration of polybutadiene segments in the matrix phase; $\bar{M}_e$ = the molecular weight between crosslinks.

Results for $\bar{M}_e$ are 34, 25, 17 and 14 kg/mol for concentrations of 12, 20, 30 and 40 wt%, respectively. Hence, the reciprocal values of $\bar{M}_e$ are proportional to the concentration. The authors conclude that entanglements play a primary role in the value of the modulus, because $\bar{M}_e$ is much smaller than the molecular weight of 64 kg/mol of the polybutadiene block. As the polybutadiene chains are anchored in the semi-rigid polystyrene domains, their lifetimes are relatively long and, according to Fig. 127, much longer than 100 s. This system behaves like a semi-permanent network as long as the stresses are not high enough to exceed the strength of the semi-rigid domains and as long as the frequencies are not too low.

At higher temperatures the semi-rigid domains become less and less rigid: the system becomes viscoplastic, which means that the dynamic moduli become strongly strain dependent (see Fig. 129). At 119 °C the system behaves like an elastic liquid with less strain dependence. In Fig. 130 the various rheological regions are shown as a kind of phase diagram.

8.4 Ethylene oxide-Propylene oxide-Ethylene oxide Triblock Copolymers

Poly(ethylene oxide) (PEO) is a predominantly hydrophilic polymer, and in aqueous solutions the polymer chain is highly swollen, with up to approximately

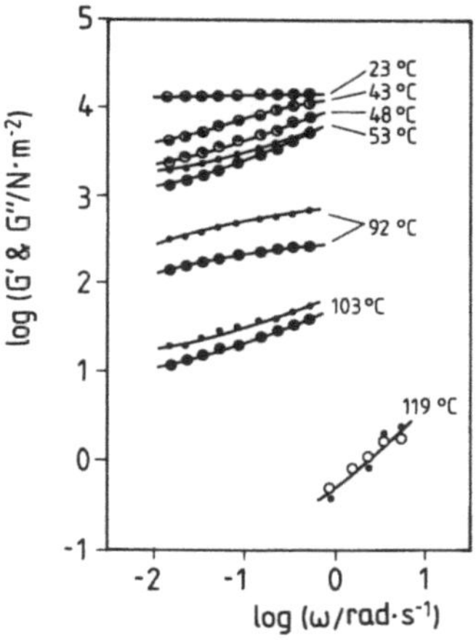

Fig. 129. Frequency dependence of the storage moduli of 20 wt% *n*-tetradecane solutions of SBS Kraton 1101 at various temperatures, as indicated; the strain dependence is also shown in this figure as the amplitude of cup oscillation: *small symbols*: 1.1°; *large symbols*: 2°. Reproduced from J Rheol [Ref. 369] by the courtesy of The American Institute of Physics, Woodbury, NY

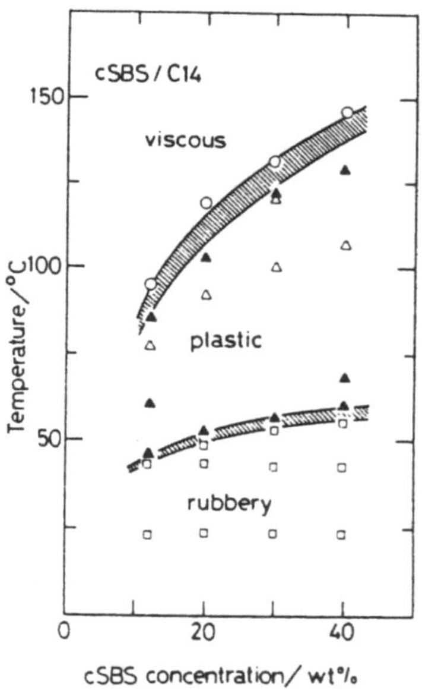

Fig. 130. Temperature-concentration classification of SBS Kraton 1101 solutions in *n*-tetradecane: (□) linear and rubber-like; (○) linear and viscous; non-linear, due to yield stress or plasticity with (▲) and without (△) thixotropy. Reproduced from J Rheol [Ref. 369] by the courtesy of The American Institute of Physics, Woodbury, NY

two water molecules per ethylene oxide unit, although this number also depends on temperature [370, 371]. It has a lower critical solution temperature (LCST) of approximately 100 °C and an upper critical solution temperature (UCST) of approximately 5 °C, both values depending on its molecular weight of course. However, far below its LCST and even at room temperature in aqueous solutions, large, loose and thermodynamically reversible association complexes can be formed. This presence of micelles or aggregates was confirmed by Polik and Burchard [372] by careful analysis of their light scattering experiments on dilute poly(ethylene oxide) solutions at temperatures varying from 20 to 90 °C. At an intermediate temperature of 60 °C, the molecular weight was maximal and larger than that at 20 °C by a factor of 100. Forces that are involved in the aggregation of PEO are hydrogen bonding, the structure of water and the hydrophobic effect [371]. It is clear that the addition of strong electrolytes also influences the aggregation behaviour of PEO, because the order of the water structure is changed, and also as a result the hydrophobic interaction [372–374]. For example, the cloud point of an aqueous PEO solution decreases by 50 °C on addition of potassium fluoride up to a concentration of 1 mol/l [374].

On the other hand, poly(propylene oxide) (PPO) is hydrophilic at low temperatures and hydrophobic at high temperatures. Thus, water is a good solvent for PPO below approximately 15 °C, whereas at higher temperatures aggregates are formed in aqueous solutions. With blocks of PEO and blocks of PPO combined into single polymer chains, one can therefore expect amphiphilic characteristics and aggregation phenomena. It has to be emphasized that, for this triblock copolymer, the outer PEO blocks are more water soluble than the inner PPO blocks, and thus the mechanism of gel formation will be different from the above mentioned triblock copolymers and is still a matter of discussion.

The triblock copolymer $EO_nPO_mEO_n$ is soluble in water and forms micelles above a critical polymer concentration and above a characteristic temperature. This was shown by Zhou and Chu [352, 375, 376]. These authors affirmed that micelle formation is temperature dependent because the hydrophilic/hydrophobic characteristics of these polymers can be easily modified by changing the temperature. They mentioned the presence of three temperature regions: a unimer region, a micelle region and a transition region where unimers and micelles coexist. The micelles seem to be monodisperse in size, even if the polymer itself is polydisperse, their radii being independent of concentration but dependent on temperature [377].

Dynamic light scattering [352, 375, 376, 378–381] and small-angle neutron scattering [377, 379, 382, 383] revealed the existence of structures at different temperatures and concentrations. In aqueous solution unimers are, in general, present only at low temperatures and low concentrations. At higher temperatures, micelles are formed. The micellar liquid undergoes a first order phase transition to a simple cubic crystal when the volume fraction φ of the swollen micelles becomes 53%. For the polymer P85, $EO_{25}PO_{40}EO_{25}$, this is the case at 25 wt% at 27 °C. In Fig. 131 a schematic representation of interacting micelles in a 25 wt% solution of P85 at different temperatures is shown. The various volume fractions needed for cubic or hexagonal packing are obtained by swelling of the micelles by water. From these values, one can calculate the number of water molecules perturbed per EO monomeric unit to be approximately 4. This is in agreement with results obtained by Malmsten et al. [384–386], who investigated the gelation behaviour of aqueous solutions of $EO_{99}PO_{65}EO_{99}$ by means of GPC and NMR self-diffusion studies. They concluded that 2 to 5 water molecules are perturbed per EO unit, the real

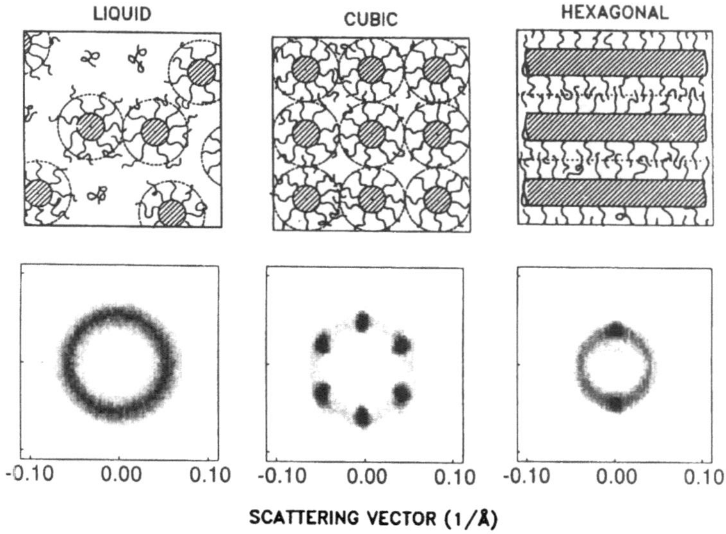

Fig. 131. *Top*: schematic illustrations of interacting micelles in a 25 wt% aqueous P85 solution at 25 °C (*left*), 27 °C (*middle*) and 68 °C (*right*). *Bottom*: two-dimensional scattering functions as obtained perpendicular to the shear plane. Reproduced from Progr Colloid Polym Sci [Ref. 383] by the courtesy of the author and of Steinkopff Verlag Darmstadt, FRG

number depending on temperature. The above-mentioned value of 53%, where micellar liquid transforms into a cubic crystal, is just the volume fraction of spheres in a closest simple cubic packing [387], although the SANS pattern shows that the ordering is a body centered cubic packing [379, 382, 383, 388] with a closest packing of $\varphi = 68\%$ [387].

On the basis of SANS, Wanka et al. [379] found that gels consist of large domains of cubic liquid crystalline regions. The micellar size in the gels seems to be roughly the same as in dilute solutions. Similar conclusions were reached by Mortensen et al. [382]. Additionally Rassing et al. [389, 390] studied the gelation of aqueous $EO_{99}PO_{65}EO_{99}$ solutions by ^{13}C NMR and found indications of conformational changes in the PPO block at higher temperatures, which were interpreted as a dehydration of the micellar core. A similar conclusion, based on DSC experiments, was drawn by Wanka et al. [379].

The formation of gels is associated with close packing of the micelles with their weak interparticle interactions in a body centered cubic lattice. This close packing in an ordered lattice is possible because the micelles are uniform in size [377]. At higher temperatures, the micelles become rod-like and pack in a hexagonal structure (simple packing $\varphi = 0.785$, hexagonal packing $\varphi = 0.907$). From their self diffusion experiments on $EO_{99}PO_{65}EO_{99}$ solutions, Malmsten et al. [384–386] concluded that the average residence time of a single polymer molecule in the aggregates is extremely long (~hours). This is of course the reason for the gel-like behaviour as long as the frequency is not too low. At

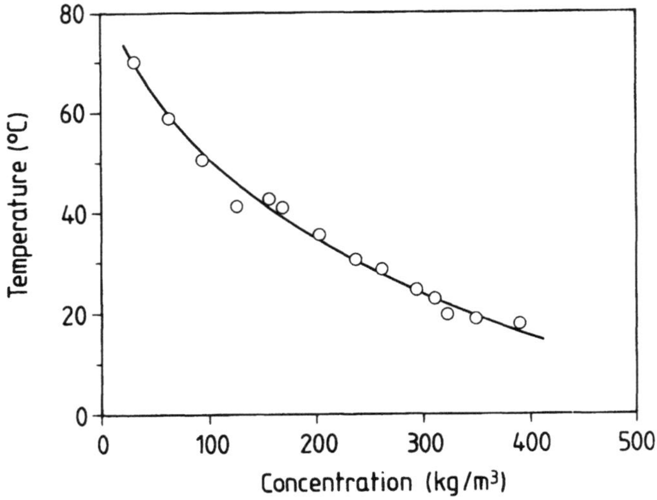

Fig. 132. Gelation temperature of aqueous P-85 solutions, determined from dynamic mechanical shear measurements, as a function of P-85 concentration. Reproduced from J Phys Chem [Ref. 378] by the courtesy of the authors and of The American Chemical Society

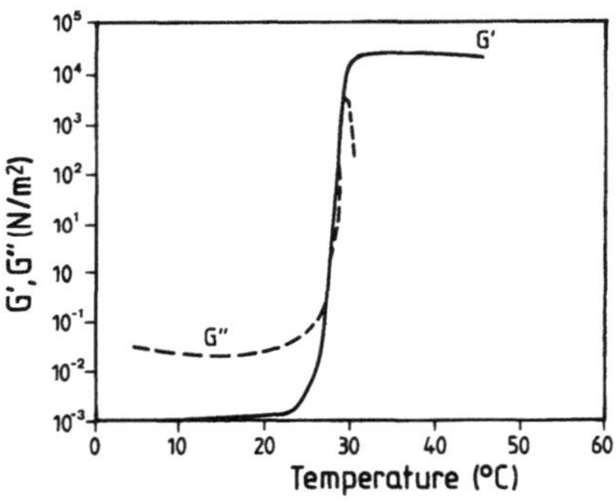

Fig. 133. Dynamic moduli, G′ and G″, of a 26 wt% aqueous p-85 solution, measured at an angular frequency of $\omega = 1$ rad/s, as a function of temperature. Reproduced from J Phys Chem [Ref. 378] by the courtesy of the authors and of The American Chemical Society

concentrations higher than 10%, the aggregates are interconnected and thus gels may be formed at temperatures depending on concentration. This is shown in Fig. 132, where the gelation temperature determined from dynamic mechanical measurements is chosen as the temperature where $G' = G''$. Below this temperature the system is liquid-like and above it the system behaves like a gel, where $G' \gg G''$: at these temperatures G′ is almost independent of temperature (see Fig. 133).

From their SANS experiments on $EO_{25}PO_{40}EO_{25}$, Mortensen and Pedersen [377] concluded that the micelles are spherical particles (with a diameter of 4–5 nm, depending on temperature) composed of a central core of dense poly(propylene oxide) and with an outer shell of hydrated poly(ethylene oxide) units. The intermicellar interactions are, in spite of the flexibility of the water dispersed PEO blocks, basically as for hard spheres, and one even observes a hard sphere crystallization as the micelle concentration approaches a critical volume fraction of 0.53 [382, 383, 388].

From hard core potential modeling of the interaction between the micelles, Wanka [379] and Mortensen and Pedersen [377, 388] concluded that the PEO mantles anchor the micelles. In general, the micelles form a polycrystalline cubic phase in a transparent paste-like material. If the polycrystalline material is exposed to shearing forces (in a Couette-instrument for example, the polycrystal abruptly transforms into a single crystal. Above temperatures of about 60–70 °C, the spherical micelles transform into a liquid with large prolate ellipsoids [377]. This liquid phase is a transition region between the cubic phase and a crystalline mesophase of hexagonally ordered rod-like micelles. It might be concluded that at room temperature a Newtonian liquid with Gaussian chains is present, which changes at higher temperatures into a solid-like gel, which again dissolves at still higher temperature. This sol gellifies again with a further increase in temperature, ultimately leading to phase separation at the cloud point. From oscillatory mechanical measurements it has become clear that the gels are thermoreversible [378, 379, 391]. At about 95 °C, large aggregates of polymers ordered in lamellae structures are formed, leading to an opaque suspension. From SAXS experiments by Glatter et al. [392], it appeared that the form factor of the micelles changes with concentration, so that the hard sphere interaction model is only a simplified model. A phase diagram relating to this is shown in Fig. 134. The drawing of this diagram is based on a number of different measuring techniques.

Viscoelastic properties, i.e. the dynamic moduli G′ and G″, were measured by Hvidt et al. [378, 380, 381, 393], Wanka et al. [379], Wang and Johnston

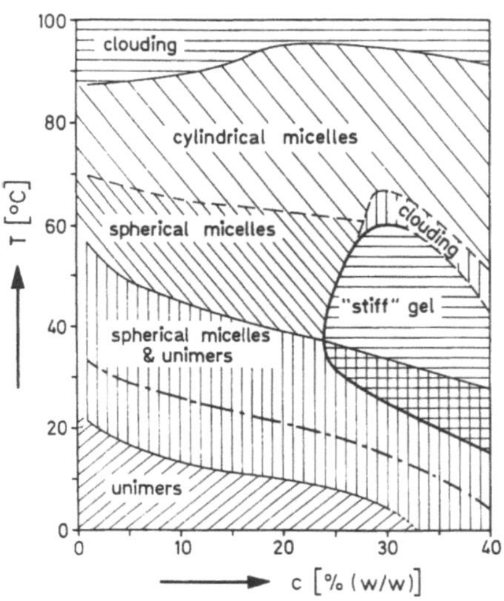

Fig. 134. Phase diagram for aqueous p-85 solutions. Reproduced from Macromolecules [Ref. 392] by the courtesy of the authors and of The American Chemical Society

[391] and Bahadur and Pandya [394]. Hvidt et al. determined the gel point of aqueous $EO_nPO_mEO_n$ systems by measurement of their dynamic moduli as a function of temperature at constant frequency. At low temperatures the systems behaved like an elastic liquid with $G'' \gg G'$. A sol-gel transition was defined as the temperature where $G' = G''$. The transition was unusually sharp. Above that temperature, $G' \gg G''$. This kind of behaviour was also reported by other authors [376, 379]. Wanka et al. [379] have shown that in the gel region a real rubber-like plateau exists: log G′ plotted vs log ω gives an almost horizontal plateau over more than three frequency decades; in this region G″ is approximately ten times smaller than G′. They were also able to determine the gel temperature by extrapolating the yield values of the gel to zero. Whatever Hvidt's definition of the gel temperature, its deviation is not more than a few degrees. For Pluronic 85 (i.e. $EO_{25}PO_{40}EO_{25}$) the gel temperature is already shown as a function of concentration in Fig. 132. Comparison with the phase diagram of Fig. 133 reveals that it just coincides at the highest concentrations with the transition from the region of "spherical micelles & unimers" into the "stiff gel" region. For concentrations lower than 25% the gelation curve passes the region of "spherical micelles" and here it is less clear what mechanism causes the system to gellify. In fact, the G′-G″ cross-over appeared to be due to drying and not to gelation [395]. At higher temperatures, the gel becomes a liquid again and thus the storage modulus decreases to very small, unmeasurable values. It is clear that the gelation and degelation temperatures and their course depend on the kind of copolymer and especially on the proportion of PO and EO blocks. In Fig. 135, the dynamic moduli are shown as a function of

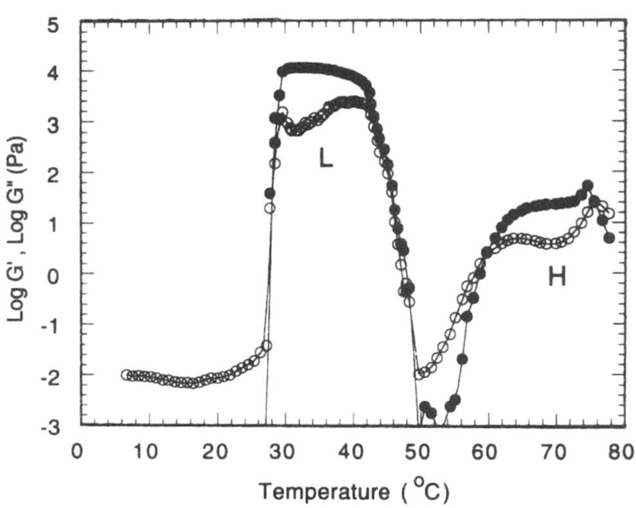

Fig. 135. Temperature dependence of the storage and loss moduli, G′ (●) and G″ (○), of a 25.5 wt% aqueous p-94 solution, measured at ω = 0.314 rad/s; L and H point to the low and high temperature gel regions. Reproduced from J Non Cryst Solids [Ref. 393] by the courtesy of the authors and of Elsevier Science-NL, Sara Burgerhartstraat 25, 1055 KV Amsterdam, The Netherlands

temperature for a 25.5 wt% solution of Pluronic 94 (i.e. $EO_{21}PO_{47}EO_{21}$). It appears that at still higher temperatures a gel is formed again, but with a lower modulus of elasticity (different by a factor of 100); the low temperature gel was called a strong or hard (H) gel and the high temperature gel a weak or soft (S) one. On a further temperature increase the gel dissolves again. The strong dependence on concentration, as shown in Fig. 136, is interesting. At concentrations of 25 and 25.5 wt% a gelation is clearly perceptible in the low temperature range, whilst a concentration of 24 wt% only shows the high temperature gelation. This high temperature gelation is not as concentration dependent as the low temperature gelation. The gap between the low and high temperature gelations decreases with increasing concentration [393]. At concentrations above 27 wt%, there is no longer a gap: only a minimum and a maximum value are present and not more than a shoulder at a concentration of 37 wt% [381]. A new maximum in G′ at 27 wt%, marked by a T (i.e. two phase turbid gel), develops between the hard and the soft gel. In Fig. 137 the sol-gel transitions of Pluronic 94 solutions are shown as a function of concentration, and in Fig. 138 their gelation and two phase regions are shown.

These results are quite similar to the results obtained for Pluronic P85 in Fig. 134, except for a shift in temperature. The hard gel region for Pluronic 94 is qualitatively present at concentrations and temperatures quite similar to a cubic crystalline region observed for Pluronic 85. The soft gel region at higher temperatures for Pluronic 94 corresponds qualitatively to the temperatures where a hexagonal crystalline region was observed by SANS on Pluronic 85, except that the gel states of Pluronic P94 are observed at concentrations as low as 1%.

From their DSC studies on aqueous solutions of poly(propylene oxide) and Pluronic 94, Hvidt et al. [393] concluded that endotherms are present in both cases at around the same temperature, whereas both endothermic effects amount to 98 ± 8 J/(g of PPO). This shows that the PPO blocks in Pluronic 94 are involved in the micelle formation, whereas the poly(ethylene oxide) blocks

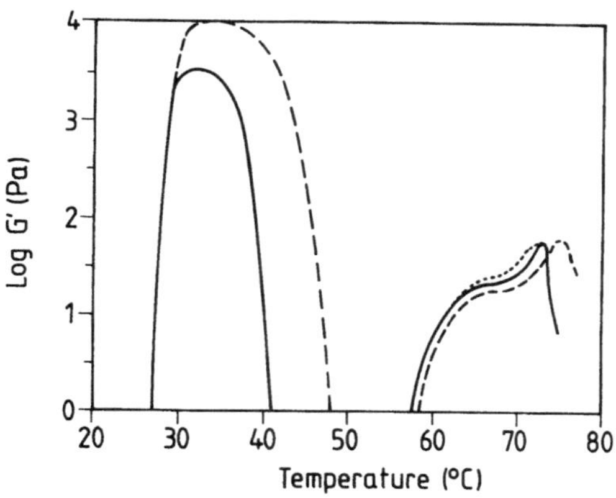

Fig. 136. Temperature dependence of the storage modulus at ω = 0.314 rad/s for aqueous p-94 solutions of different weight concentrations: (·····) 24.0%; (——) 25.0% and (-----) 25.5%. Reproduced from J Non Cryst Solids [Ref. 393] by the courtesy of the authors and of Elsevier Science-NL, Sara Burgerhartstraat 25, 1055 KV Amsterdam, The Netherlands

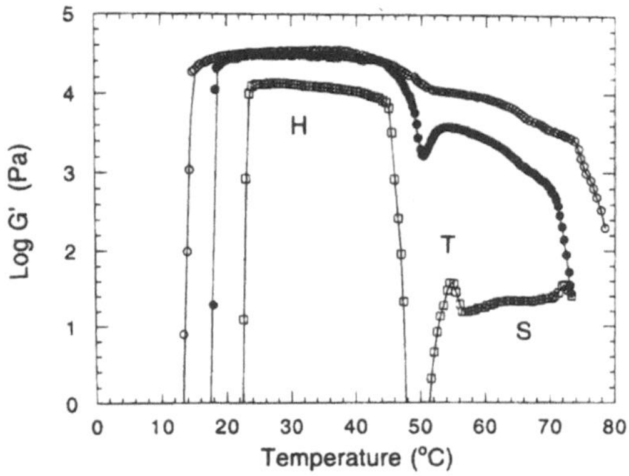

Fig. 137. Temperature dependence of the storage modulus, measured at $\omega = 0.314$ rad/s, for aqueous p-94 solutions of various concentrations in wt%: (□) 27.0; (●) 32.0; (○) 37.0; H, S and T denote hard, soft and turbid gels, respectively. Reproduced from J Phys Chem [Ref. 381] by the courtesy of the authors and of The American Chemical Society

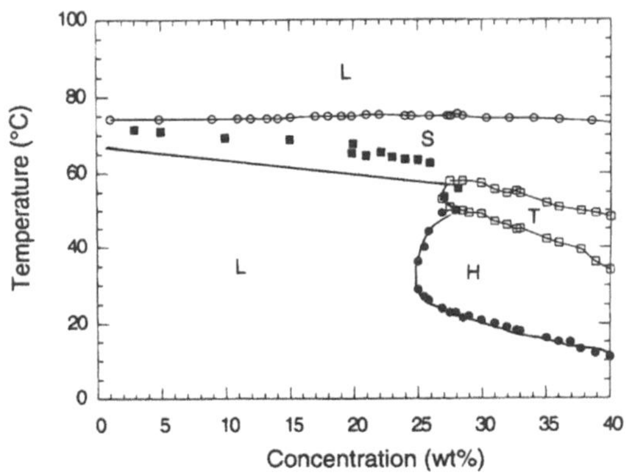

Fig. 138. Gelation and two-phase regions for aqueous p-94 solutions. *Filled symbols* refer to temperatures where glass spheres become unable to move. *Heavy curves* are traces of gelation curves from viscoelastic measurements. *Open symbols* and *fine curves* denote temperatures where a transition between the two phases and a single phase is observed. H, S and T refer to hard, soft and turbid gel, respectively; L is a liquid state. Reproduced from J Phys Chem [Ref. 381] by the courtesy of the authors and of The American Chemical Society

keep the micelles in solution. Gelation cannot be originated by entanglement coupling of the PEO blocks because they are too short. The authors proposed an affine deformation mechanism by which the spherical micelles deform into ellipsoids, with a higher surface area exposed to water. The elasticity should be related to the thermodynamically unfavourable interactions between the hydrophobic PPO chains and water. However, it has been shown before that micelles and aggregates are formed in aqueous PEO solutions at temperatures above 25 °C. In this case a network could be formed by the interactions of the PEO blocks. This might also be the origin of elasticity of the EO-PO-EO copolymer/water systems. However, on the other hand, it was shown by Hvidt et al. [381] that an aqueous 29.7 wt% solution of PEO-6000 (with higher $\bar{M}_w$ than the PEO blocks in Pluronics) only shows a small value for the loss modulus between 5 and 80 °C and no detectable storage modulus at 0.3 rad/s.

8.5 Propylene oxide-Ethylene oxide-Propylene oxide Triblock Copolymers

The triblock copolymer $PO_nEO_mPO_n$ is the reverse of $EO_nPO_mEO_n$. In aqueous solutions the outer blocks form dense, swollen PPO domains interconnected by the more flexible, water swollen PEO chains. Mortensen et al. [354] studied the aqueous (D_2O) solution behaviour of Pluronic-25R8 $PO_{15}EO_{156}PO_{15}$ by SANS, dynamic light scattering, small amplitude dynamic mechanical rheology and viscosity measurements. Again, depending on concentration and temperature, the presence of regions with unimers, randomly cross-linked network, micellar network, micellar crystals and elastic lamellar gel was affirmed. A phase diagram of the mentioned system is shown in Fig. 139. In a rather small part of the phase diagram a micellar mesophase exists: only between 50 and 70 wt% and at temperatures below 40 °C is a solid-like elastic

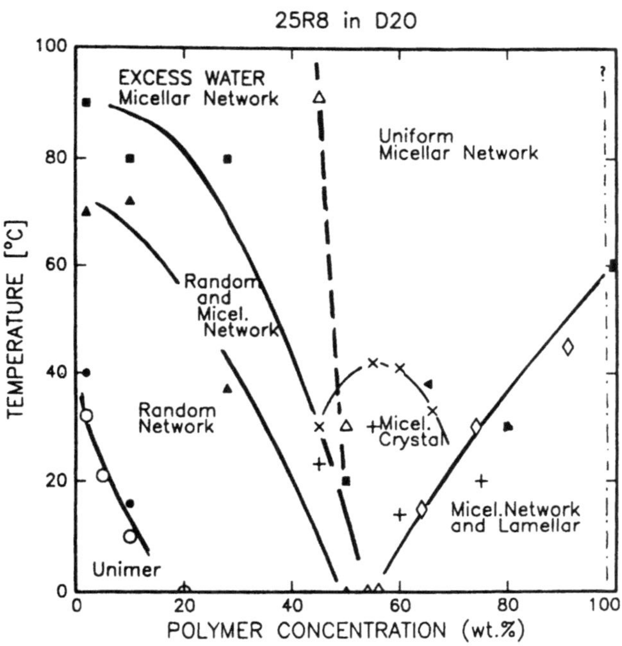

Fig. 139. Phase diagram of the D_2O-25R8 system; *closed symbols* refer to neutron scattering results, *open symbols* to light scattering and *crosses* to rheological data. Reproduced from Macromolecules [Ref. 354] by the courtesy of the authors and of The American Chemical Society

gel present. This became clear from rheological measurements. The viscosity as a function of concentration of this system is shown in Fig. 140 at 40, 45, 60 and 80 °C. At around 55 wt%, the viscosity passes through a broad maximum and at 40 °C the 55 and 60 wt% systems are gels. A gel temperature is again defined as the temperature where G′ equals G″. Around the gel temperature the change in G′ is rather sharp, just as for the reverse copolymer $EO_nPO_mEO_n$. The result is shown in Fig. 141: two transition temperatures are present in the micellar crystal region. In the gel region linear viscoelasticity is only present at strain amplitudes of 0.0005 and less.

8.6 Multiblock Copolymers

He et al. [98, 396–398] studied the properties of the multiblock copolymer $(A_{n1}B_{n2})_n$ where A = dimethyl siloxane and B = 1-(dimethylsilyl)-4-(dimethylvinylsilyl) benzene or COSP: crystalline organosilicic polymer (see Fig. 142). The polydimethyl siloxane blocks are flexible at room temperature, the glass

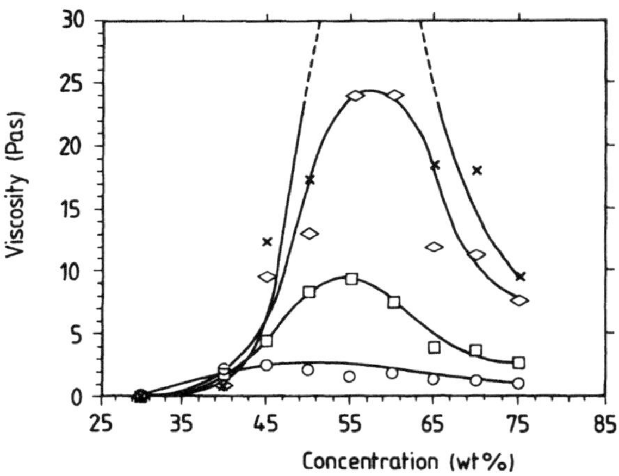

Fig. 140. Viscosity of aqueous 25R8 solutions, measured at various temperatures: (×) 40 °C; (◇) 45 °C; (□) 60 °C; (○) 80 °C. Reproduced from Macromolecules [Ref. 354] by the courtesy of the authors and of The American Chemical Society

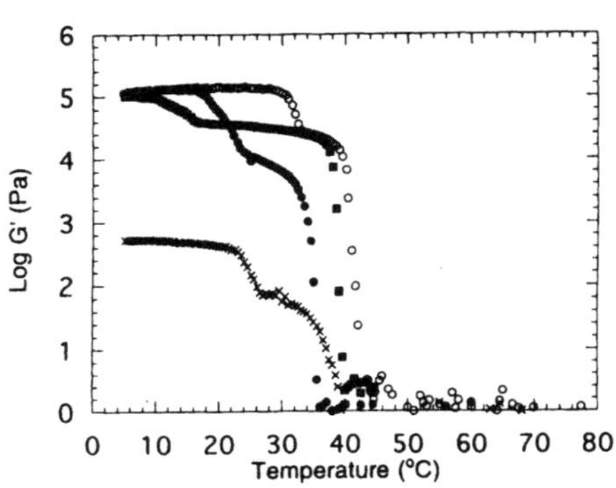

Fig. 141. Shear storage modulus, measured at ω = 0.314 rad/s, of aqueous 25R8 solutions, vs temperature for various concentrations: (×) 45; (○) 55; (■) 60 and (●) 75 wt%. Reproduced from Macromolecules [Ref. 354] by the courtesy of the authors and of The American Chemical Society

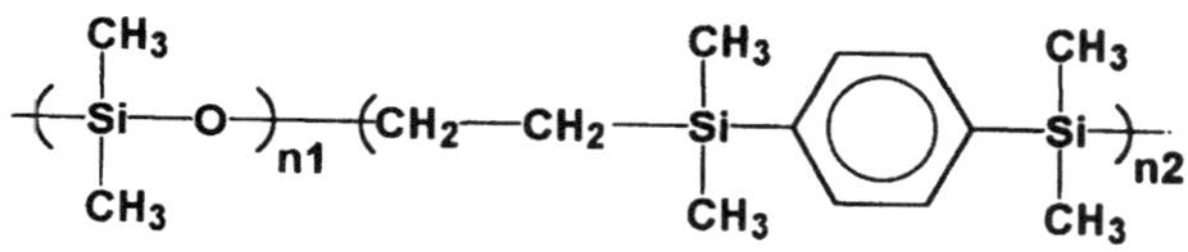

Fig. 142. The molecular structure of the multiblock copolymer, studied by He et al. [2, 98, 396–398]

Table 6. Characteristics of the multiblock copolymer used in the study by He et al. [98, 396–398]

Name	$\bar{M}_{n,A}$	$\bar{M}_{n,B}$	X_B	$\bar{M}_v$	$\bar{M}_n$	n	n_1	n_2
Copo10	17.1	2.0	0.11	200	90	5	219	9
Copo20	17.1	4.0	0.19	120	67	3	219	18
Copo50	9.7	9.7	0.50	290		10?	131	44

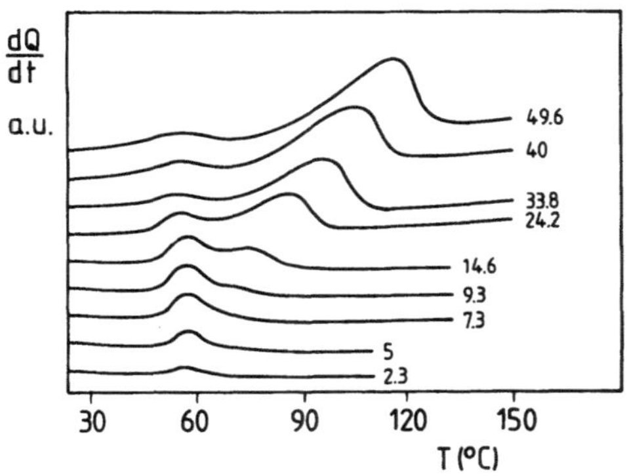

Fig. 143. DSC thermograms from Copo50 gels quenched at −18 °C for 24 h, then kept at 30 °C for a week. Heating rate 20 °C/min; concentration as indicated in wt%. Reproduced from Macromolecules [Ref. 98] by the courtesy of the authors and of The American Chemical Society

transition temperature being −127 °C, whilst the B-polymer consists of associated crystalline hard blocks with a glass transition temperature varying from −2 to 24 °C and a melting temperature varying from 165 to 189 °C, both values depending on molecular weight [2] (see Table 6). Copo10 means a polymer with approximately 10 wt% of COPS. The PDMS and COPS blocks are incompatible.

Gels were made from solutions in *trans*-decalin, bromobenzene and 1-phenyl-dodecane. In order for a gel to be formed, the solutions were quenched to low temperatures (−18 °C in the case of *trans*-decalin and bromobenzene and 20 °C in the case of 1-phenyldodecane) for 24 h to complete gelation and then kept for a week at 30 °C prior to any measurement. DSC thermograms for Copo50 gels are shown in Fig. 143. There is a small endothermic effect at about 50 °C, independent of concentration. At higher temperatures another endothermic effect is present, but its amount, and the temperature of the maximum, increase with concentration. Approximate phase diagrams could be constructed from these thermograms, as is shown, e.g., for Copo20 in *trans*-decalin in Fig. 144 (for an explanation the reader is referred to the book by Guenet [2] on thermoreversible gels). A monotectic transition is present at 50 °C and 17 wt%, which corresponds to the DSC thermograms, and a liquidus line corresponding to the endothermics at higher temperatures. The coexistence curve on the left hand side of T_M could not be detected, but should be present. Its presence has been detected clearly for Copo10 in 1-phenyl dodecane. The height of the monotectic transition depends on the composition of the polymer and on the solvent used: it increases with increasing amount of crystallizable B-blocks and for 1-phenyl dodecane the monotectic temperature is much higher than for the other two solvents.

Stress relaxation experiments at room temperature showed almost ideal rubber elastic behaviour (Fig. 145), at least up to 200 min: the slopes, $d(\log \sigma)/d(\log t)$, vary from -0.01 to -0.03. The compression modulus of the gels is constant up to the temperature of the monotectic transition. Above that temperature gels with $c < c_M$ become liquid-like, whilst the compression modulus of gels with $c > c_M$ gradually decreases; this decrease is faster if the concentration is lower. In the case of gels in 1-phenyldodecane this dependence is less clear.

The composition and concentration dependence of the compression modulus, measured at constant temperature, is shown in Figs. 146a and b. In Fig. 146a $\log E$ is plotted vs $\log c$ for the three copolymers dissolved in *trans*-decalin: straight lines are obtained from which it follows that E is proportional to $c^{4.5}$. By dividing the compression modulus by the fraction of the crystallizable component one master curve is obtained (see Fig. 146b). It may be concluded that E is proportional to X^2_{COSP}. Hence [397]

$$E\ (N/m^2) = (1.24 \pm 0.07) \cdot 10^8 X^2_{COSP} C^{4.5}. \tag{28}$$

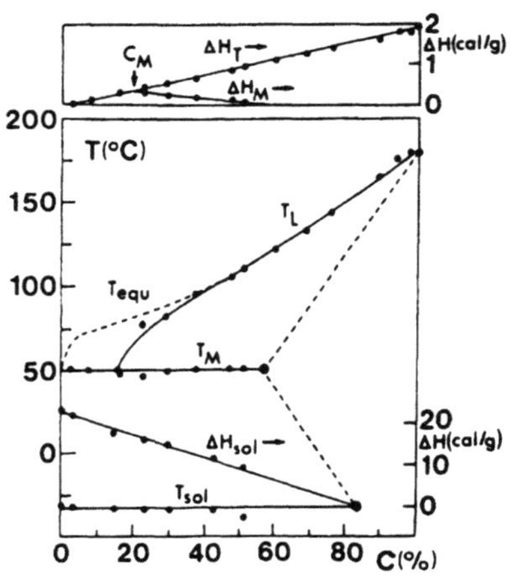

Fig. 144. Temperature-concentration phase diagram for Copo20 solutions in *trans*-decaline. Tamman's diagram (ΔH vs c) is also given in the phase diagram. Reproduced from Macromolecules [Ref. 98] by the courtesy of the authors and of The American Chemical Society

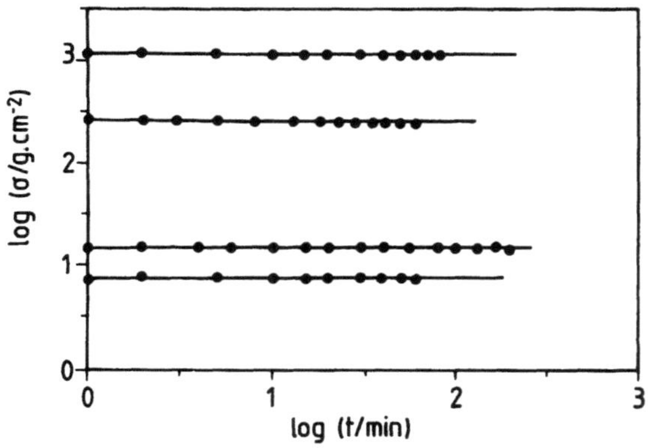

Fig. 145. Stress relaxation curves for $\lambda = 0.9$ for various CoPo systems. From *top* to *bottom*: Copo50 c = 42 wt%; Copo50; c = 15 wt%; Copo10; c = 36 wt% and Copo20; c = 16 wt%. Reproduced, with permission of the author and publishers, from [2]

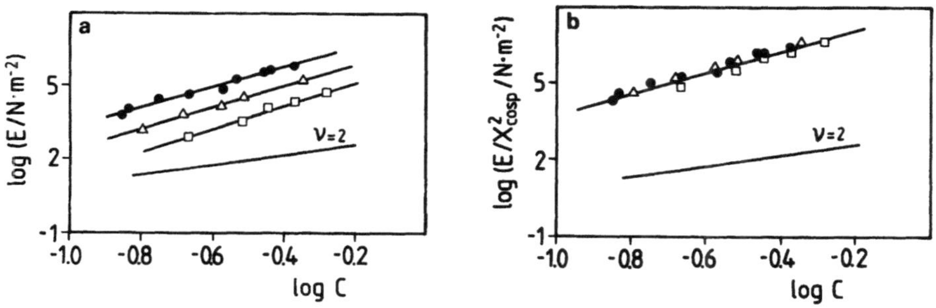

Fig. 146. **a** Young's modulus vs concentration. **b** Reduced Young's modulus vs concentration: (●) Copo50; (△) Copo20; (□) Copo10. Reproduced from Macromolecules [Ref. 98] by the courtesy of the authors and of The American Chemical Society

This shows that the copolymer concentration is the determining factor. According to the authors, this strong concentration dependence is due to a combination of the concentration effect and an undercooling effect present in gels with crystalline crosslinks. The concentration effect is always present in chemically crosslinked networks and gives, for good solvents, a proportionality between E and $C^{2.25}$. The undercooling effect leads to an extra concentration dependence. The strong concentration dependence of the modulus is often observed in gels with (semi-) crystalline crosslinks.

8.7 Conclusions

In solutions of triblock, ABA, or multiblock, $(AB)_n$, copolymers gelation may occur in a selective solvent due to the tendency of one kind of block to form aggregates. In general, the A-blocks are subject to aggregation, thus forming micelles, with flexible B-blocks in between. In this way a micellar network will be formed if the concentration is high enough, and there is a more or less random network at low concentrations (see Fig. 122). The network is a permanent network with infinite relaxation times if the temperature is so low that the swollen aggregates are in their glassy state. If the temperature is higher, then a temporary network is obtained with finite relaxation times. The longest time constants of the dynamics of the swollen A-blocks are not only a function of temperature, but also of the molecular weight of the A-block polymer, which affects their value far more.

Much work has been done by Mijnlieff et al. in their study of viscoelasticity of styrene-isoprene-styrene triblock copolymer solutions in *n*-tetradecane, although more investigations are needed to elucidate the structure and the dynamics of the gels formed.

Aqueous systems of ethylene oxide-propylene oxide-ethylene oxide triblock copolymers have been the subject of many studies, primarily in the determination of the phase diagrams. In this case the solvent is a precipitant for the middle block, and a solvent for the outer blocks. Poly(ethylene oxide)-water systems show lower and higher critical solution temperatures, however, so that associates are formed in an extremely complex way with the formation of spherical and cylindrical micelles with various ways of ordering, depending on concentration and temperature. Moreover, gels are formed in a special region of the phase diagram. Results obtained from phase diagrams agree more or less with relatively scarce studies of viscoelastic properties, most of them by Hvidt et al. Thus this subject is still an underdeveloped area of research.

For the rest, some work has been done on organosilicic multiblock copolymers and on styrene-butadiene-styrene and propylene oxide-ethylene oxide-propylene oxide triblock copolymers.

It will be clear that in the field of block copolymers, fallow areas are present to a high degree. The difficulty of the measurement of viscoelastic properties is the sensitiveness of the gels to (shear) deformations: linear viscoelasticity is, in general, only measurable at very small deformations, so that very small stresses have to be measured. Special sensitive instruments are needed for these studies.

9 Liquid Crystalline Polymers

9.1 Introduction

In solutions of liquid crystalline polymers (LCPs), both lyotropic and thermotropic, gel formation may take place by interaction of the mesogenic groups of different polymer molecules. As a result, thermoreversible gels develop by the formation of thermoreversible, physical crosslinks. Although the study of LCP solutions is an interesting subject, little has been published as yet. Papers concerning LCP solutions discuss time and concentration dependent viscosities. In recent years, other techniques have also been used to study the behaviour of solutions and gels, e.g. static and dynamic light scattering, microscopy and differential scanning calorimetry and rheometry.

Gelation is an important phenomenon in LCP solutions manifested in fibre spinning and, for example, in many biological systems. In the solution spinning process gelation occurs in the coagulation bath. The strength of the fibres depends on the gelation conditions, whereas phase transitions in gels can be used as chemical sensors [399].

Although in many studies the viscoelastic behaviour of thermotropic and lyotropic LCPs in bulk, with their good mechanical properties and processability, has been described, only a few publications deal with the viscoelastic properties of their thermoreversible gels because, in general, it is difficult to dissolve them on a molecular scale, and immediate applications are unlikely.

The problem of the ordering of rigid rods in solutions of low concentrations was discussed by Onsager [400] in 1949. He concluded, on the basis of the calculation of the second virial coefficient of solutions of rigid rod particles, that a thermodynamic phase transition occurs when the particle concentration increases above a critical concentration. It is well-known that the viscosity of such solutions of stiff and semi-flexible polymers increases with increasing concentration up to the concentration where phase separation takes place: at that point the viscosity decreases dramatically, after which it steadily increases again with concentration. From his lattice theory for semi-flexible polymer chains, Flory [401, 402] predicted a phase diagram consisting of three distinct phases, depending on temperature. In Fig. 147, the result is shown for a phase diagram of rigid, impenetrable rods with an axial ratio of 150 [402–405]. The solution is in the isotropic state at low polymer concentrations and high temperatures and in an anisotropic state at high polymer concentrations and relatively high temperatures. In between, there is a narrow, chimney-like region where both phases coexist. At lower temperatures, the chimney widens strongly and there is a broad region where both phases coexist. The form of this phase diagram was experimentally verified by Miller et al. [403–408] for solutions of the semi-flexible poly (γ-benzyl-α,L-glutamate) in various solvents (see Fig. 148). The theoretical narrow biphasic region moves towards higher concentrations

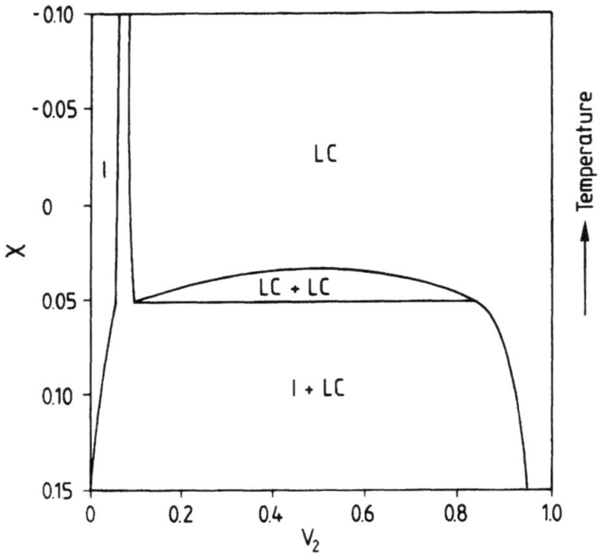

Fig. 147. Lattice model phase diagram for rigid, impenetrable rods of axial ratio 150, indicating the three phases: isotropic (I), liquid crystalline (LC) and the biphasic region (I + LC) (see [402–405]). Reproduced from Liquid Crystals and Ordered Fluids, Vol 2 [Ref. 405] by the courtesy of Plenum Publishing Corporation, New York

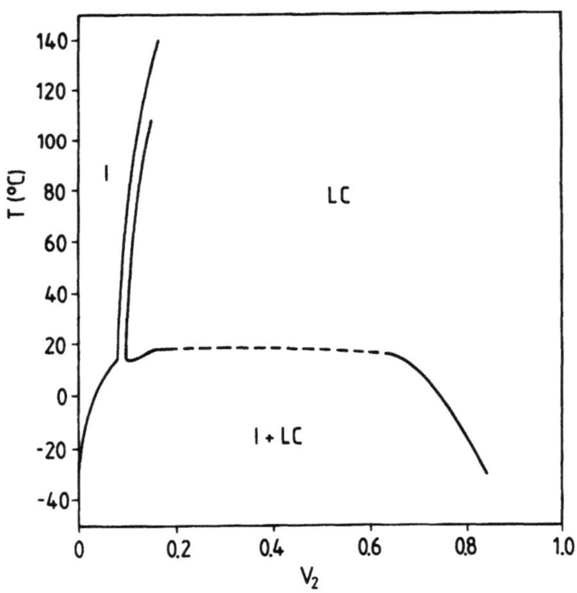

Fig. 148. Experimental phase diagram for PBLG (M = 310 kg/mol) in DMF. The *dashed line* indicates the area of insufficient data. Reproduced from Liquid Crystals and Ordered Fluids, Vol 2 [Ref. 405] by the courtesy of Plenum Publishing Corporation, New York

as the axial ratio is reduced. Apparently, this is the case in PBLG as the biphasic region turns to higher concentrations when the temperature increases [405, 406]. This is in agreement with the strong decrease of the intrinsic viscosity with increasing temperature [405]. According to the phase rule, the chimney

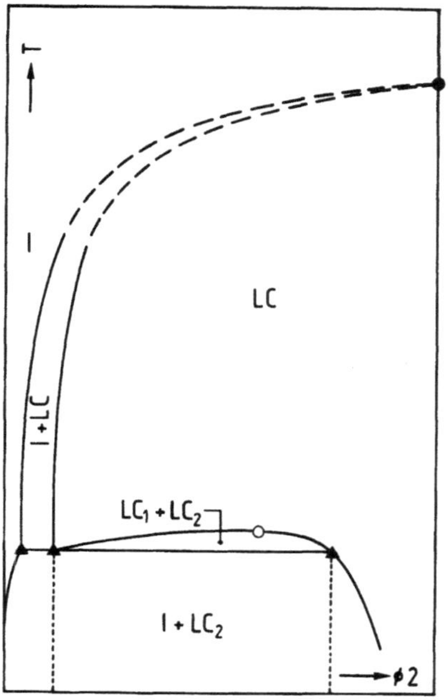

Fig. 149. Schematic phase diagram of solutions of rigid-rod molecules according to the phase rule. Reproduced from Makromol Chem [Ref. 409] by the courtesy of the authors and of Hüthig & Wepf Verlag Publishers, Zug, Switzerland

must eventually turn towards the axis of the solvent free polymer [409] (see Fig. 149). Moreover, the phase rule prescribes that the narrow chimney and the broad biphasic area are separated by a three-phase line. In general, this is a shallow region and for that reason it easily escaped direct observation [404, 409].

9.2 SCLCPs: Side Chain Liquid Crystalline Polymers

In a series of papers by Shukla [399] and Muthukumar et al. [410–413] the gelation behaviour of poly-(γ-benzyl-α,L-glutamate) gels in benzyl alcohol is described: rheology [399, 410, 411], thermodynamics and kinetics [399], and static and dynamic light scattering [411–413]. The authors made use of a PBLG (see Fig. 150) with a molecular weight of 345 kg/mol and studied the gelation behaviour of 1–15 wt% solutions in benzyl alcohol. From the determination of the Mark Houwink exponent ($a = 1.46$), the authors conclude that the polymer chain is semi-flexible, because for flexible chains in a good solvent $a \approx 0.8$, whereas $a = 2$ for rigid rods [414]. As has been mentioned before [405, 406], the flexibility increases with increasing temperature. A PBLG with a molecular weight of 100 kg/mol exists as an α-helical rod with a contour length of 68.5 nm (see e.g. [416]). A solution of this polymer is semidilute above a concentration of 0.5%. The side-chain groups of PBLG have a tendency to associate in a head-to-tail side-by-side aggregation and thus aggregation will occur in quite dilute

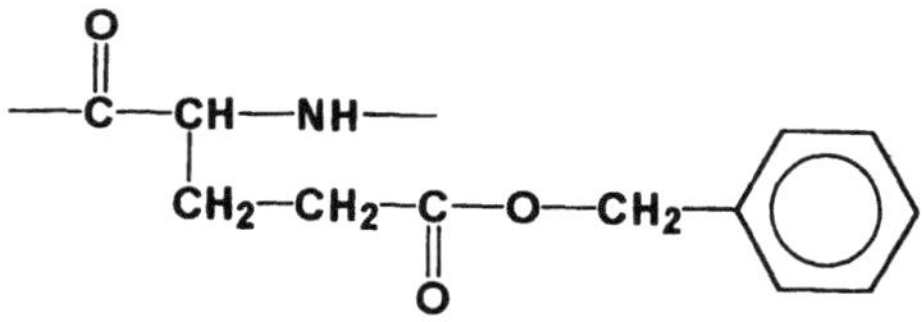

Fig. 150. Monomeric unit of poly(γ-benzyl-α,L-glutamate)

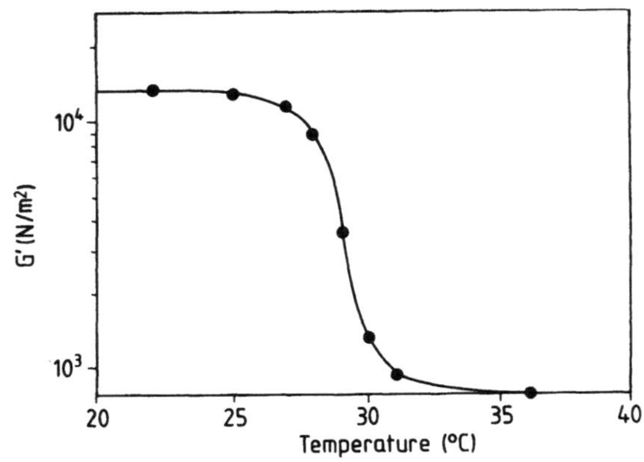

Fig. 151. Dependence of the storage modulus of a 1 wt% PBLG solution in benzyl alcohol on (decreasing) temperature (ω = 1 rad/s, strain amplitude = 1%) from the isotropic solution into the gel phase. Reproduced from Polym Eng Sci [Ref. 410] by the courtesy of the authors and of The Society of Plastics Engineers, Brookfield, UK

solutions. Upon sudden cooling of an isotropic solution to a temperature at which the ordered phase is thermodynamically stable, the rate of the phase transition is dramatically dependent on temperature; a time of the order of 100 min, on average, is needed before an equilibrium state is reached. However, the return to the isotropic state takes only a few minutes [406] (crystallization vs melting). By varying the polarity of the solvent from polar to non-polar the association tendency of the polymer increases considerably, even in dilute solution [415]. However, the solvent polarity does not significantly affect the phase boundaries of the narrow biphasic region. The effect of the decrease of the polarity is not only the increase of the association tendency of the polymer, but also the increase in temperature where the narrow and wide biphasic regions intersect [406]. The gelation temperature of a solution of a certain concentration was determined by Shukla et al. [410] by measurement of the storage modulus on going from high to low temperatures. The result of such a measurement is shown in Fig. 151 for a 1% solution: the gelation temperature is approximately 30 °C. The change in the storage modulus at the transition is more dramatic on going from an isotropic solution on the left hand side of the chimney than on going from the anisotropic phase on the right hand side of the chimney. The temperature of this transition is dependent on concentration. In this way the phase diagram could be drawn approximately (see Fig. 152). This phase diagram is in agreement with the one measured by Sasaki et al. [416, 417]. The chimney itself was difficult to determine. However, the result is in qualitative agreement with Flory's model, although he expected the isotropic and

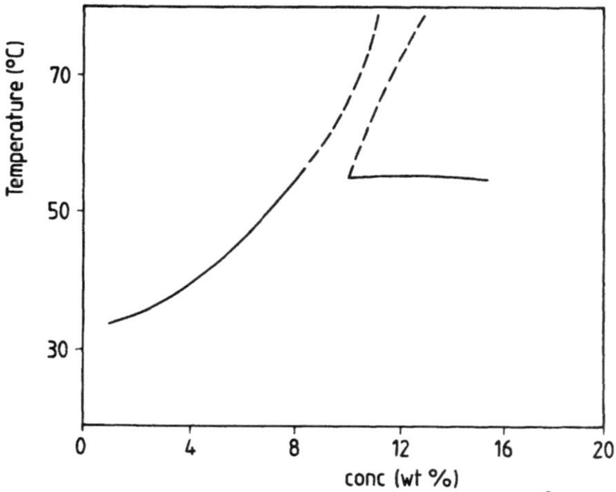

Fig. 152. Phase diagram for the PBLG-benzyl alcohol system, obtained from gelation temperature measurements. Reproduced from Polym Eng Sci [Ref. 410] by the courtesy of the authors and of The Society of Plastics Engineers, Brookfield, UK

anisotropic phases to coexist instead of the occurrence of gelation. According to Miller et al. [406], ordered spherulites develop in the narrow biphasic region, which suggests a nucleation mechanism. Hence, it cannot be determined by measurement of the gelation temperature. Gelation will only occur in the wide biphasic region, whether coming from the isotropic state, the ordered state or from the narrow biphasic region. The minimum volume fraction of PBLG (M = 310 kg/mol) needed for gelation in DMF at $-40\,^{\circ}C$ has the extremely low value of 0.05 vol% [406].

Sasaki et al. [416, 417] reported the shear moduli for a number of gels in benzyl alcohol of PBLGs, varying in molecular weight from 40 to 600 kg/mol. Results are shown in Figs. 153 and 154, where the dependences on temperature and molecular weight for various concentrations are shown. Again, it is clear that the moduli have extremely high values, taking into account the low concentrations. An example of dynamic properties is shown in Fig. 155 for a 9 wt% solution at 28 °C. Both dynamic moduli, G' and G'', show a weak increase with frequency (slopes: $n \approx 0.07$), whereas the complex viscosity η^* decreases with frequency, following a straight line with a slope equal to -0.94. These values are in agreement, because from the theory of linear viscoelasticity it follows that the sum of the absolute values of the two slopes should be equal to 1. Moreover, for the loss angle it follows $\tan\delta = \tan(n\pi/2) = \tan 0.11 = 0.11$, which is in agreement with the results of G' and G'' of Fig. 155. It has to be emphasized that these gels are very fragile and that linear viscoelastic behaviour can be measured only at very small strain amplitudes (less than 3%). The frequency behaviour of both G' and G'' is in close agreement with the results of stress relaxation experiments: $\log G(t)$ plotted vs log time yields curves with slopes that are approximately equal (negative) to those of the storage modulus obtained from dynamic mechanical measurements under the same conditions. A comparison of both results is shown in Fig. 156 for a 9 wt% solution.

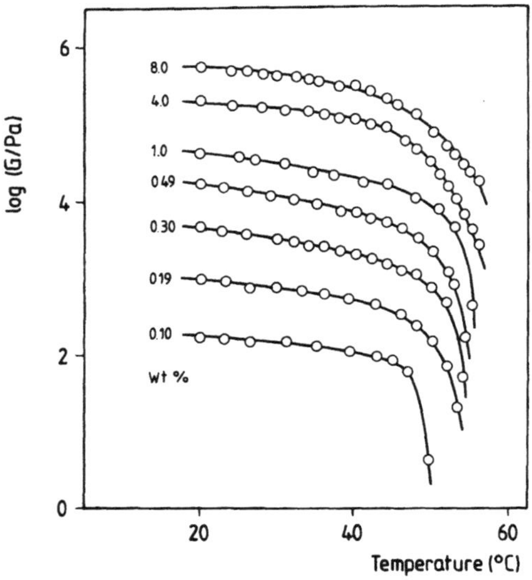

Fig. 153. Temperature dependence of the shear modulus of PBLG (M = 150 kg/mol)-benzyl alcohol gels of various concentrations, as indicated. Reproduced from Polym J [Ref. 416] by the courtesy of the authors and The Society of Polymer Science, Japan

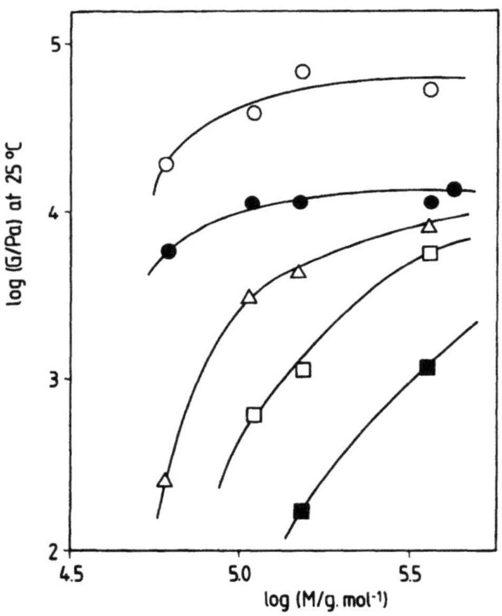

Fig. 154. Molecular weight dependence of the shear modulus at 25 °C of PBLG-benzyl alcohol gels of various concentrations in wt%: (○) 2.0; (●) 0.49; (△) 0.30; (□) 0.20; (■) 0.10. Reproduced from Polym J [Ref. 416] by the courtesy of the authors and The Society of Polymer Science, Japan

The results shown in Fig. 157 are very interesting, where the slope d [log G(t)]/d (log t) of the stress relaxation experiments is plotted vs concentration for various temperatures: it appears that the negative slope increases with both temperature and concentration. Although the height of the rubber plateau increases with concentration (up to 8 wt%) the gels show more ideal rubber

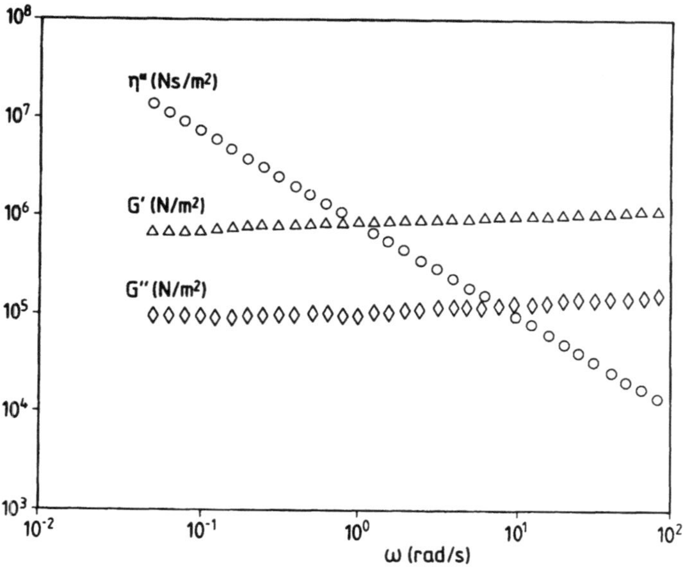

Fig. 155. Dynamic moduli, G′ and G″, and complex viscosity, η^*, vs angular frequency on double logarithmic scales for a 9 wt% PBLG-BA gel at 28 °C and 1% strain. Reproduced from Polym Eng Sci [Ref. 410] by the courtesy of the authors and of The Society of Plastics Engineers, Brookfield, UK

elasticity behaviour at low concentrations than at high concentrations. On the other hand, creep measurements show that a 2 wt% gel is more liable to creep than a 6 wt% gel [410].

The storage modulus plotted vs concentration at different temperatures shows a sharp maximum around a concentration of 8 wt%. At high temperatures this is just where the chimney in the phase diagram reaches the broad biphasic region. In general, the viscosity shows a steep maximum around the isotropic-anisotropic phase transition. According to Hermans [418] this concentration, where the viscosity is maximal, corresponds with the narrow biphasic region in the phase diagram of a rigid rod polymer system. In Fig. 158 a shoulder is present on the left hand side of the maximum. Kiss and Porter [419] also noted in the PBLG/m-cresol system a shoulder on the right hand side of the maximum. However, they made use of a measurement geometry of eccentric plates: by rotation of one of the disks an oscillation is superimposed on the steady shear flow. One has to be very careful in measuring the viscoelastic properties of these very fragile LCP gels in this way, as it is known that they cannot withstand shear deformations above a few percent: mechanical degradation of the network is the result of too high deformations. Nevertheless, shoulders on both sides of the maximum point to the existence of a chimney-like region at temperatures above the broad two-phase region. From dynamic light scattering experiments, combined with microscopy, Shukla et al. [411,413] concluded that the formation of the gel phase is a nucleation free, continuous,

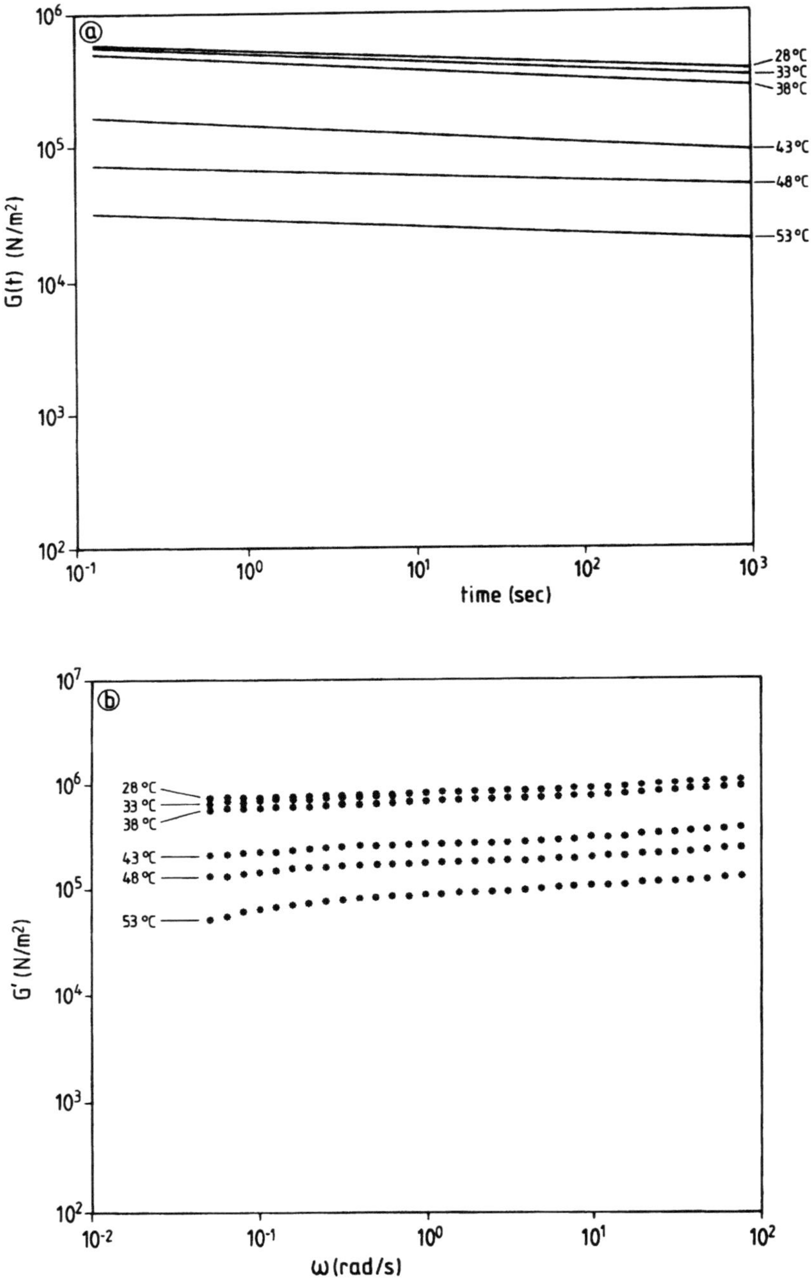

Fig. 156. a Shear moduli vs time for a 9 wt% PBLG-BA gel at different temperatures. **b** storage moduli vs angular frequency for a 9 wt% PBLG-BA gel at different temperatures. Reproduced from Polym Eng Sci [Ref. 410] by the courtesy of the authors and of The Society of Plastics Engineers, Brookfield, UK

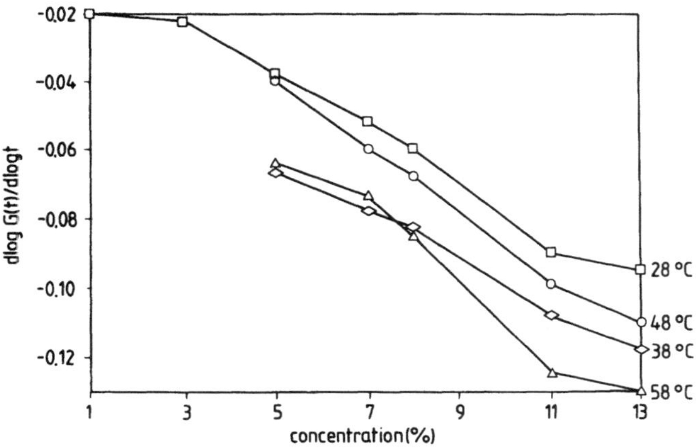

Fig. 157. Slope d log G(t)/d log t at 1% strain vs concentration of PBLG-benzyl alcohol gels for different temperatures. Reproduced from Polym Eng Sci [Ref. 410] by the courtesy of the authors and of The Society of Plastics Engineers, Brookfield, UK

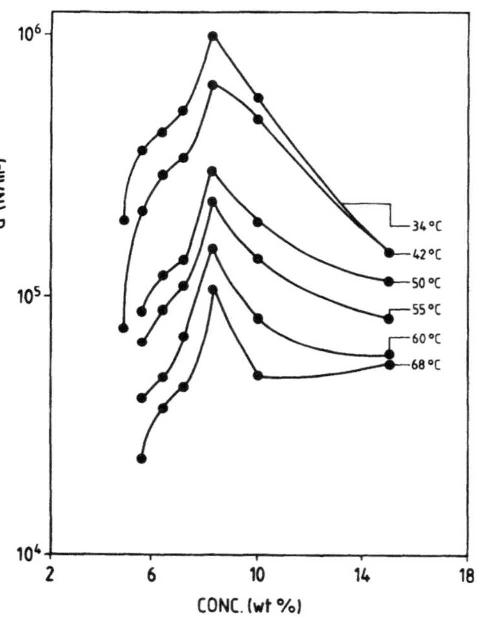

Fig. 158. Storage modulus vs concentration for a PBLG-BA gel, at different temperatures. Reproduced from Polym Eng Sci [Ref. 410] by the courtesy of the authors and of The Society of Plastics Engineers, Brookfield, UK

kinetic process. It is not the result of spinodal decomposition, but of a diffusion controlled aggregation process, with an Avrami exponent of 1.5, leading to the formation of fibrils. In dilute solutions gels are formed, due to aggregation to inhomogeneously distributed clusters of microfibrils (see Fig. 159). On the other hand, in concentrated solutions aggregation below the spinodal curve is

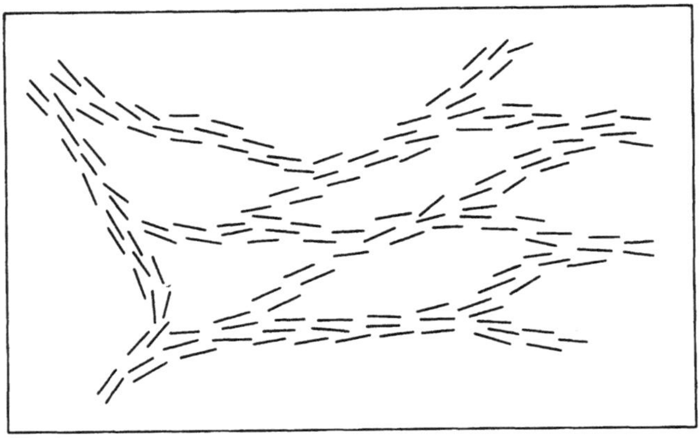

Fig. 159. Model for network formed by aggregation of PBLG chains. Reproduced from Polymer [Ref. 399] by the courtesy of the author and Elsevier Science Ltd, The Boulevard, Langford Lane, Kidlington 0X5 1GB, UK

followed by spinodal decomposition. This results in a bicontinuous system of interpenetrating, ordered regions of low and high polymer concentrations, followed by crystallization of the polymer in the concentrated phase. Gels formed in concentrated solutions in the area between the binodal and spinodal curves are similar to gels formed in dilute solutions.

It should be noted that, according to Miller et al. [406], in the PBLG-DMF system gelation takes place roughly 10–15 °C above the temperature where birefringence is observed. Presumably, birefringence is present only below the spinodal curve as a consequence of crystallization.

9.3 MCLCPs: Main Chain Liquid Crystalline Polymers

Guezala et al. [420] studied the viscoelastic behaviour of solutions of the thermotropic main chain liquid crystalline X7G (see Fig. 160), i.e. a copolyester ($\bar{M}_n \approx 20$ kg/mol) of poly(hydroxy benzoic acid) (60%) and poly(ethylene terephthalate) (40%) in *m*-cresol: the material used is non-random with respect to chain structure and domains of poly(hydroxy benzoic acid) have been reported. Solutions were studied, varying in concentration from 2.5 to 45 wt%. Special care was taken to avoid water absorption, because in its presence degradation and/or transesterification might occur.

The steady shear flow behaviour could be described by a Casson model:

$$\sigma^{1/2} = \sigma_o^{1/2} + \eta_\infty^{1/2}\left(\frac{d\gamma}{dt}\right)^{1/2} \qquad (29)$$

where σ_o is a yield stress and η_∞ the plastic viscosity. The Casson model is normally able to describe the flow behaviour of two-phase systems, e.g. suspensions. According to this phenomenon, the authors conclude that the copolymer

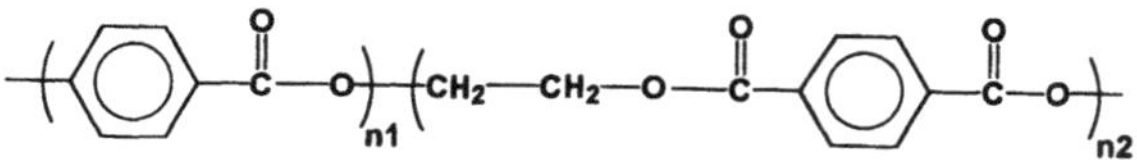

Fig. 160. Molecular structure of X7G

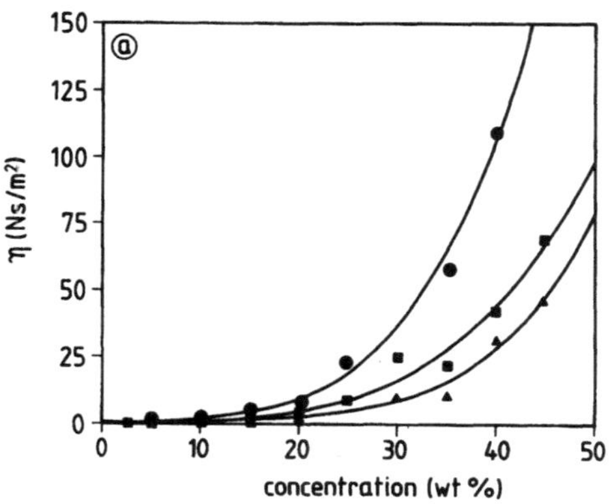

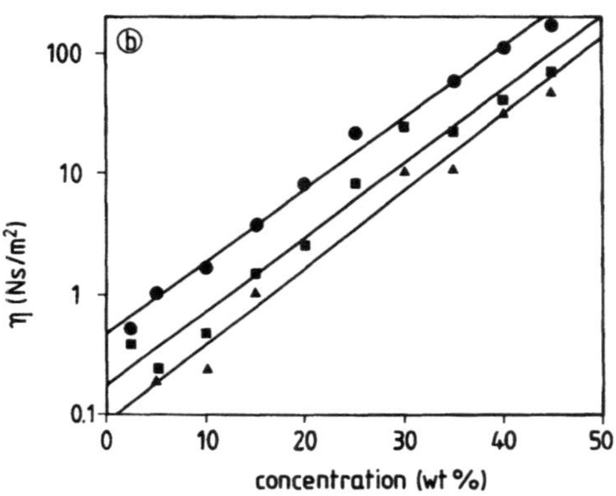

Fig. 161a, b. Plastic viscosities (Casson model) vs concentration of X7G solutions in *m*-cresol; temperatures 70 °C (●), 100 °C (■) and 120 °C (▲): **a** linear scale; **b** semi-logarithmic scale (constructed from data presented in Table 1 of [420])

system shows two-phase morphology in a *m*-cresol solution. In general, a yield stress only appears when phase separation has occurred. By plotting $\sigma^{1/2}$ vs $(d\gamma/dt)^{1/2}$, straight lines are obtained, yielding the values of σ_o and η_∞. The yield stress increases at first with concentration, subsequently it decreases dramatically to a minimum value at 25 wt%, after which it increases again and gelation starts. The viscosity increases with concentration, but to a high extent only above 20 wt%. This is also in agreement with the concentration where gelation just starts. The authors state that the viscosity is constant up to 20 wt%. However, from an inspection of their tabulated results, one has to conclude that in this concentration range the viscosity increases by a factor of 10. Results of viscosities are shown in Figs. 161a and b, linear and semi-logarithmic, respectively.

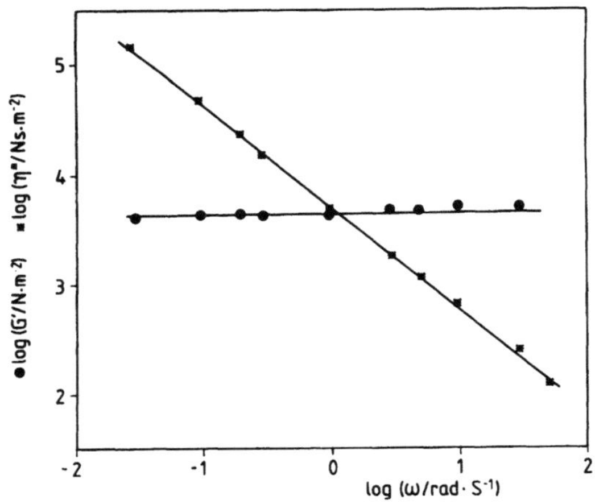

Fig. 162. Storage modulus, G′, and complex viscosity, η*, vs frequency for a 30% solution of X7G in *m*-cresol at 21 °C. Reproduced from Polymer [Ref. 420] by the courtesy of the authors and Elsevier Science Ltd, The Boulevard, Langford Lane, Kidlington 0X5 1GB, UK

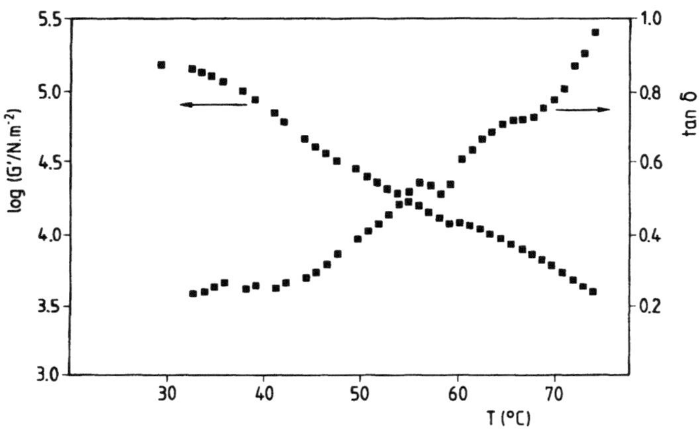

Fig. 163. Storage modulus, G′, and loss tangent, tan δ, measured at 1 Hz, vs temperature for a 40% solution of X7G in *m*-cresol. Reproduced from Polymer [Ref. 420] by the courtesy of the authors and Elsevier Science Ltd, The Boulevard, Langford Lane, Kidlington 0X5 1GB, UK

Dynamic mechanical properties are shown in Figs. 162 and 163.

The storage modulus of a 3 wt% solution shows almost rubber elastic behaviour with a slope of approximately 0.035. The slope of the η* curve, i.e. 0.94, is in reasonable agreement with the slope of G′. The rubber plateau decreases gradually (almost monotonically) with increasing temperature, although a weak transition is perceptible at about 40 °C, where the slope changes little.

Greiner et al. [421] investigated the gelation behaviour of 0.13–44 wt% solutions of a polyester in 3-phenoxy toluene (see Fig. 164). Even at a concentration of only 0.13% gels are formed at temperatures below 60 °C (see Fig. 165), where $G' \gg G''$ ($\tan\delta \approx 0.05$ for $T < 55$ °C). Above 55 °C G′ and G″ decrease

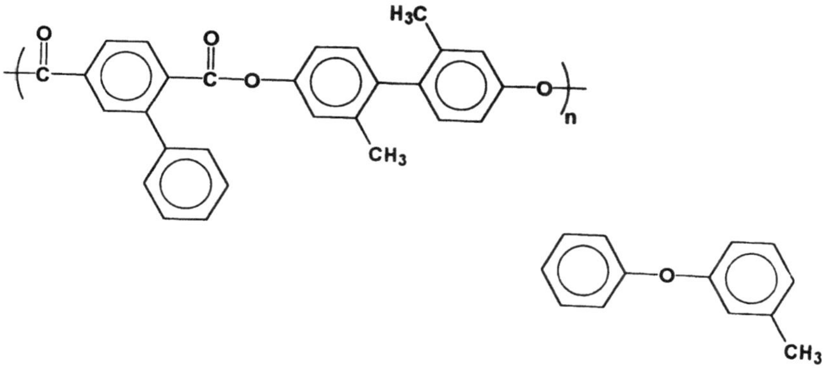

Fig. 164. Molecular structure of the copolyester and of the solvent 3-phenoxytoluene mentioned in [421]

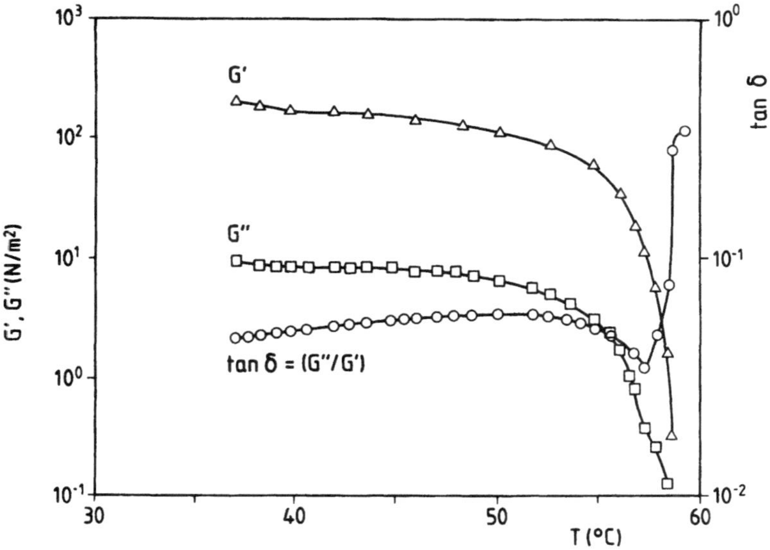

Fig. 165. Dynamic moduli, G′ and G″, and tan δ vs temperature for a 0.13 wt% mixture of the polyester of Fig. 164 in 3-phenoxytoluene; ω = 1 rad/s, strain amplitude = 25%. Reproduced from Makromol Chem [Ref. 421] by the courtesy of the authors and of Hüthig & Wepf Verlag Publishers, Zug, Switzerland

dramatically, whereas tan δ increases. Results for a 7 wt% solution at 50 and 60 °C are shown in Fig. 166. At 50 °C, the behaviour is gellike (G′ ≫ G″), whereas at 60 °C the system looks like being somewhere around the gel point: G′ ≈ G″, with a slope ≈ 0.5. DSC measurements reveal that a transition takes place around 57 °C, nearly independent of concentration (0.13–44 wt%). The origin of the endothermic transition at this temperature is not fully understood.

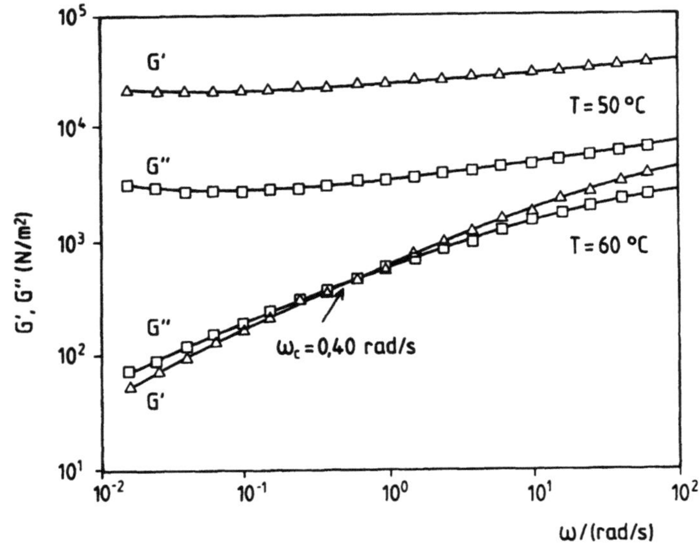

Fig. 166. Dynamic moduli, G′ and G″, vs angular frequency for a 7 wt% mixture of the polyester, shown in Fig. 164, in 3-phenoxytoluene. The *upper curves* are measured at 50 °C, the *lower curves* at 60 °C. Reproduced from Makromol Chem [Ref. 421] by the courtesy of the authors and of Hüthig & Wepf Verlag Publishers, Zug, Switzerland

The gelation threshold at this temperature coincides with disappearance of birefringence in the sample. This could be attributed to the formation of micro-crystallites of the polyester during gelation or to the formation of crystal solvates.

9.4 Conclusions

Gelation in solutions of liquid crystalline polymers is the result of interactions between the mesogenic groups in the polymers. In SCLCPs the mesogenics in the side chain form (liquid) crystalline regions, or at least show strong interactions. These regions of increased interactions act as crosslinks of unknown functionality (see Fig. 3G), especially if the main chain is a flexible chain, or lead to networks of aggregated rods (see Fig. 159). On the other hand, in MCLCPs mesogenics of parts of the main chain will interact. In general, the rodlike mesogenics form regions of strong interactions, thus also resulting in the formation of networks of aggregated rods, with crosslinks of high functionality (see also Fig. 159).

Investigation of the viscoelastic properties of LCP-gels is an extremely difficult subject, because the gels, in general, cannot even withstand low shear forces. Hence, instruments are needed with which very low shear stresses can be measured. Notwithstanding the interesting subject, few studies have been reported. Although linear viscoelasticity is difficult to study, determination of the transition from linear to non-linear viscoelasticity (i.e the break-down of the network) is also of importance in order to determine the strength of the network junctions. Much work has still to be done to elucidate the network properties of LCP gels.

10 Gelatin

10.1 Introduction

Gelatin may be defined as a protein made soluble by hydrolysis of collagen derived from the skin, white connective tissue and bones of animals. It is distinguished from animal glue by the way it is isolated from the raw material. For this purpose more purification is needed than with the preparation of glue. In this way a product of high quality is obtained [422].

A more scientific definition is given by Veis [423]: "The gelatins are a class of proteinaceous substances that have no existence in nature, but are derived from the parent protein collagen, by any one of a number of procedures involving the destruction of the secondary structure of the collagen and, in most cases, some aspects of the primary and tertiary structures. Collagen is the principle proteinaceous component of the white fibrous connective tissues, which serve as the chief, tensile stress-bearing elements for all vertebrates, whereas related proteins are found in any of the lower phyla".

Collagen is a protein with an unusual amino acid composition. It has high contents of glycine and of the imino acids proline and hydroxyproline (both with a pyrrolidine ring) and contains very small amounts of aromatic and sulfur-containing amino acids. The amino acid composition [424] of mammalian and avian collagens is remarkably constant. For these kinds of collagen glycine accounts for about 33% of the total number of amino acid residues, proline and hydroxyproline for about 20% and alanine for about 11% (see Fig. 167). Thus, these four amino acids account for two-thirds of the amino acid residues in collagen. However, fish collagens show a wider variation in composition, especially in the content of the imino acids. This content is lower, the lower the temperature of the water the fishes live in (see Table 7). The composition of reptile and amphibia collagens is rather variable, too, but generally between that of mammals and fishes. The shrinkage temperature or denaturation temperature of collagen fibres is primarily dependent on the content of imino acids, this temperature being lower if the imino acid content is smaller [424–430] (see Table 8 and Fig. 168). These data show that the stability of the structure of collagen is dependent on the imino acid content. From partial hydrolysis studies [424] a widespread distribution of glycine and frequent occurrences in the polypeptide chain of the sequences GLY-PRO-Y, GLY-X-HYP and GLY-PRO-HYP, where X and Y can be any amino acid, have been demonstrated (see e.g. [430–433]).

At first it was assumed that the combination GLY-PRO-HYPRO was of primary importance for the stability of collagen. In that way, the intramolecular bond between the carboxyl group of glycine and the hydroxyl group of hydroxyproline would be responsible for the stability. From measurements of the heat of denaturation and of the shrinkage temperature of collagen, however, one

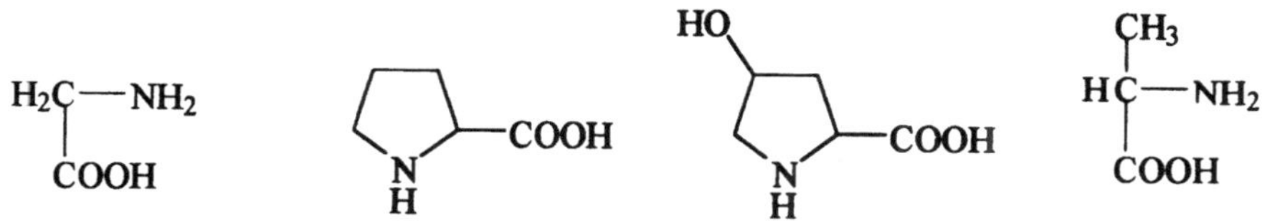

glycine **proline** **hydroxyproline** **alanine**

Fig. 167. Chemical structures of the most important amino acids of collagen and gelatin

Table 7. Number of imino acid residues per 1000 total residues of skin collagen of various animals [423]

Rabbit	Crocodile	African lungfish	Pike	Cod
237	221	207	199	155

Table 8. Dependence between imino acid content and the shrinkage temperature of collagen [423]

Source of collagen	Imino acid content per 1000 residues	Shrinkage temperature °C
Rat	226	40.8
Pike	199	30.6
Cod	155	20.0

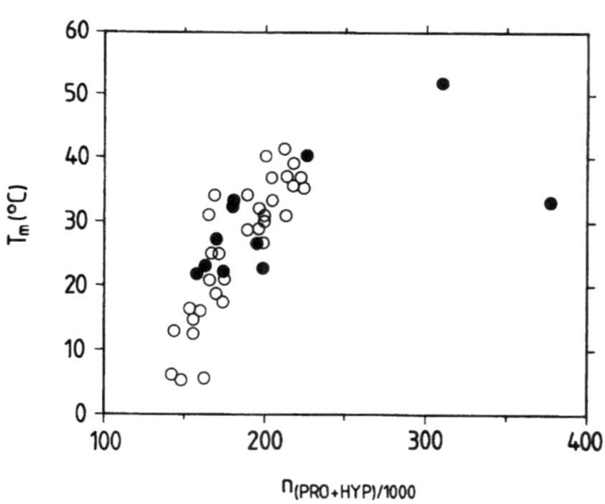

Fig. 168. The collagen melting temperature in salt-free solution with pH 3.7 vs total imino acid content per 1000 residues: (○) vertebrates; (●) invertebrates; constructed after data presented in [430]

had to conclude that hydrogen bonds do play a part in the stabilization of the collagen structure, but not the most important one. McClain and Wiley [435] concluded from the heat and entropy of denaturation that the tertiary super helix structure of the collagen molecule is stabilized primarily by the steric restrictions imposed by the pyrrolidine residues. Consequently, both the transition enthalpy and the transition temperature increase with increasing collagen stability. Their mutual dependence is shown in Fig. 169, where $\Delta H_{res/1000}$, i.e. the average transition enthalpy per 1000 amino acids present, is plotted vs the denaturation temperature T_m. It appears that, for mammalian collagens, $\Delta H_{res/1000}$ is approximately equal to 6 kJ/mol.

Privalov et al. [427, 436] supposed that the regular water structure near the macromolecules also plays an essential role in stabilizing the collagen structure. The presence of immobilised and oriented water was proved by Berendsen and Migchelsen [437, 438] with the aid of broad band ^{2}H-NMR. On the basis of calorimetric experiments, Piez and Sherman [439] concluded that hydroxyproline as such cannot be considered to be the only candidate to effect the stabilization of collagen, but that, among other things, a tripeptide GLY-PRO-Y could also be responsible for the collagen structure. These authors supposed that the formation of a collagen type helix only requires the presence of the repeating tripeptide sequence $(GLY\text{-}X\text{-}Y)_n$, where X and Y can be any amino acid, provided there is an average of at least one residue of proline or hydroxyproline every other triplet.

In a comprehensive review by Privalov [430], the role of proline and hydroxyproline in the stabilisation of the collagen triple helix was discussed. It was shown that for the stabilisation the tripeptide sequences GLY-PRO-Y, GLY-X-HYP and GLY-PRO-HYP are needed. From circular dichroism of model peptides it was shown [440] that the transition temperature of $(GLY\text{-}X\text{-}HYP)_n$ is higher than that of $(GLY\text{-}PRO\text{-}Y)_n$. In vertebrates the PRO and HYP residues are in general present on the X and Y positions, respectively. Moreover,

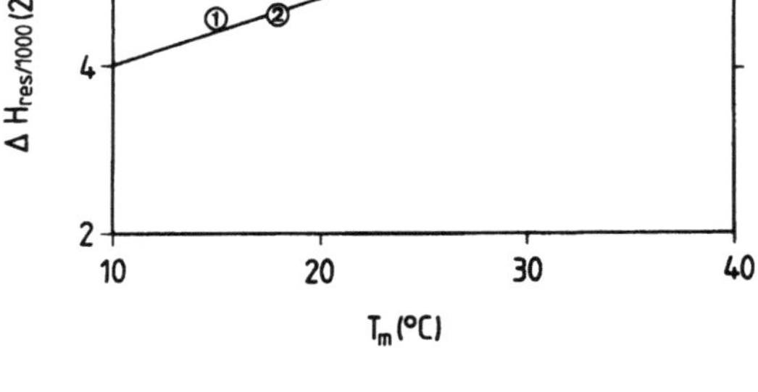

Fig. 169. Average residual transition enthalpy vs transition temperature for collagens of different origin: (1) cod skin; (2) halibut; (3) frog skin; (4) pike skin; (5) carp swim bladder; (6) rat skin; (7) sheep skin. Reproduced, with permission of the author and publishers, from [430]

4-hydroxyproline is a better stabilising imino acid than 3-hydroxyproline, this being present only exceptionally. Furthermore, the enthalpic contribution and the melting temperature of collagens correlate best with the content of 4-hydroxyproline. This is shown in Figs. 170 and 171, where $\Delta H_{res/1000}$, i.e. the average transition enthalpy per 1000 amino acids present, is plotted vs the total imino acid content and vs the 4-hydroxyproline content and T_m vs the 4-hydroxyproline content and vs the proline content. It is interesting to note that $\Delta H_{res/1000}$ extrapolated to $n_{HYP/1000} = 0$ coincides with the

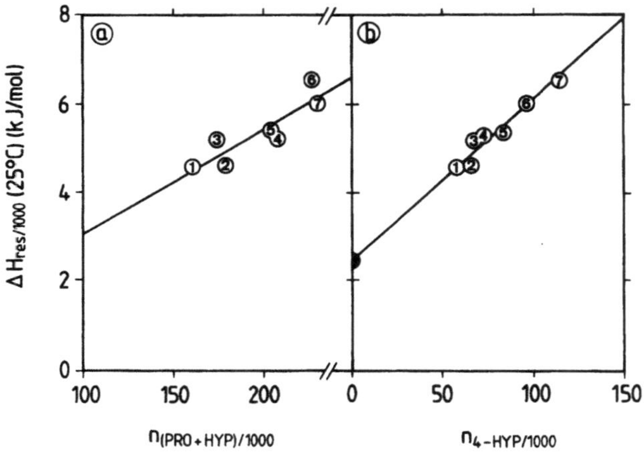

Fig. 170a, b. Average residual transition enthalpy reduced to 25 °C vs: **a** total imino acid content; **b** hydroxyproline content, both per 1000 residues, in the helical part of various collagens (numeration see Fig. 169; (●) is the value obtained for the synthetic (PRO-PRO-GLY)$_n$). Reproduced, with permission of the author and publishers, from [430]

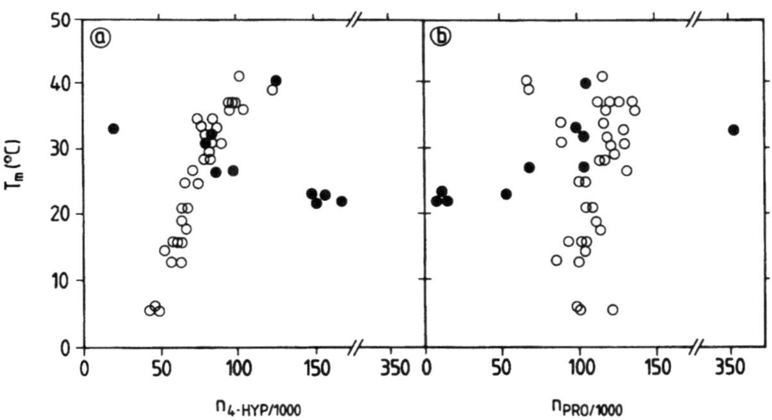

Fig. 171a, b. Transition temperature vs: **a** 4-hydroxyproline content; **b** proline content, both per 1000 residues, in various collagens: (○) vertebrates; (●) invertebrates; constructed after data presented in [430]

value for the synthetic hydroxyproline-less $(GLY\text{-}PRO\text{-}PRO)_n$. It appears that the collagens of invertebrates and vertebrates differ in their imino acid dependence of the quantities shown. It has to be emphasized that in vertebrates the 4-hydroxyproline residues are primarily present in the Y position, but in invertebrates in both positions with almost equal probability.

The transition entropy, from the random coil into the triple helix conformation, is negative, not only because of the decreased entropy of the polypeptide chain itself, but primarily due to the entropy decrease of the water molecules, because in the triple helix state the water molecules are immobilised and oriented in the axis direction in a structured water mantle. The presence of hydrogen bridges was presumed on energetic grounds [430] and was proved recently by high resolution X-ray crystallography on model triple helices [441]. Interchain hydrogen bonding occurs between the peptide groups of different chains [430, 442].

Because of the negative entropy contribution, the contribution of the transition enthalpy to the Gibbs energy has to be considerably negative. A great part of the negative enthalpy is the result of the above-mentioned presence of hydrogen bonds between the three helices [430, 431, 441, 443], and the formation of ion pairs [442, 444, 445] is also thought to play an important role. The ionizable residues (LYS, ARG, GLU and ASP) constitute 15–20% of all residues in fibrillar collagen and approximately 40% of all GLY-X-Y triplets contain at least one charged residue [443]. Negatively charged residues (GLU and ASP) are predominantly present in the X-position and the positively charged residues (LYS and ARG) in the Y-position.

The structure of collagen is mainly derived from detailed X-ray diffraction and electron microscopy studies. By stretching collagen, Cowan et al. [446] were able to improve the X-ray diffraction pattern considerably. Similar structures were proposed by three groups of researchers [447–450]. At present the following model is generally accepted. The collagen molecule, a rod of 300 nm, consists of three separate peptide chains, each with a molecular weight of about 100 kg/mol, coiled along a left-handed threefold axis with three residues in one turn [425, 451, 452]. These three helices complete their turn in about 0.9 nm. They are arranged parallel to each other so that, when viewed from above, their axes are set at the corners of an equilateral triangle of side 0.5 nm. Furthermore, a slight right-handed super helix is formed about a common axis, with an overall repeat distance of about 8.6 nm and a translation of 2.86 nm from a residue on one chain to the corresponding residue on another one. In this model, two features have to be emphasized: the steric hindrance of rotation, as imposed by the high content of pyrrolidine residues, enhancing the rigidity, and the occurrence of glycine at every third residue, permitting the chains to approach each other closely enough for the formation of hydrogen bonds. It is interesting to note that substitution of glycine in a tripeptide GLY-X-Y by alanine, causes a subtle alteration of the conformation, with a local untwisting of the triple helix, due to the impossibility of close packing of the chains near the central axis [441]. Close packing requires that the small glycine molecules occupy every third position. This is the reason for the generation of the GLY-X-Y repeating sequence. The change in structure by the substitution of GLY by another amino acid residue is responsible for hereditary diseases, like the Ehlers–Danlos syndrome IV, that are attended with a weakness of the skin, connective tissues, bones etc. [441, 442, 453, 454].

The number of hydrogen bonds (one or two) that are involved in the triplets containing only one pyrrolidine residue have puzzled many authors. One hydrogen bond is formed between the amino group of glycine and the carboxyl group of a residue in the X-position of an adjacent helix [430, 442]. The distance between donor and acceptor in the second hydrogen bond is too long, namely 0.51 nm [455]. Thermal denaturation and rates of deuterium and tritium exchange have, however, produced results, in favour of the double-bonded structures [424]. For this reason it was suggested that these groups are connected by one or two water molecules [430]. In this way adjacent helices are also linked. Hydroxyproline residues play an important role in these second hydrogen bonds. The gradual right-handed twist of the individual strands allows the side groups of amino acids of variable sizes to come into the structure [456]. The triple helix structure with nearly the same parameters is observed in synthetic polytripeptides [457] of variable length, like $(GLY\text{-}PRO\text{-}PRO)_n$ or $(GLY\text{-}PRO\text{-}HYPRO)_n$.

The denaturation process of collagen involves the melting of the ordered hydrogen bonded structure. In this process the triple helix structure is destroyed to produce one, two or three random chains of gelatin molecules. In aqueous solutions the gelatin molecules are believed to be in the coil conformation at temperatures above 40 °C. When these solutions are cooled to temperatures below 30 °C a reversible coil-helix transition takes place, which can be detected by optical rotation, mainly due to the left-handed single strand helix formation [458, 459]. A mechanism for the partial renaturation of gelatin, imposed by Harrington & Rao [460], is shown in Fig. 172.

At very low concentrations (< 0.1%) intramolecular bonds are formed preferentially by back folding of the single chains. At concentrations above 1% the helix growth induces chain association and three-dimensional network formation. For that purpose two subsequent processes are involved: a) the formation of single helix nuclei and b) aggregation of these single helices to a triple helix. It appears that the extent and manner of reversion to the collagen fold (triple helix) structure is dependent on solvent, temperature and concentration.

As early as 1904, Leick [461] investigated the elasticity of gelatin gels. He concluded that Young's modulus was nearly proportional to the square of the concentration of gelatin up to 25 wt%. In 1915, Arisz [462] published his comprehensive studies of the sol and gel state of gelatin solutions. One of his findings, which puzzled himself, was the remarkable dependence of the sol-gel transition on the thermal history. In 1925, Poole [463] and in 1932, Hatschek [464] also concluded that the modulus of elasticity of gelatin gels is proportional to the square of the concentration. However, in 1921, Sheppard and Sweet [465] did not report a similar relation in the concentration range from 10 to

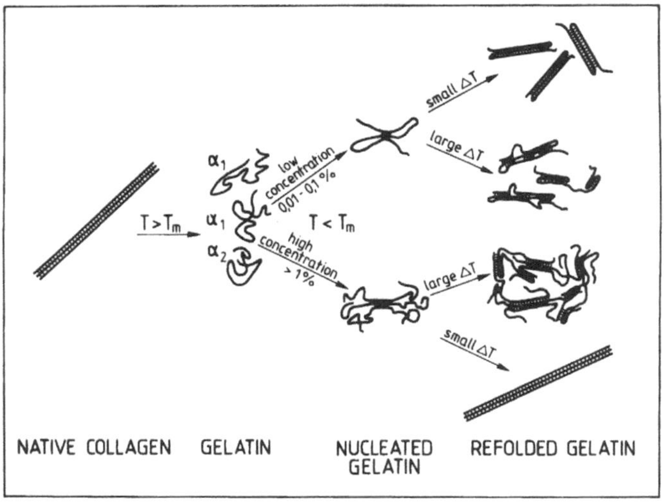

Fig. 172. Schematic view of the renaturation process of gelatin. Reproduced from Biochemistry [Ref. 460] by the courtesy of The American Chemical Society

45%. In his review on protein gels in 1948, Ferry [466] paid much attention to gelatin gels. From then on many investigations on gelatin have been published. Many of these have been concerned with the viscoelastic behaviour of gelatin gels. Special attention has been paid to the dependence of the shear modulus, G, on molecular weight, concentration and temperature.

10.2 Time and Temperature Dependences of Dynamic Moduli

10.2.1 Plain Ageing

An example of the gelation process in gelatin solutions is shown in Fig. 173 for a 1.95 wt% aqueous gelatin solution at temperatures varying from −1.2 to 20.6 °C [11, 12, 23]. The angular frequency is the same in all cases, namely $\omega = 0.39$ rad/s (in fact it appears that the storage modulus is almost independent of frequency around 0.39 rad/s). The ageing process is much faster at low temperatures than at high temperatures. After some hours of ageing an equilibrium value of the storage modulus seems to be reached. However, from Fig. 174, where the storage modulus is plotted vs log ageing time, it appears that even after more than 100 h no tendency to reach an equilibrium value is perceptible.

Another interesting feature is the presence of what is called by Te Nijenhuis [11, 23, 24] an induction time, or a gel time, before which the storage modulus has no measurable value. For low temperatures this gel time is very short (less than 1 min), whereas for a temperature of 25 °C this value appears to be about

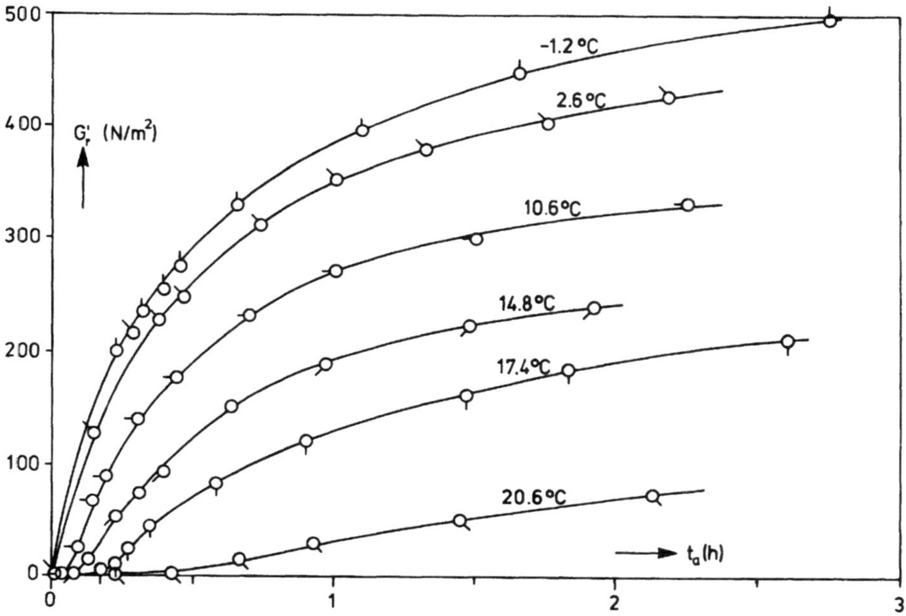

Fig. 173. Storage modulus of aqueous 1.95 wt% gelatin ($\bar{M}_w = 70$ kg/mol) vs ageing time for several ageing temperatures; angular frequency $\omega = 0.393$ rad/s. Reproduced from Colloid Polym Sci [Ref. 23] by the courtesy of Steinkopff Verlag Darmstadt, FRG

8 h. A maximum gelation temperature was calculated to be around 27 °C [11, 24, 467].

In Fig. 175 the storage modulus of this solution at 14.8 °C is plotted on double logarithmic scales vs the angular frequency, for different ageing times. (Note that the values, especially those after 6 min of ageing, were obtained by an interpolation procedure, where the storage modulus is plotted vs time). Even after 6 min of ageing (i.e. 1 min after the gel time) the storage modulus is almost independent of frequency. It means that a real rubber network is formed, whose crosslink density increases with time.

Even more pronounced is the independence of the equilibrium modulus of the frequency in Fig. 176. Here the different curves belong to different ageing temperatures after 120–170 h of ageing. Over about five decades in the frequency, the storage modulus has an almost constant value, the slope being about 0.01.

From these results, no relaxation at lower frequencies is perceptible. However, Ross–Murphy and Higgs [8, 9] measured the creep behaviour of gelatin gels (2.4–15 wt%) at room temperature. The gels show a typical creep behaviour of polymer solutions with equilibrium compliances, J_e, of the order of 10^{-5}–10^{-3} m^2/N and with rather high viscosities of the order of 10^8 Ns/m^2. For a 4 wt% solution at room temperature, a retardation time of approximately

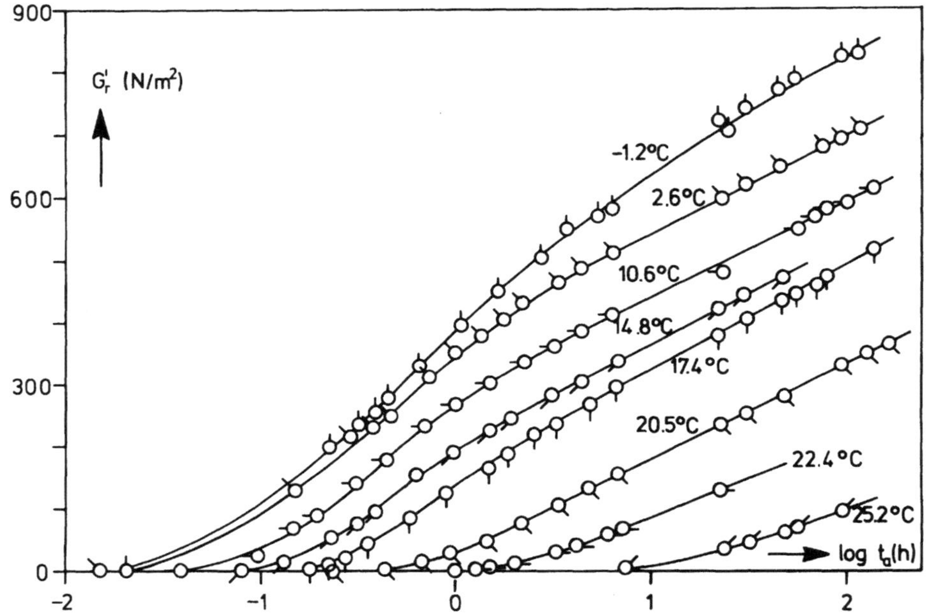

Fig. 174. Reduced storage modulus of aqueous 1.95 wt% gelatin ($\bar{M}_w = 70$ kg/mol) plotted vs log ageing time for several ageing temperatures; angular frequency $\omega = 0.393$ rad/s; reference temperature $-1.2°C$. Reproduced from Colloid Polym Sci [Ref. 23] by the courtesy of Steinkopff Verlag Darmstadt, FRG

3×10^4 s might be calculated. Similar results were obtained by Kramer [468]. These results urged the present author to carry out a careful inspection of his own results [11, 23, 24], as his results, depicted by the storage moduli as a function of frequency (see Figs. 175 and 176), did not show a tendency to a relaxation process at low frequencies. From the measurements by Ross–Murphy and Higgs [8, 9], it appears that for a 4 wt% solution at room temperature a relaxation process should be expected at frequencies around 3×10^{-5} rad/s. A result is shown in Fig. 177 where, for a 2 wt% solution, both dynamic moduli, G′ and G″, measured at 2.6 °C are plotted vs angular frequency. Around a frequency of 10^{-2} rad/s there is a minimum in the loss modulus. Such a result betrays that at much lower frequencies a relaxation mechanism should be present, attended with a maximum in the loss modulus and a decrease in the storage modulus (depicted by dotted lines). This result is in agreement with the results of creep measurements. It shows that gelatin gels show no rubberlike behaviour up to very low frequencies.

In the literature, viscoelastic measurements before the gel point of gelatin solutions are largely unreported. As far as the author knows, Djabourov et al. [469] were the first to report measurements of viscoelastic properties of aqueous gelatin solutions before the gel point (down to $G' = 7 \times 10^{-4}$ N/m²) of a 4.7% aqueous solution. In about 3 h the system gradually changes from a Newtonian

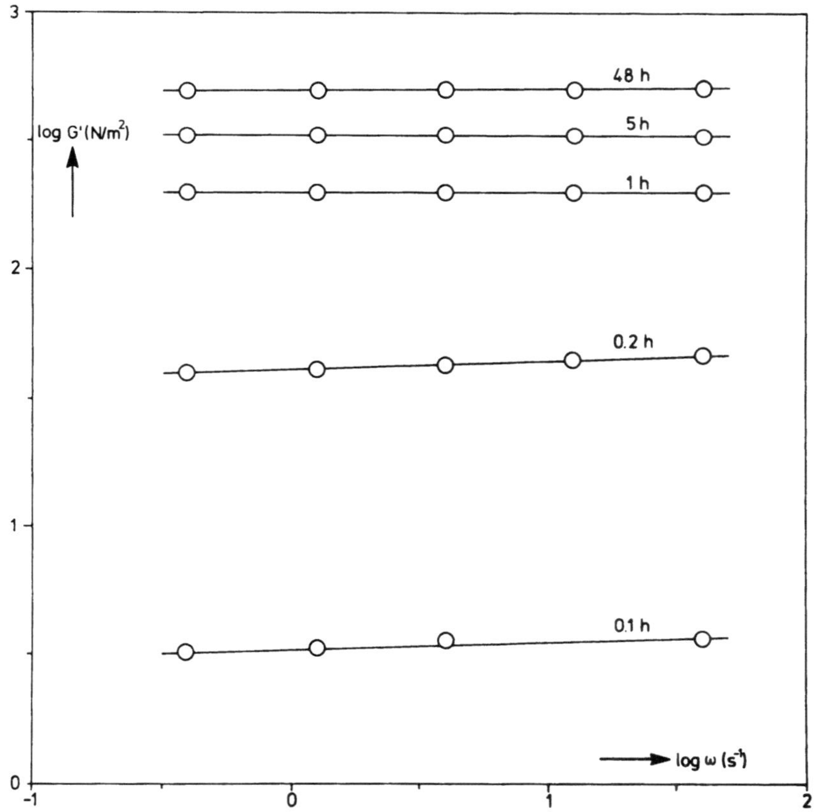

Fig. 175. Storage modulus of aqueous 1.95 wt% gelatin ($\bar{M}_w = 70$ kg/mol) vs angular frequency for several ageing times; ageing temperature 14.8 °C. Reproduced from Colloid Polym Sci [Ref. 23] by the courtesy of Steinkopff Verlag Darmstadt, FRG

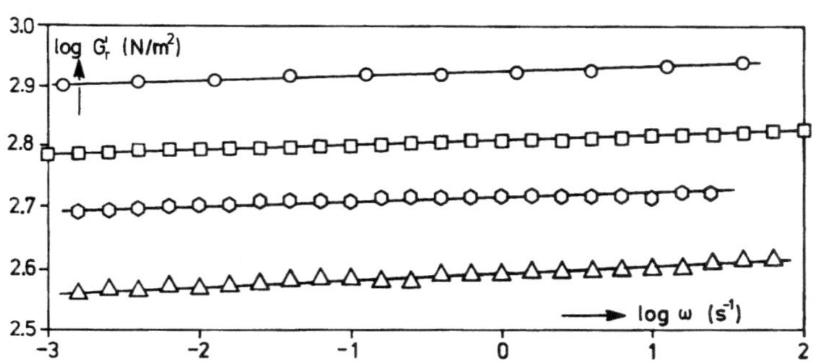

Fig. 176. Reduced storage modulus of aqueous 1.95 wt% gelatin ($\bar{M}_w = 70$ kg/mol) vs angular frequency for several amounts of ageing; reference temperature −1.2 °C; (○) 120 h at −1.2 °C; (□) 145 h at 10.6 °C; (⬡) 146 h at 17.3 °C; △ 170 h at 20.5 °C. Reproduced from Colloid Polym Sci [Ref. 23] by the courtesy of Steinkopff Verlag Darmstadt, FRG

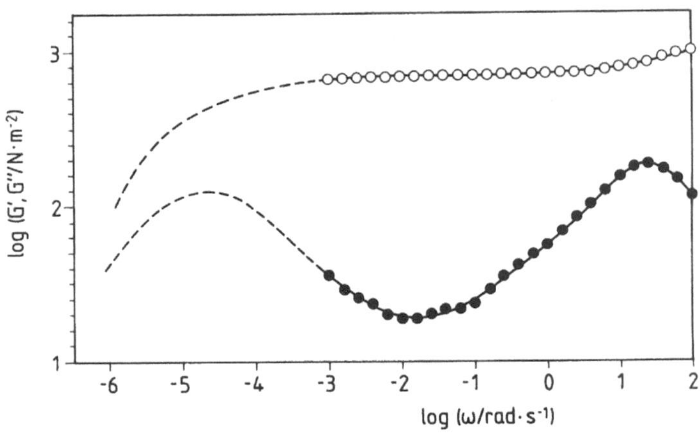

Fig. 177. Frequency dependence of the dynamic moduli, (○) G′ and (●) G″, of aqueous 1.95 wt% gelatin ($\bar{M}_w = 70$ kg/mol) gel, after 45 h of ageing at 2.6 °C (unpublished measurements of the present author)

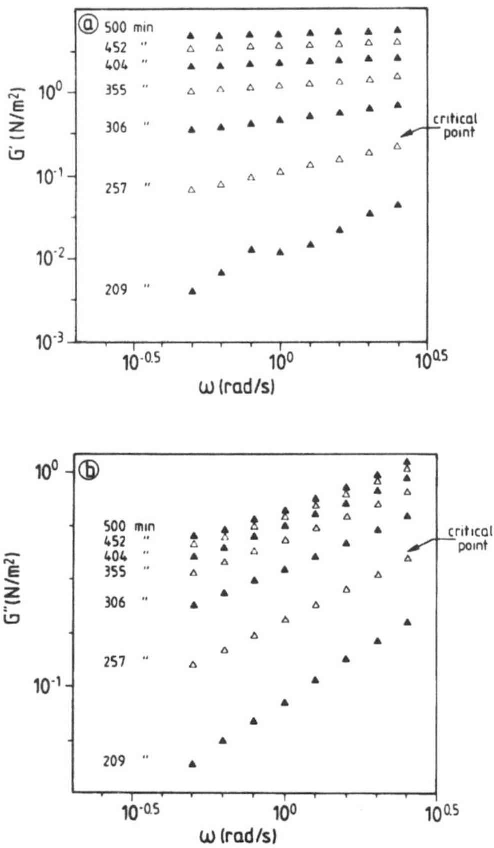

Fig. 178. **a** Storage moduli vs angular frequency. **b** Loss moduli vs angular frequency; 5 wt% aqueous gelatin ($\bar{M}_w = 151$ kg/mol) at 26 °C for ageing times varying from 209 min (i.e. before the gel point) to 500 min (i.e. after the gel point). Reproduced from Rheol Acta [Ref. 470] by the courtesy of the authors and of Steinkopff Publishers Darmstadt, FRG

solution into a rubber network with a storage modulus of about 0.8 N/m^2. Meanwhile, more results of pregel viscoelastic measurements have been reported [470–473], and most of them were also used for the determination of the gel point with the aid of the Winter-Chambon method. An example is given in Fig. 178 for a 5 wt% solution at 26 °C [470]. The critical point, where tan δ is independent of frequency, is reached after 257 min of ageing. The slopes, n, at the critical point and the value of n calculated from tan δ agree very well: n = 0.69. This is also in agreement with results obtained by many authors [470–473]: in general, a value $n = 0.68 \pm 0.08$ has been reported.

10.2.2 Special Temperature History

A remarkable phenomenon arises if a gelation process is started at a certain temperature and is continued at a lower temperature. The pregelation at the higher temperature has a dramatic effect on the rate of the ageing processes at the lower temperature: after the temperature quench the storage modulus increases very fast and rises (far) above that obtained by plain ageing at this lower temperature. The existence of this effect was reported by Pouradier [474], Stainsby [475], and later by Ledward [476], and has been studied systematically by Te Nijenhuis [11, 24]. A result of this study is shown in Fig.179.

The effect increases if the preageing has lasted a longer time, but it seems to reach an equilibrium state after a couple of days of preageing [11]. It is worth mentioning that preageing at 17.4 °C even for a short time, i.e. within the induction period at that temperature, causes the storage modulus to rise above the plain ageing level. The level of the preageing temperature also affects the rate of the excess ageing at −1.2 °C: a maximum effect is observed at 16 °C [11, 24].

The excess ageing is annulled completely by increasing the temperature again. This is shown in Fig. 180 [11, 24]: the storage modulus ultimately obtained is equal to the modulus obtained without an intermediate temperature decrease. It is concluded that the excess ageing is caused by an enormous growth of existing nuclei resulting in crosslinks of relatively small triple helices. During the low temperature period the formation of more stable triple helices also continues: these crosslinks will increase in number. This also causes an increase in the storage modulus. However, this formation is no faster at the lower temperature; this follows from the modulus ultimately obtained which is no higher than that without the intermediate low temperature period. The conclusion is, first, that the rate of formation of triple helices of a certain stability is independent of temperature and, second, that the heat of activation of such processes is very small, e.g. smaller than 3 kJ/mol. This conclusion is in agreement with that of Eagland et al. [477]. On the basis of their combined measurements of optical rotation and viscosity in the temperature range from 5 to 15 °C on very dilute gelatin solutions, these authors conclude that the reaction rate constant of the rate determining step in the ageing process of gelatin is almost independent of temperature. Hence, the fast increase of the rate of the gelation

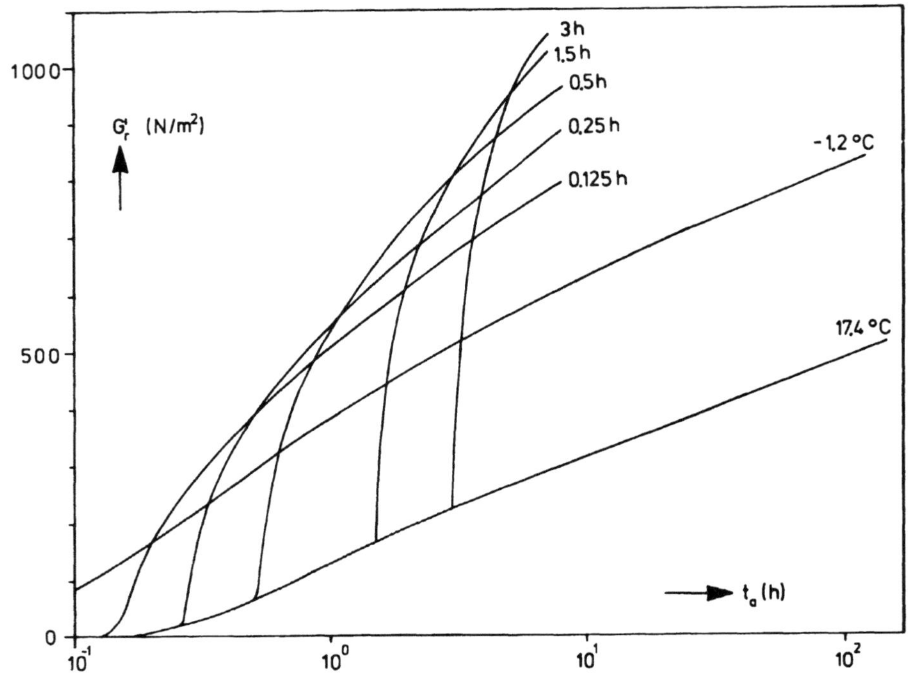

Fig. 179. Reduced storage modulus of aqueous 1.95 wt% gelatin ($\bar{M}_w = 70$ kg/mol) of several ageing histories plotted vs log ageing time; angular frequency $\omega = 0.393$ rad/s; reference temperature 17.4 °C. Reproduced from Colloid Polym Sci [Ref. 24] by the courtesy of Steinkopff Verlag Darmstadt, FRG

process of gelatin solutions is not caused by an increase in the reaction rate constant, but by an increased availability of less stable crosslinks. Similar conclusions were reported by Busnell et al. [478] on the basis of viscoelasticity combined with differential scanning calorimetry and optical rotation experiments (see Sect. 5).

Another view is expressed by De Wolf (private communication), who, on the basis of his circular dichroism studies [479], suggested that at low temperatures monohelix formation is likely to occur. Hence, in a system that was prematured at room temperature and that was subsequently cooled down to low temperatures, the chains are stiffened by the formation of monohelices in between the triple helical crosslinks. This should mean that the increase of the storage modulus is not the result of an increased crosslink density, but primarily of chain stiffening due to the formation of monohelices. These helices melt again with increasing temperature. This model closely resembles the model suggested by Berghmans et al. [263] in the gelation process of syndiotactic poly(methyl methacrylates) at low temperatures, as discussed in Sect.4.

Two points are worth mentioning. First, the results shown by Te Nijenhuis are for a 2 wt% gelatin solution. The effects of excess ageing by lowering the

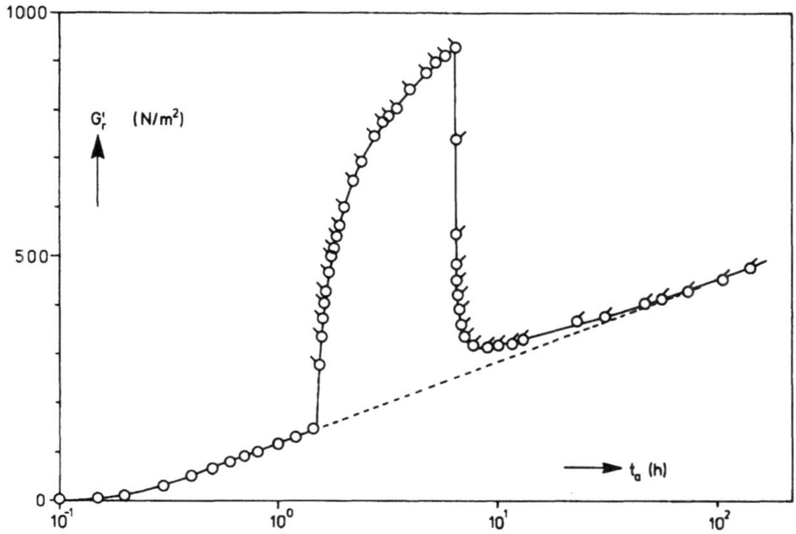

Fig. 180. Reduced storage modulus of aqueous 1.95 wt% gelatin ($\bar{M}_w$ = 70 kg/mol) with a special ageing history, plotted vs log ageing time; angular frequency ω = 0.393 rad/s; reference temperature − 1.2 °C (○) and (ර) 17.4 °C; (ර) − 1.2 °C. Reproduced from Colloid Polym Sci [Ref. 24] by the courtesy of Steinkopff Verlag Darmstadt, FRG

temperature decrease with increasing concentration: for a 5 wt% solution the effect almost disappears [11, 24], and at higher concentrations it is no longer present (unpublished results of Goossen and Te Nijenhuis, 1991). Second, every temperature increase causes a decrease of the storage modulus; eventually the modulus is equal to that obtained after plain ageing at this higher temperature. Even if the temperature increase falls within the induction period at the higher temperature, the rubber modulus disappears and appears again after the induction period. This effect was observed by Arisz in 1915 [462].

10.3 Concentration Dependence

10.3.1 Phenomenology

Near the critical concentration, c_o, below which no gelation occurs, the power law dependence of the equilibrium shear modulus, G_e, on the concentration, c, is high and variable. However, in many, rather historical, publications [461, 463, 464, 466, 480–489] it was demonstrated that the quotient G'/c^2 or G_e/c^2 has a constant value in a certain concentration region, or that G' or G_e are linear functions of c^2. For the value of G', its equilibrium value (whatever it may be) should be used. In the previous section it was shown that even after 150 h of ageing the storage modulus still linearly increases with log ageing time. Hence, it is not surprising that the values of G'/c^2 reported show a rather large scattering. Sheppard and Sweet [465] did not find such a quadratic relationship at high concentrations (10–45 wt%). Saunders and Ward [486] report that for concentrations below 2 wt% the storage modulus is a linear function of the concentration. In his doctoral thesis Te Nijenhuis [11] shows that, for 2–5 wt% aqueous gelatin solutions, plots of G' vs c^2 are linear for every value of the ageing time from 0.3 to 100 h. However, linear extrapolation to $G' = 0$ did not lead to $c = 0$, but to a positive value c_o, which decreases with increasing gelation time. This means that only after prolonged ageing is G' really proportional to c^2. This is also clear from Fig. 181, where G'/c^2 is plotted vs log ageing time for concentrations of 1–5 wt%. It appears that only after a long ageing times (>40 h) and only for concentrations of 2–5 wt% are the curves more or less coincide. Moreover, at a concentration of 1 wt% this dependence does not hold at all. Apparently intramolecular crosslinks play an important role at these low concentrations. Unpublished measurements of Goossen and Te Nijenhuis (1991) show that the c^2 dependence is not completely satisfied in the concentration range of 10–50 wt%, and especially not above 40 wt%; however, the deviations are small.

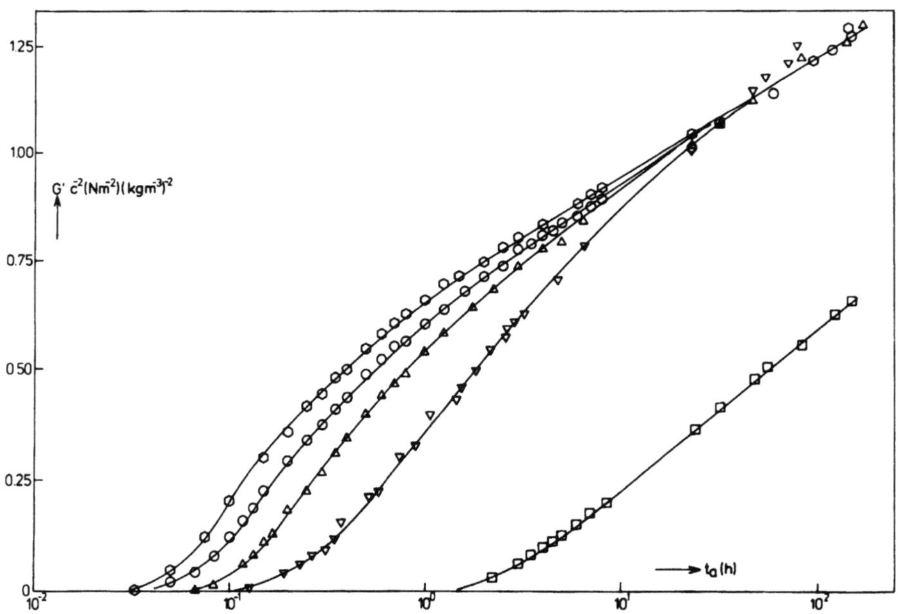

Fig. 181. G'/c^2 of aqueous gelatin ($\bar{M}_w = 70$ kg/mol) of several concentrations plotted vs log ageing time; ageing temperature 17.4 °C; angular frequency $\omega = 0.393$ rad/s; gelatin concentrations (wt%): (□) 1.00%; (▽) 1.95%; (△) 3.00%; (○) 4.00%; (⬡) 5.00%. Reproduced from Colloid Polym Sci [Ref. 23] by the courtesy of Steinkopff Verlag Darmstadt, FRG

10.3.2 Theoretical Considerations

Clark and Ross-Murphy [1, 487] calculated the concentration dependence of the equilibrium shear modulus of gels by making use of Hermans' treatment [488], in which the association of primary chains is described by a monomer-dimer equilibrium. The crosslinks are treated as dimer interactions and, hence, their functionality is 4. From that consideration they were able to calculate the equilibrium shear modulus as a function of the reduced concentration, i.e. c/c_o, where c_o is the minimal concentration needed for network formation.

A slightly different approach is given in the Appendix, where it is shown that, provided the maximum number of possible crosslinks is much larger than their actual number (i.e. $\bar{\gamma}_{w,max} \gg \bar{\gamma}_w$), the relationship between the equilibrium shear modulus and the concentration is given by

$$G_{er} = \frac{4}{f}\left[\frac{c}{c_o}\right]^{0.5f-1}(1 - w_s^{0.5f}) - 2(1 - w_s). \tag{30}$$

For relatively high concentrations the sol fraction tends to zero in the equilibrium state. In that case the expression for the reduced equilibrium shear modulus will be

$$G_{er} = \frac{4}{f}\left[\frac{c}{c_o}\right]^{0.5f-1} - 2 \approx \frac{4}{f}\left[\frac{c}{c_o}\right]^{0.5f-1}. \tag{31}$$

Hence, the reduced modulus is proportional to $(c/c_o)^{0.5f-1}$, whereas the equilibrium shear modulus G_e is proportional to $c^{0.5f}$. Results of calculations are shown in Fig. 182, where G_{er} is plotted vs c_r on double logarithmic scales for

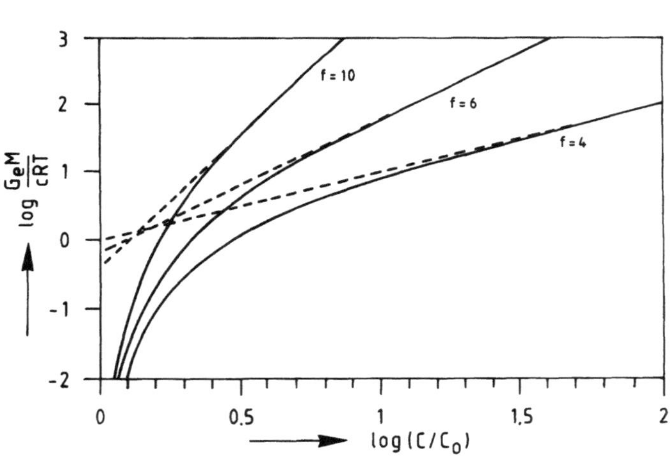

Fig. 182. Reduced equilibrium shear modulus, as calculated from Eq. (30), plotted vs reduced concentration for crosslink functionalities f = 4, 6 and 10; the *broken lines* are the asymptotes for high concentrations

crosslink functionalities of 4, 6 and 10. At high values of the reduced concentration the slope is equal to 0.5f − 1, as shown by the asymptotes. The asymptotic values are reached for reduced concentrations of about 2.8, 7 and 80 for crosslink functionalities of 4, 6 and 10, respectively. In the case of gelation of gelatin solutions the value of c_o is about 1 wt% (see Sect. 10.4.2 and [487]), which means that the asymptotic values would be reached at a concentration of 7 wt%.

A probably more realistic situation is a non-infinite value of $\bar{\gamma}_{max}$. In that case G_e cannot be expressed easily as a function of w_s and c/c_o. The reduced equilibrium shear modulus can be calculated as a function of the reduced concentration on substituting certain values of $\bar{\gamma}_{max}$ into Eq. (A7) and making use of Eqs. (A9) and (A10). In Fig. 183 plots of $\log G_{er}$ vs $\log c_r$ are shown for various values of $\bar{\gamma}_{max}$ (4, 10, 50 and ∞) and crosslink functionalities f = 4 and 6. It appears that an equilibrium value (independent of concentration) is reached for high values of the reduced concentration, which means that in those cases the equilibrium shear modulus is proportional to the concentration:

$$G_e = \frac{2cRT}{M}\left[\bar{\gamma}_{max}\frac{f-2}{f} - 1\right]. \tag{32}$$

For aqueous gelatin solutions (with $c_o \approx 1$ wt%) these equilibrium values are reached for unrealistically high concentrations, however. It is not possible to find a quadratic relationship between G_e and c in the concentration range of 2 to 40 wt% and not even in the range of 2 to 5 wt%.

In order to get an idea of the values of $\bar{\gamma}_{max}$, an example is given for a 5 wt% aqueous solution of gelatin with a weight average molecular weight of 70 kg/mol; in this case the equilibrium shear modulus at room temperature

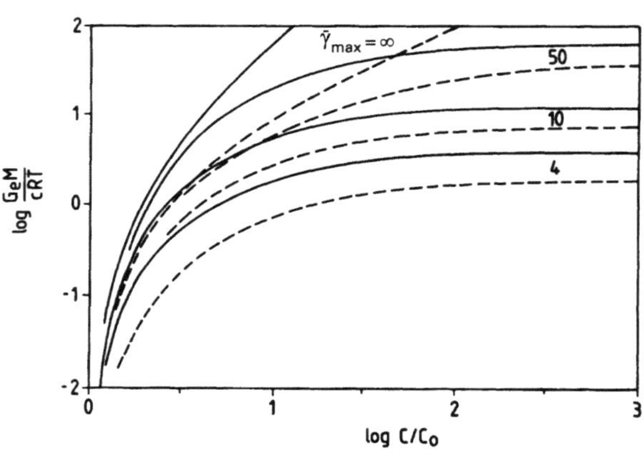

Fig. 183. Reduced equilibrium shear modulus, as calculated from Eqs. (A7), (A9) and (A10), plotted vs reduced concentration for crosslink functionalities f = 6 (*full lines*) and f = 4 (*broken lines*) and various maximum numbers of potential junction zones per polymer molecule: $\bar{\gamma}_{max}$ = 4, 10, 50 and ∞

is given by

$$G_e(N/m^2) = 2400\,\bar{\gamma}_{w,max} - 3600. \qquad (33)$$

For values of $\bar{\gamma}_{w,max}$ of 5, 10 and 20, the equilibrium shear modulus is equal to 8.4, 20.4 and 44.4 kN/m^2, respectively. These values are much higher than those measured after 150 h of ageing (at 17.4 °C $G_e \approx 3$ kN/m^2 [11, 23]). From this comparison it is clear that the actual concentration dependence is determined more by the kinetics of the gelation process than by thermodynamics.

In this respect it is interesting to mention the work of Peniche–Covas et al. [17]. These authors made measurements of Young's modulus and of the optical rotation as a function of time for a 5.7 wt% aqueous gelatin solution and applied the cascade theory of network formation [17] in order to treat their data quantitatively. In Fig. 184, E is plotted vs the relative conversion of the reaction α/α_c, as calculated from optical rotation measurements (α_c = the reaction conversion at the gel point). The solid lines depict the cascade theory calculations. At high conversions, curve C, calculated for hexafunctionality of the triple helix crosslinks and the maximum number of potential crosslink sites being eight, gives the best fit, whereas, for lower conversions, tetrafunctionality seems to give a better fit. The dashed lines, i.e E vs $\bar{\gamma}/\bar{\gamma}_g$ (where $\bar{\gamma}_g$ is the crosslinking index at the gel point and equal to $2/(f-2)$), are calculated by the present author, making use of tetra-, hexa- and octafunctional crosslinks and assuming uniform polymer with a molecular weight of 105 kg/mol. In his model [39–44], the potential number of crosslinks is unlimited. These calculations also show that,

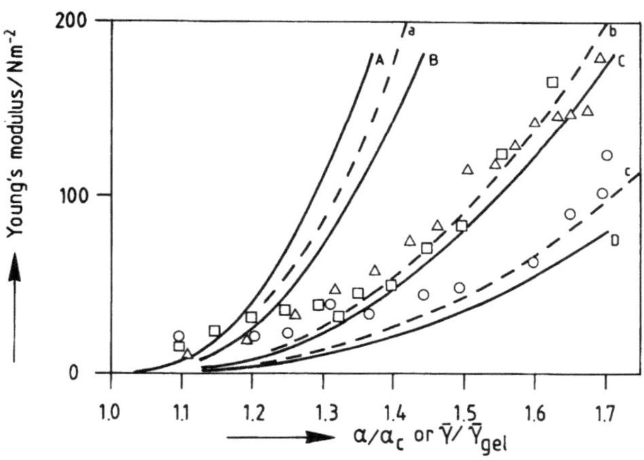

Fig. 184. Young's modulus as a function of relative conversion α/α_c (from optical rotation) for aqueous gelatin solutions (in triplo); ageing temperature 26.9 °C. *Solid lines*: theoretical predictions based on the cascade theory of network formation for various crosslink functionalities and maximum number of potential junction zones; curve C, which gives the best fit, $f = 6$ and maximum number of potential junction zones = 8 [17, 489]. *Broken lines* (a, b, c) results of calculations with the network model [39–44] for crosslink functionalities $f = 4$, 6 and 8, plotted vs $\bar{\gamma}/\bar{\gamma}_{gel}$

for high conversions, hexafunctionality of the crosslinks gives the best fit and, for low conversions, tetrafunctionality.

10.3.3 Gel Melting Temperature

According to Eldridge and Ferry [160] the relationship between the melting temperature T_m of gels with tetrafunctional crosslinks and the concentration is given by

$$\left[\frac{d \ln c}{dT_m}\right]_{\bar{M}_w} = \frac{\Delta H_m^o}{RT_m^2} \tag{34}$$

where c = polymer concentration in the gel,
T_m = melting temperature of the gel,
$\bar{M}_w$ = weight average molecular weight of polymer,
ΔH_m^o = heat of melting of the crosslinks,
R = gas constant.

In the following way, this relationship can be extended to crosslink functionalities f. We have the equilibrium

$$0.5f \text{ potential junctions} \rightleftarrows 1 \text{ f} - \text{functional crosslink} - \Delta H^o$$

with the equilibrium constant

$$K = \frac{n_{cl}}{n_l^{0.5f}} \tag{35}$$

with n_{cl} = the number of f-functional crosslinks,
n_l = the number of potential junctions still present.

In working out this relationship in the way shown by Eldridge and Ferry for tetrafunctional crosslinks, one obtains

$$\left[\frac{d \ln c}{dT_m}\right]_{\bar{M}_w} = \frac{\Delta H_m^o}{(1 - 0.5f)\ RT_m^2} \tag{36}$$

or

$$\left[\frac{d \log c}{d(1/T_m)}\right]_{\bar{M}_w} = - \frac{\Delta H_m^o}{2.303 R(1 - 0.5f)}\,. \tag{37}$$

Assumptions made in the derivation of these relationships are as follows. a) The fraction of the total number of junctions in the gel that contribute to the elasticity, i.e. the fraction of the number of elastically active crosslinks, is a function of concentration and not of temperature. This assumption is questionable, since the degree of intramolecular crosslinking will depend on the concentration, becoming smaller with increasing concentration. However, the variation in T_m with concentration is relatively small, so that the question of the strict validity of this assumption is rather irrelevant.

b) The number of still unused junctions per unit of gel volume is proportional to the concentration. This is only true if the number of junctions already occupied in crosslinks is small with respect to the original number of potential crosslinks. In other words, this holds if the ageing process has not proceeded very far; at the melting temperature and the maximum gelation temperature this demand will generally be satisfied.
c) The number of potential junctions still available is not a function of the temperature. This assumption is rather questionable.

10.3.4 Determination of Maximum Gelation Temperature

Many methods for the determination of the gel melting point are based on a slow temperature increase, and determine the temperature at which the gel does not exist any more. This is not a good method because the melting process of a gelatin gel is too slow and depends on the amount of previous ageing. It even happens that a gel will melt at a certain temperature, and during prolonged maturation at that temperature the system will gel again (see e.g. [462]). Another, possibly better, method to determine the gel melting temperature is to determine the induction period of gelation as a function of temperature and extrapolate this period to infinite induction time, or, better, extrapolate t_{ind}^{-n} to zero ($n > 0$). This temperature is called T_{gel} [11, 24]. However, T_{gel} cannot be identified with T_m as, in general, the temperature at which gelation just takes place is lower than the melting temperature of an already formed gel (see e.g. [458, 490]). For a further discussion see Sect. 10.4.

10.3.5 Melting Enthalpy ΔH_m^o

Measurement of T_{gel} as a function of concentration and subsequently plotting ln c vs $1/T_{gel}$ should yield a straight line with a slope equal to $-\Delta H_m^o/[R(1-0.5f)]$. Results of these measurements are shown by Te Nijenhuis [11, 24], from which it appears that $\Delta H_m^o/(1-0.5f) = -221$ kJ/(mol of potential junction regions). Eldridge and Ferry [160] report values varying from -206 to -924 kJ/mol, depending on the molecular weight and on T_m: $-\Delta H_m^o$ also decreases with decreasing molecular weight.

The absolute value of the melting enthalpy is also smaller when the gels are formed at lower temperatures (on going from 0 to 15 °C, $-\Delta H_m^o$ increases by a factor of 2 to 3). Apparently larger numbers of big structures are formed at high than at low ageing temperatures, although the heat of activation for growth is small; apparently the formation of nuclei at 15 °C is faster than at 0 °C. Nevertheless, the equilibrium shear modulus is higher at 0 °C than at 15 °C. This is a consequence of the larger number of small, less stable structures (i.e. crosslinks). A schematic representation of the presence of different structures is shown in Fig. 185. The consequence is that more time is needed before these gels

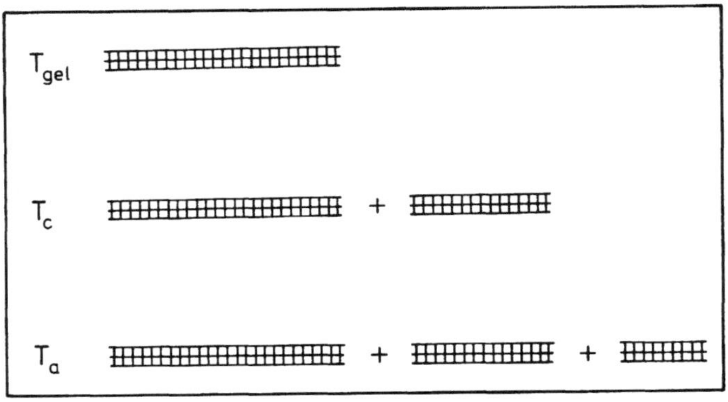

Fig. 185. Schematic view of the stability of collagen-like structures as a function of temperature: $T_a < T_c < T_{gel}$; besides large stable structures smaller ones will become stable with decreasing temperature [11]

melt; if the temperature increase would be infinitely slow and if no growth would occur, then both gels would melt at the same temperature.

From experiments by Eldridge and Ferry [160], it appears that gels formed at 15 °C melt at higher temperatures than those formed at 0 °C. The height of the melting point is determined by the number and size of the crosslinks. When the melting temperature is lower, the structures that melt are smaller, so that the values of $-\Delta H_m^o$ are lower; this means that the slope of curves of ln c vs $1/T_m$ becomes less negative with decreasing temperature. Results from Eldridge and Ferry are in agreement with these conclusions (see Fig. 186). Hence, one should assume an average value of ΔH_m^o.

Another important parameter determining the melting point is the molecular weight. Low molecular weight polymers need more crosslinks per unit volume to form a gel than high molecular weight polymers. With the same number of relatively large structures a low molecular weight gel will melt easier than a high molecular weight gel. This conclusion is in agreement with measurements reported by e.g. Eldridge and Ferry [160].

In his review, Ferry [466] tabulated literature values of ΔH_m^o, which vary from -260 to -315 kJ/mol. Tarr [491] reports values varying from -252 to -317 kJ/mol, obtained with the method used by Eldridge and Ferry for aqueous gelatin in the presence of salts and surfactants. From measurements by Flory and Garrett [492] one can derive a value of -149 kJ/mol. Note that these values are all equal to $\Delta H_m^o/(0.5f - 1)$. Harrington and Rao [460] report values varying from -170 to -340 kJ/mol for ΔH_m^o corrected for $f = 6$, whereas Te Nijenhuis [11, 24] found a value of -442 kJ/mol. For the values of ΔH_m^o, mentioned before, corrected for hexafunctionality of the triple helix crosslinks, results are shown in Table 9.

From their measurement of the temperature dependence of optical rotation of very dilute α-gelatin solutions, Eagland et al. [477] conclude that ΔH^o of

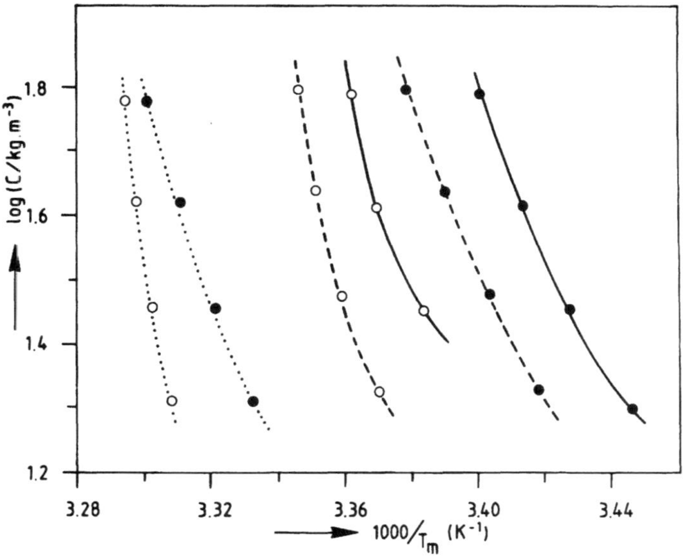

Fig. 186. Results reported by Eldridge and Ferry: log c plotted vs $1/T_m$ for gelatins of various molecular weights: (——), (-----) and (·····) $\bar{M}_w = 48$, 52.7 and 72 kg/mol, respectively; ageing temperature 0 °C (●) and 15 °C (○) (constructed after data presented in [160])

Table 9. Values of ΔH_m^o of gelatin gels, corrected for hexafunctionality of the crosslinks

Authors	$\bar{M}_n$ (kg/mol)	$-\Delta H_m^o$ (kJ/mol)	Method
Ferry [466]	29–45	520–630	T_m (c)
Eldridge & Ferry [160]		412–1848	T_m (M and c)
Tarr [491]		512–634	T_m (c)
Flory & Garrett [492]		298	
Harrington & Rao [460]		340–680	
Te Nijenhuis [11, 24]	35	442	T_{gel} (c)

a folding process has a value of −18.8 kJ/mol. Oth et al. [493] decided on a value of − 17.6 kJ/mol and Flory and Garrett [492] on − 18.8 kJ/mol. In fact, measurements of the optical rotation yield a value of ΔH^o corresponding with the formation of a repeating unit of a turning for the formation of a mono helix. The average value is − 18.3 kJ/mol. Hence, for one turning of the three monohelices in the triple helix, the average value of ΔH_m^o is 55 kJ/mol. This is in agreement with results obtained by Privalov [430], who concluded that the hydroxyproline content dependence of the average residual enthalpy is given by (see Fig. 170)

$$\Delta H_{res/1000} = 2.400 + 0.0358 n_{4-HYP/1000}\,(\text{kJ/mol}). \tag{38}$$

Table 10. Number of helix turnings in a triple helical crosslink

Authors	Temp. °C	Number	Method
Eldridge & Ferry [160]	0	8–11	M dependence
Eldridge & Ferry [160]	15	18–34	M dependence
Tarr [491]	?	9–12	Solvent dependence
Harrington & Rao [460]		6–13	
Te Nijenhuis [11, 24]	26–30	8	T_{gel} (t_{ind})

For the collagen of mammals $n_{4-HYP/1000}$ is approximately 100, and thus $\Delta H_{res/1000} \approx 6$ kJ/mol. Consequently, the three turnings in a triple helix are attended with a transition enthalpy of approximately 54 kJ/mol. Hence, the average number of turnings of triple helices that the crosslinks consist of can be calculated from the experimental values of ΔH_m^o. Results, shown in Table 10, are in fair agreement with the assumption by Harrington and Rao [460] that at low temperature 7 to 11 turnings are needed for the formation of a stable helix, whereas at higher temperatures this number will be considerably higher.

10.3.6 Crosslinking Enthalpy ΔH^o

All values of ΔH^o, mentioned in the foregoing section, are calculated from measurements of the gel melting point. However, it is also interesting to know the values at lower temperatures, i.e. during the gelation process. As a starting point for these calculations, use is normally made of the expression of the equilibrium constant

$$K = \frac{n_{c1}}{n_1^2} \tag{39}$$

for tetrafunctional crosslinks. For f-functional crosslinks this constant is given by

$$K = \frac{n_{c1}}{n_1^{0.5f}} = K^o e^{-\Delta H^o/RT}. \tag{40}$$

Hence,

$$\frac{d \ln K}{dT} = \frac{\Delta H^o}{RT^2} = \frac{d \ln n_{c1}}{dT} - 0.5f \frac{d \ln n_1}{dT}. \tag{41}$$

According to Pouradier [500] and Eldridge and Ferry [160] the concentration of crosslinks might be written as

$$n_{c1} = \frac{G_e}{0.5 fRT} \tag{42}$$

where it is assumed that the network is ideal. This results in

$$\frac{d\ln K}{dT} = \frac{\Delta H^\circ}{RT^2} = \frac{d\ln G_e}{dT} - \frac{1}{T} - 0.5f\frac{d\ln n_1}{dT}. \tag{43}$$

If the dependence of n_1 on temperature is neglected, then one obtains

$$\frac{d\ln K}{dT} = \frac{\Delta H^\circ}{RT^2} = \frac{d\ln G_e}{dT} - \frac{1}{T}. \tag{44}$$

However, in this way only the elastically effective crosslinks are counted, whereas crosslinks in the sol fraction and other elastically ineffective crosslinks also contribute to the crosslinking enthalpy. Hence, a better method might be to make a substitution [127]:

$$n_{c1} = \frac{\bar{\gamma}_n}{0.5f}\frac{c}{\bar{M}_n} = \frac{\bar{\gamma}_w}{0.5f}\frac{c}{\bar{M}_w} \tag{45}$$

where $\bar{\gamma}_n$ and $\bar{\gamma}_w$ are the number- and weight-average crosslinking indices, respectively, in the gel as a whole (i.e. sol fraction + network fraction). In that case we arrive at

$$\frac{d\ln K}{dT} = \frac{\Delta H^\circ}{RT^2} = \frac{d\ln \bar{\gamma}_n}{dT} - 0.5f\frac{d\ln n_1}{dT} = \frac{d\ln \bar{\gamma}_w}{dT} - 0.5f\frac{d\ln n_1}{dT}. \tag{46}$$

From reference data collected by Pouradier [500], gelatin AP ($\bar{M}_n = 47$ kg/mol, 4 wt% solution) is chosen as an example, together with results reported by Te Nijenhuis [11, 23, 24] ($\bar{M}_n = 35$ kg/mol, 2 wt% solution). Shear and storage moduli are shown as a function of temperature in Fig. 187. If it is assumed that both polymers have a Flory distribution ($\bar{M}_w/\bar{M}_n = 2$), then results are obtained as shown in Fig. 188. In fact, these values of $-\Delta H^\circ$ are too high, because n_1 is certainly dependent on temperature: $d(\ln n_1)/dT < 0$. Extrapolation of Pouradier's data to the melting temperature (i.e. to $\bar{\gamma}_w = 0.5$) yields $\Delta H^\circ_m = -785$ kJ/mol at $T_m = 28.3\,^\circ$C; Te Nijenhuis' data similarly give $\Delta H^\circ_m = -288$ kJ/mol at $T_m = 29.0\,^\circ$C. It was found before that $T_{gel} \approx 27\,^\circ$C. This discrepancy between T_m and T_{gel} is in agreement with the above-mentioned difference between the melting and crystallization temperatures. It should be emphasized that the extrapolation procedure is rather inaccurate, and hence one should not set much value on these ΔH°_m data. At low temperatures, the increase in G' with decreasing temperature is relatively small. This results in rather low values of ΔH° (approximately -10 kJ/mol in the temperature range from 2 to 15 °C), even smaller than the value of 55 kJ/mol for the transition enthalpy for one turning in the triple helix. In Te Nijenhuis' case, at temperatures below 10 °C $-\Delta H^\circ$ rises again. This is not understandable physically, but it results from an s-shaped curve obtained on plotting the storage modulus vs temperature [127].

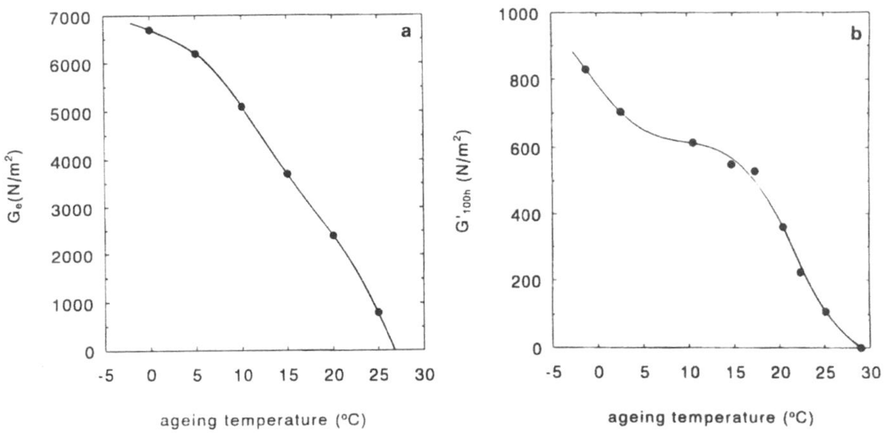

Fig. 187. **a** Shear modulus of aqueous gelatin ($\bar{M}_n = 47$ kg/mol, c = 4 wt%; Pouradier [500]. **b** Storage modulus of aqueous gelatin ($\bar{M}_n = 35$ kg/mol, c = 2 wt%; Te Nijenhuis [11,23,24]. Reproduced from Polym Gels Networks [Ref. 127] by the courtesy of Elsevier Science Ltd, The Boulevard, Langford Lane, Kidlington 0X5 1GB, UK

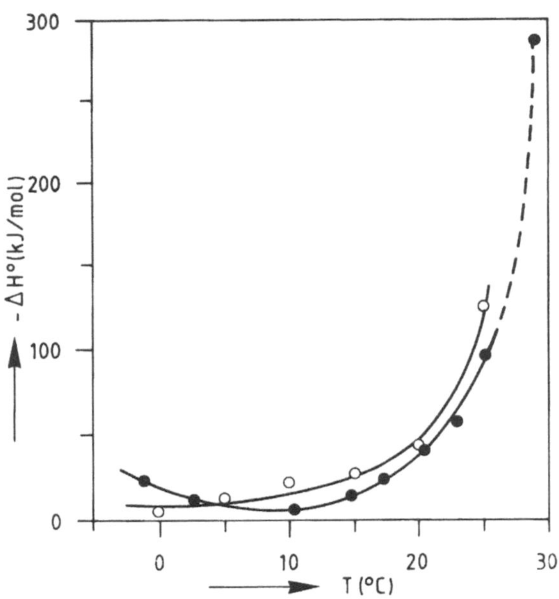

Fig. 188. Crosslinking enthalpy of gelatin as a function of temperature; calculated with the aid of Eq. (46) from measurements reported by (○) Pouradier [500] and by (●) Te Nijenhuis [11,23,24]. Reproduced from Polym Gels Networks [Ref. 127] by the courtesy of Elsevier Science Ltd, The Boulevard, Langford Lane, Kidlington 0X5 1GB, UK

The number of crosslinks of certain sizes is determined by the matching equilibrium constant

$$K_i = \frac{n_{c1,i}}{n_{1,i}^{0.5f}} = K_{o,i} e^{-\Delta H_i^\circ/RT} \tag{47}$$

so that

$$n_{c1} = \sum_i n_{c1,i} = \sum_i K_{o,i} e^{-\Delta H_i^\circ/RT} n_{1,i}^{0.5f}. \tag{48}$$

In a first approximation, ΔH_i° will be proportional to the size of the triple helices.

The lower boundary of the summation (i) depends on temperature: with decreasing temperature, more kinds (read: sizes) of structures will cooperate in the network formation. Hence, two parameters play a role as follows.

a) ΔH_i° of every kind (size) of triple helix. It might be assumed that at lower temperatures the tendency for crosslink formation increases for every size; hence, the rubber equilibrium shear modulus increases with decreasing temperature, not as a consequence of an increasing rate of crystallization of that size (the activation energy is only small), but by the increased number of stable structures.

b) With decreasing temperature, smaller structures become stable, so they can contribute to network formation.

10.4 Critical Gelation

10.4.1 Temperature Dependence

The induction time or critical gelation time was defined by Te Nijenhuis [11, 23, 24] as the time needed for the development of a measurable increase in the storage modulus. It appears that this time increases strongly with ageing temperature. At the maximum gelation temperature, T_{gel}, the critical gelation time becomes infinite so that its reciprocal value tends to zero. It was found empirically by Te Nijenhuis [11, 24] that the $t_c^{-1/2}$ vs temperature plot yields a straight line for a 1.95 wt% aqueous gelatin solution (see Fig. 189); extrapolation to $t_c^{-1/2} = 0$ gives the maximum gelation temperature as 26.3 °C. Ross-Murphy [467] made use of these data to fit them into his model, based on the random percolation simulation [494]:

$$t_c = K_T \times \left(1 - \frac{T}{T_c}\right)^q. \tag{49}$$

His calculations resulted in $q = -2.22$ and $T_c = 27.24$ °C. At this moment it is not clear what is the significance of the exponent -2.22 (or -2 in Te Nijenhuis' empirical case). However, it has to be emphasized that the value of q is very sensitive to the value of choice of T_c. Results of linear regression are given in Table 11. It is clear that it is difficult to decide what is the exact value of the exponent q.

Amis et al. [471, 472] determined the gelation time for 1.75 wt% solutions as a function of temperature with the aid of the Winter-Chambon method and tabulated their results for temperatures varying from 22 to 27 °C. Application of

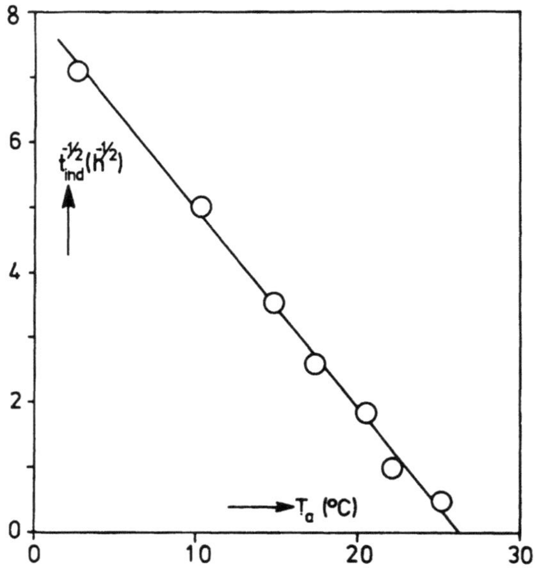

Fig. 189. Reciprocal square root of critical gelation time vs temperature [11,24]. Reproduced from Colloid Polym Sci [Ref. 24] by the courtesy of Steinkopff Verlag Darmstadt, FRG

Table 11. Results of calculations of the parameters of Eq. (49) for measurements by Te Nijenhuis [11,24]

T_c	Correlation coefficient	q	K
27.67	− 0.9962	− 2.41	4.33×10^{-5}
27.24	− 0.9959	− 2.22	6.64×10^{-5}
26.65	− 0.9936	− 2.00	1.32×10^{-4}

Ross-Murphy's model yields $q = -2.20$ and $T_c = 28.51$ °C, close to the results from Te Nijenhuis' measurements. A comparison of both results is shown in Fig. 190.

10.4.2 Concentration Dependence

According to the gelation kinetics as proposed by Oakenfull et al. [495–497], the critical gelation time is given by

$$t_c \approx \frac{1}{kc^n} \tag{50}$$

where k is the rate constant, c is the polymer concentration and n is the number of crosslink sites which form a junction zone or crosslink. The Oakenfull plot is defined as a plot of $\log(1/t_c)$ vs $\log c$; its slope gives the reaction order, which is

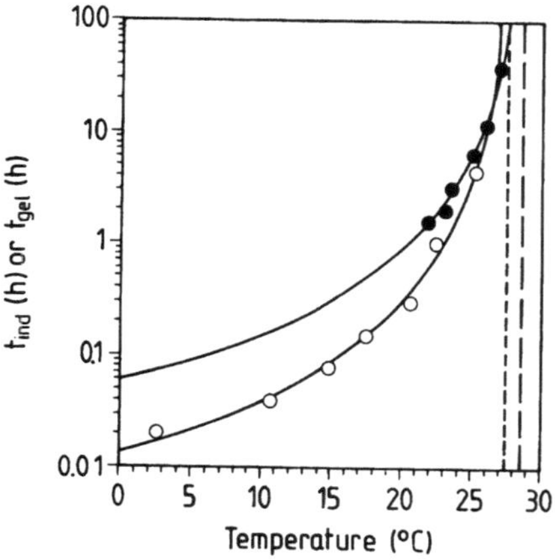

Fig. 190. Critical gelation time of measurements by Te Nijenhuis [11, 23, 24] and by Amis et al. [471, 472], compared with calculations from Ross-Murphy's model [467] with the aid of Eq. (49). (○) Te Nijenhuis' results: $q = -2.41$, $T_c = 27.67\,°C$; (●) Amis et al. results: $q = -2.20$, $T_c = 28.51\,°C$

assumed to be the number of polymer chains participating in the formation of a junction zone [498].

Ross-Murphy has criticised the model on a number of grounds [498, 499] and most particularly on a) the assumption of high order kinetics with $n \gg 2$ ($n = 4.5$ for iota carrageenan and 12.5 for kappa carrageenan) and b) the absence of a critical gel concentration c_o. He derived an alternative equation

$$t_c = \frac{k}{[(c/c_o)^{n'} - 1]^p} \tag{51}$$

where n' is the true kinetic order of the reaction, k is a rate constant and p is an exponent ($0 < p < 3$). For high values of c/c_o the equation reduces to the Oakenfull equation with $n = pn'$. Results are given by Ross-Murphy for measurements which obey both equations, and are thus for concentrations $c \gg c_o$. From measurements by the present author [11, 23] on aqueous gelatin solutions, results are shown in Fig. 191 (Oakenfull plots for 17.4 and 25.2 °C) and in Fig. 192 (Ross-Murphy plot for 17.4 °C). It appears that the Oakenfull plot holds for concentrations of 2–5 wt%, whereas the 1 wt% sample deviates from this plot. The slopes are $n = 1.67$ (17.4 °C) and $n = 3.11$ (25.2 °C). This should mean that the reaction order at 25.2 °C is 3, so that nuclei of mono helices are formed quickly, followed by the rate determining step of the formation of triple helices. At 17.4 °C, on the other hand, both steps (nuclei formation and triple helices formation) are rate determining. One could expect that the formation of nuclei is rate determining ($n = 1$) at lower temperatures; however, for those temperatures the gelation process is so fast that it is impossible to determine the critical gelation time accurately enough. Ross-Murphy plots of the 17.4 °C measurements are shown in Fig. 192 for $n' = 3$ as well as for $n' = 2$. For both values of n', the differences between experimental and calculated results are imperceptible. The values of p are 0.536 and 0.756 respectively, so

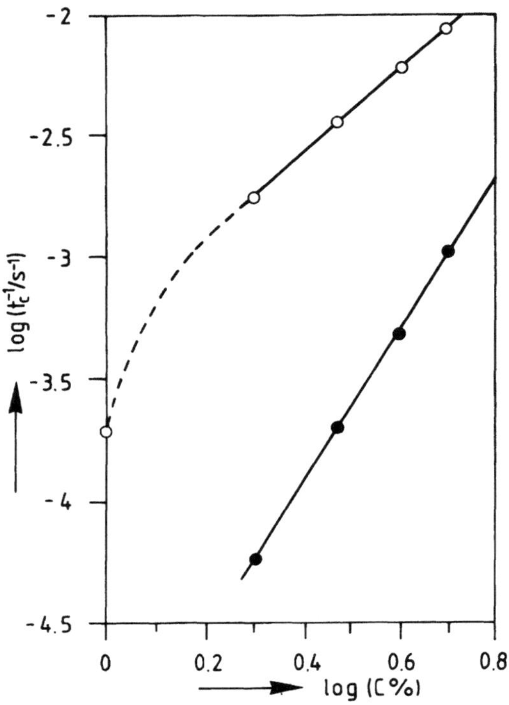

Fig. 191. Oakenfull plots for measurements reported by Te Nijenhuis [11, 23, 24]; concentrations 2–5 wt%; ageing temperatures: (○) 17.4 °C; (●) 25.2 °C

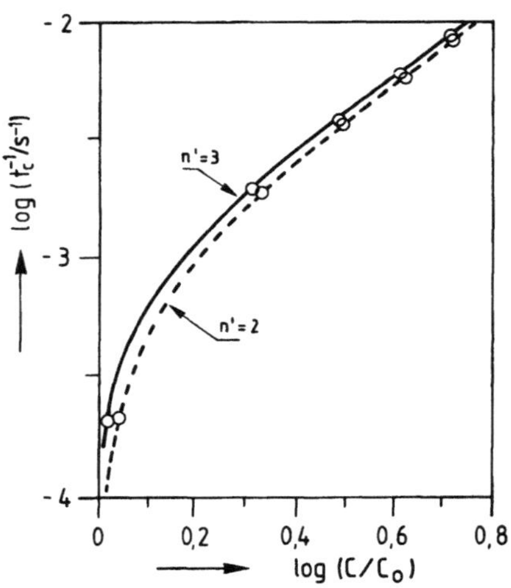

Fig. 192. Ross-Murphy plots for measurements reported by Te Nijenhuis [11, 23, 24], for cross-link functionalities f = 4 and 6; for the parameters used, see text

that pn′ = 1.51 and 1.61 respectively, rather close to the slope of the Oakenfull plot (n = 1.67). The critical concentrations for the best fit are $c_o = 0.96$ wt% (n′ = 3) and $c_o = 0.91$ wt% (n′ = 2), respectively.

10.5 Viscoelasticity and Optical Rotation

Investigation of the change of optical rotation has been the subject of many studies of structural changes during gelation of gelatin solutions [17, 458–461, 466, 469, 478, 485, 501–513]. In many of these studies the change in optical rotation was compared with changes in viscosity, NMR spectroscopy etc.

In their 1987 review, Clark and Ross-Murphy [1] described the measurement of optical rotation of gelatin solutions as a valuable tool to study the structural changes during gelation, consequent on the large differences between the specific rotation of the helix conformation and the random coil. In the early stages of gelatin aggregation, substantial changes in optical rotation are expected, whereas the viscosity changes are only small. On the other hand, later processes (i.e. after the gel point) have less effect on optical rotation, but they cause considerable changes in viscoelasticity.

In his 1948 review, Ferry [466] described the rigidity of gelatin gels as a function of concentration and temperature and paid some attention to optical rotation, but the two phenomena were not interrelated. In the work of Ferry and Eldridge [485], both phenomena were only mentioned without any interrelation. Only in 1961 did Todd [459] show a linear relationship between the specific optical rotation of a 5.5 wt% gelatin solution and the square root of the equilibrium shear modulus. The work of Peniche-Covas et al. [17] is an interesting study (see Sect. 10.3.2). These authors combined optical rotation and viscoelasticity measurements with the cascade theory of network formation [489] to calculate the potential number of crosslinking sites in gelatin networks.

Djabourov et al. [458, 469, 508–510] determined the relationship between viscoelastic properties and optical rotation. From optical rotation the weight fraction, χ, of helices present in an aqueous gelatin system could be calculated, because of the big difference in optical rotations in the coiled and the helical (collagenlike) states:

$$\chi = \frac{[\alpha]_\lambda^{\text{exp}} - [\alpha]_\lambda^{\text{coil}}}{[\alpha]_\lambda^{\text{collagen}} - [\alpha]_\lambda^{\text{coil}}} \,. \tag{52}$$

For a wavelength $\lambda = 436$ nm it was found that $[\alpha]^{\text{collagen}} = -800 \pm 10\,^\circ$ and $[\alpha]^{\text{coil}} = -256 \pm 5^\circ$. From their optical rotation studies, Djabourov et al. [458, 510, 511] calculated χ as a function of ageing time. Results are shown in Fig. 193 for 4.7 wt% aqueous gelatin ($\bar{M}_w = 206$ kg/mol) solutions at various temperatures. The fraction of helices increases slowly with ageing time and the curves χ vs $\log t_a$ closely resemble the curves G' vs $\log t_a$. However, the helix content starts to increase immediately after the temperature quench and an induction period is not present. By comparing the storage modulus and the helix content as functions of ageing time, a plot of the modulus vs helix content could be constructed. In Fig. 194, G^* (which is almost equal to G') is plotted against χ for the 4.7 wt% solution of gelatin [458, 510]. It is clear that a certain level of helices is needed for the formation of a three-dimensional network. The critical

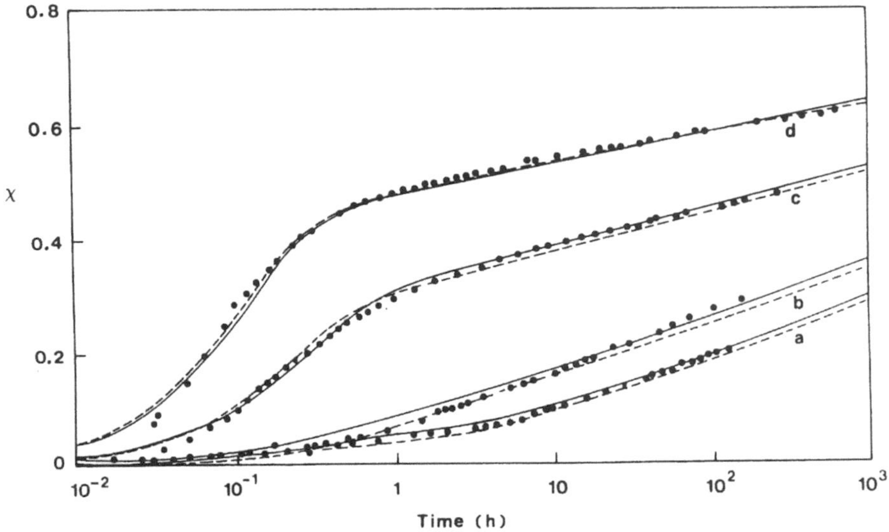

Fig. 193. Increase of the weight fraction of helices during the ageing process of aqueous 4.7 wt% gelatin ($\bar{M}_w = 206$ kg/mol) solutions with ageing time measured at various temperatures: (a) 10 °C; (b) 20 °C; (c) 26 °C and (d) 28 °C; the *continuous* and *dotted lines* correspond to the phenomenological analysis proposed by Djabourov et al. Reproduced from J Phys France [Ref. 458] by the courtesy of the authors and of Les Éditions de Physique, Les Ulis, France

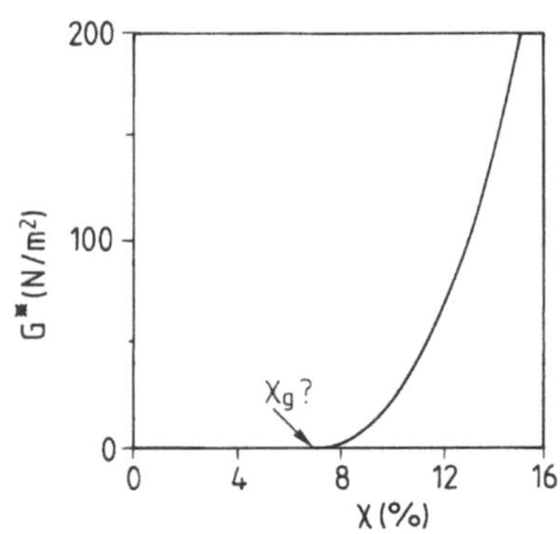

Fig. 194. Complex shear modulus G* of an aqueous gelatin gel (47 kg/m³) vs the amount of helix formation in wt% of gelatin present. Measurements by Djabourov et al. [469, 510]. Reproduced from J Phys France [Ref. 469] by the courtesy of the authors and of Les Éditions de Physique, Les Ulis, France

value at the gel point χ_g, however, is difficult to determine from this plot. From their definition of the gel point, i.e. the moment where G′ equals G″ at low frequencies (0.015 Hz), the cross-over point, they decided on an approximate critical helix content $\chi_g = 7$ wt%. From the above-mentioned results (G* vs χ), Te Nijenhuis [40] calculated the weight average crosslinking index $\bar{\gamma}_w$ as a function of the helix content (assuming a Flory distribution of the molecular weights $\bar{M}_w/\bar{M}_n = 2$ and a hexafunctionality of the triple helical crosslinks). The result is a linear relationship between $\bar{\gamma}_w$ and χ (see Fig. 195). The critical value

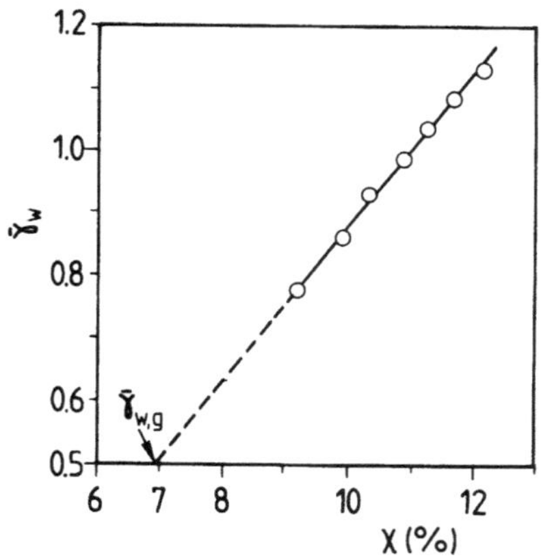

Fig. 195. Weight average crosslinking index of the gelatin gel of Fig. 194 vs the amount of helix formation in wt% of gelatin present. Reproduced from Makromol Chem, Macromol Symp [Ref. 40] by the courtesy of Hüthig & Wepf Verlag Publishers, Zug, Switzerland

of $\bar{\gamma}_w$ at the gel point is equal to 0.5 ($\bar{\gamma}_{w,g} = 2/(f-2) = 0.5$); linear extrapolation to the gel point yields the critical value of the helix content: $\chi_g = 6.9$ wt%. This value is in close agreement with the value obtained by Djabourov and Leblond [510]. According to the present author this value depends on temperature: at higher temperatures, on average, only larger structures (i.e. triple helices) are stable. This means that at the same number of crosslinks the helix content is larger at higher temperatures. However, there is no proof for this postulate at this moment. On the other hand, Clark and Ross-Murphy [1] conclude from measurements by Durand et al. [512, 513] that χ is proportional to c^{-1} at the gel point, and hence $\chi_g c$ should have a constant value independent of concentration and temperature. It is a feeling of the present author that careful measurements of optical rotation and viscoelastic properties will lead to the result that the product $\chi_g c$ is dependent on temperature and increases with increasing temperature.

Busnell et al. [478] investigated the prematuring effect on the gelation behaviour of gelatin solutions with the aid of viscoelastic experiments combined with differential scanning calorimetry and optical rotation experiments (see also Sect. 10.2.2). In Fig. 196, results are shown for the temperature dependence of the storage modulus G′ of 2 wt% gels that were quenched to 5 °C for 16 h (either directly or after holding for 24 h at 20 °C), prior to the gradual temperature increase. The prematured gels have a higher rigidity over the whole temperature range. It is interesting to note that the differentiated curves have maxima close to the temperatures where the DSC experiments show endothermic minima. In order to study the temperature dependence of the helix content in the gels, optical rotation was measured during the temperature increase. The optical rotation, differentiated with respect to the temperature, is shown as a function of temperature in Fig. 197. The difference between the prematured and non-prematured gels is very clear. In the same figure the result for a gel that was only matured at 20 °C for 24 h is also shown. It is clear that the bimodal behaviour of the prematured gel is the result of the existence of large structures that are stable

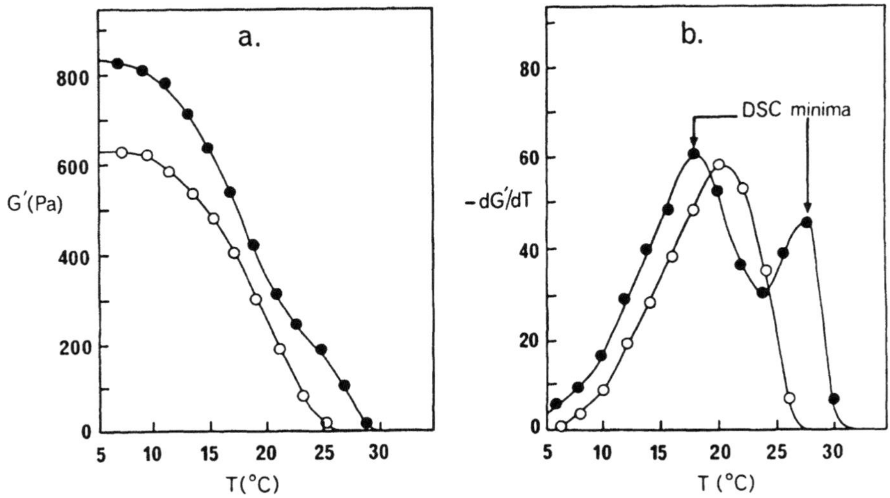

Fig. 196a, b. Temperature dependence of the storage modulus, measured at 0.5 Hz, for 2 wt% aqueous gelatin gels that were quenched to 5 °C for 16 h, either directly (○) or after holding for 24 h at 20 °C (●): **a** storage moduli; **b** change of G′ with temperature, dG′/dT, with increasing temperature. Reproduced from Gums and Stabilisers for the Food Industry, Vol 4 [Ref. 478] by the courtesy of the authors and of Oxford University Press

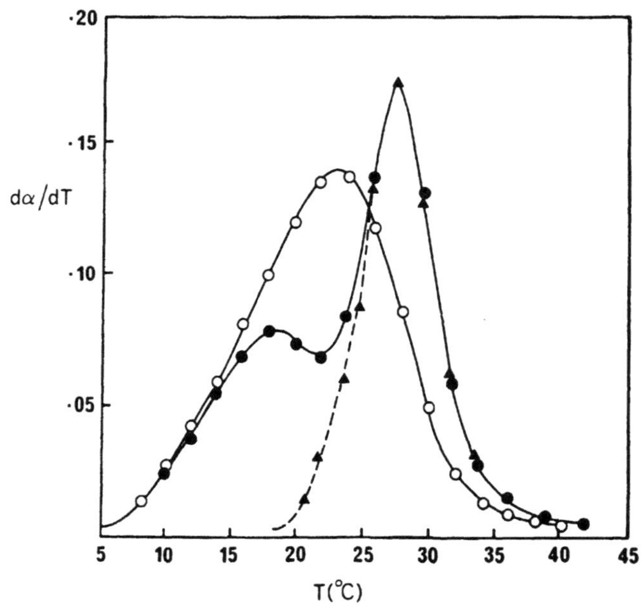

Fig. 197. Rate of change of optical rotation with temperature of the gels mentioned in Fig. 196; (○) quenched directly to 5 °C; (●) quenched to 5 °C after holding for 24 h at 20 °C; also shown are the results of a sample that was quenched to 20 °C only (▲). Reproduced from Gums and Stabilisers for the Food Industry, Vol 4 [Ref. 478] by the courtesy of the authors and of Oxford University Press

at high temperature and of small structures that are stable only at low temperatures. The behaviour of the non-prematured gels shows the existence of a normal distribution of structures with a normal size distribution. These conclusions are in agreement with those mentioned in Sect. 10.2.2.

10.6 Conclusions

Gelatin gels belong, together with poly(vinyl chloride) and poly(vinyl alcohol) gels, to the class of thermoreversible gels that have been investigated to a considerable extent. Nevertheless, research is still going on, due to the industrial importance of gelatin gels in cosmetics, pharmaceutics and photography etc, and the difficulties still present in the elucidation of the structure of the gels. The crosslinks formed are collagen-like triple helices. Many different techniques have been used for the investigation of the structure of the junction zones and of the network structure. Again, it appears, first, that viscoelasticity yields much information, due to its extreme sensitivity to changes in the structure and, second, that the combination of viscoelastic behaviour and results obtained from other analytical techniques deepens the insight into gelation properties. One of the advantages of gelatin gels is the apparently known functionality of the crosslinks, being 6. The consequence is that thermodynamics and kinetics of the crosslinking process can be rather straightforward, although one always has to bear in mind that the functionality for various reasons may differ from 6.

11 Agarose

11.1 Introduction

Agar agar, the native polysaccharide of agarose, forms thermoreversible gels when dissolved in water. It consists of two main components, agarose and agaropectin. According to Watase and Arakawa [514–516], agarose plays the major role in the mechanical behaviour of aqueous agar agar gels, as revealed from stress relaxation measurements. Agarose is a linear polysaccharide consisting primarily of β-1,3 linked D-galactose and α-1,4-linked 3,6-anhydro-α-L-galactose and contains a few ionised sulfate groups (see Fig. 198, e.g. [1]). As some of the 3,6-anhydro-α-L-galactose residues are replaced by other galactose residues, perfect ordering and subsequent precipitation of agarose chains is not likely; thus gelation can take place only when parts of the chains form ordered regions (junctions). Agarose is normally insoluble in organic solvents and cannot form gels. Apparently the gelling possibility in aqueous solutions is also a consequence of the structure of water. Adding hydrogen bond decomposing agents like urea can have a tremendously negative effect on the gelling properties.

Gelation occurs at temperatures below 40 °C, whereas the melting temperature appears to be 90 °C (see e.g. [519]). From optical rotation measurements of 0.1% aqueous solutions [1] (see Fig. 199) it appears that transitions are present at 40 and 60 °C. At these temperatures the mechanical properties of solutions/gels change. On account of these observations Tako and Namura [517] propose a gelation mechanism. At temperatures below 60 °C intramolecular hydrogen bonding takes place (see Fig. 200), which contributes to the rigidity of the molecular agarose chain. The authors believe that the agarose chains are coiled above 60 °C. In addition, below 40 °C intermolecular hydrogen bonding takes place, in which the anhydro O-3, 6 ring is involved (see Fig. 201). Both the intra- and inter-molecular hydrogen bonding are thought to be responsible for the gelation up to 60 °C. In this hydrogen bonded system water molecules are also bonded. This also contributes to the stiffness of the chains. At temperatures above 90 °C the agarose molecules are hydrated and dissociation occurs.

This model is in agreement with the models proposed by Hayashi et al. [518] and Arnott et al. [519], who, on the basis of a combination of X-ray fibre diffraction and optical rotation, decided on the presence of a double-stranded helix. Two identical, left-handed, three-fold helices might form a tertiary structure of agarose molecules in diluted, aqueous solutions and in the solid state as well. However, other techniques apparently show that the agarose network consists of junctions of substantial bundles of agarose chains. The network model obtained in this way is shown in Fig. 202. It is clear that the addition of hydrogen bonding decomposing agents like urea affects the gelation in a negative way.

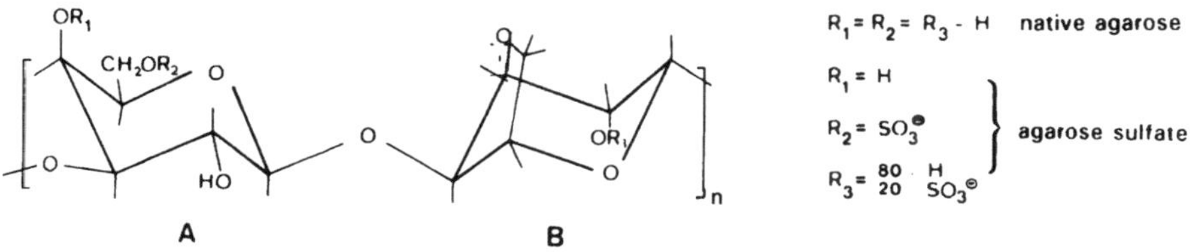

Fig. 198. Idealised AB repeat unit of agarose polymer based on 1,3-linked β-D-galactose residue (A) and α-1,4-linked 3,6-anhydro-α-L-galactose residue (B). Possible patterns of substitution involving sulfate groups are also indicated. Reproduced from Adv Polym Sci [Ref. 1] by the courtesy of the authors and of Steinkopff Publishers Darmstadt, FRG

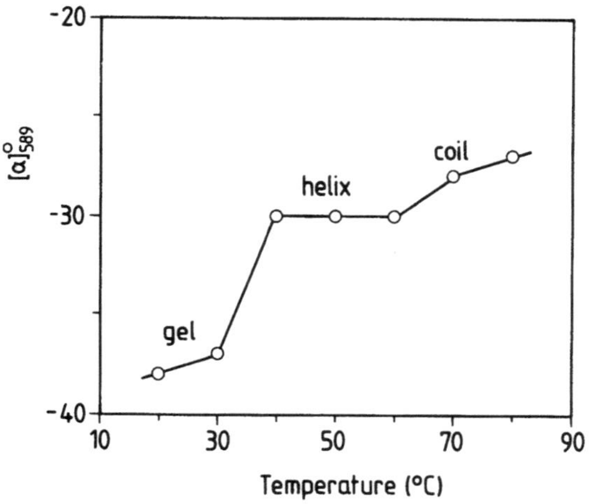

Fig. 199. Specific rotation of a 0.1% aqueous agarose solution as a function of temperature (constructed after data presented in [517])

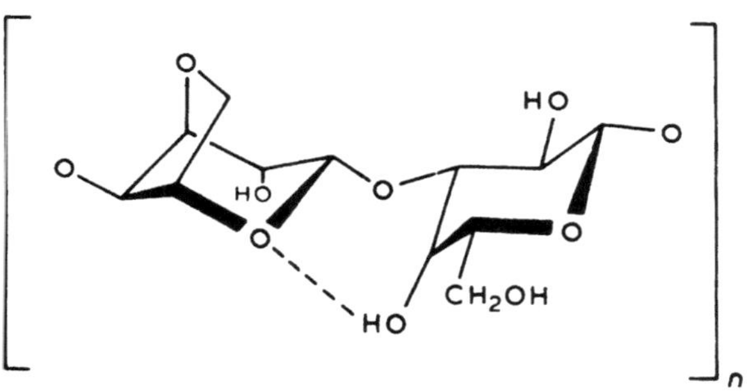

Fig. 200. Model, as proposed by Tako and Nakamura, of intramolecular hydrogen-bonding of agarose in aqueous solution; the *dotted line* refers to hydrogen-bonding. Reproduced from Carbohydr Res [Ref. 517] by the courtesy of the authors and Elsevier Science Ltd, The Boulevard, Langford Lane, Kidlington 0X5 1GB, UK

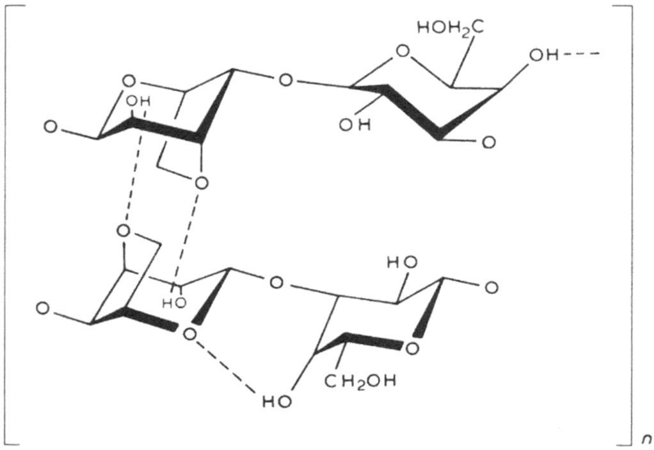

Fig. 201. Model, as proposed by Tako and Nakamura, of intermolecular hydrogen-bonding of agarose molecules in aqueous solution; the *dotted lines* refer to hydrogen-bonding. Reproduced from Carbohydr Res [Ref. 517] by the courtesy of the authors and Elsevier Science Ltd, The Boulevard, Langford Lane, Kidlington 0X5 1GB, UK

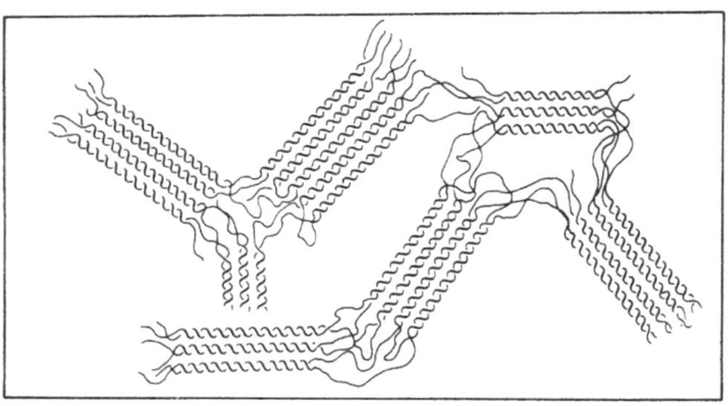

Fig. 202. Model for agarose network formation; crosslinks involve both double helix formation and substantial association of double helices to form microcrystalline junction zones. Reproduced from Adv Polym Sci [Ref. 1] by the courtesy of the authors and of Steinkopff Publishers Darmstadt, FRG

11.2 Viscoelastic Behaviour

11.2.1 Introduction

The gelling behaviour of aqueous agarose gels has been extensively studied by Nishinari et al. [166, 204, 520–529] with various measuring techniques (stress relaxation, dynamic mechanical behaviour, differential scanning calorimetry, optical rotation etc).

Agarose solutions already form gels at room temperature at very low concentrations < 0.1%). Rheological properties of agarose gels are determined by a) the degree of regularity of alternation of D-galactose and 3,6-anhydro-L-galactose residues, b) the content (dependent on the alkali treatment) and position of sulfate groups and c) the molecular weight.

11.2.2 Alkali Treatment

Agarose is generally obtained by alkali treatment of *Gracilariopsis Chorda* (i.e. agar agar). The gelling properties depend greatly on the concentration of, e.g., the sodium hydroxide solution used. The effect of alkali treatment is the formation of the 3,6 anhydro bond [521] (see Fig. 203). By deesterification, the double-helix structure is stabilised by newly created hydrogen bonds. In this way the ionic character of agarose also disappears. The result of these changes in chemical bonds is a decrease of the solubility in water and consequently an increasing tendency to form gels.

In Fig. 204, results composed from literature data [522] are shown for 3 and 5% aqueous solutions of agarose specimens that were obtained by alkali treatment with 2, 3 or 10% sodium hydroxide solution. Treatment with solutions of increasing sodium hydroxide concentration yields agaroses with increasing tendencies to form gels that, in addition, are more stable. Stress relaxation measurements at temperatures varying from 25 to 75 °C over nearly 4 decades in time (0.01 to 30 h) have been published. The authors tried to make use of a kind of time-temperature superposition principle to obtain stress relaxation curves over a larger time scale: they shifted their curves over the time axis (horizontally) in order to obtain one mastercurve. However, results obtained in this way are questionable, because the normal (WLF) time-temperature superposition is allowed only if the structure of the system is independent of temperature. Nevertheless, it appears (also from Fig. 204) that the effect of

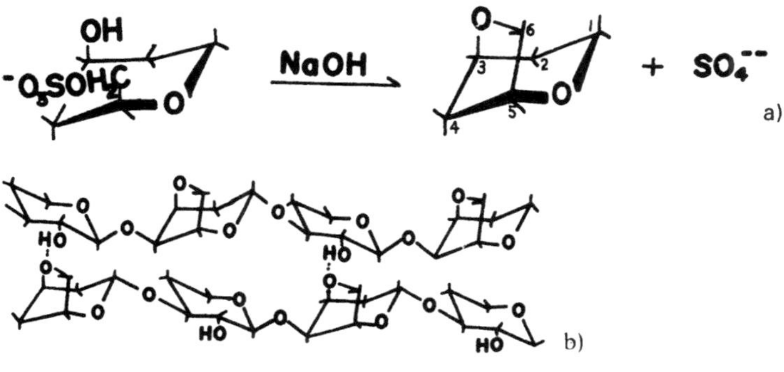

Fig. 203. a Mechanism of desulfation. **b** Formation of hydrogen bonds in helical structure of desulfated specimens. Reproduced from Rheol Acta [Ref. 521] by the courtesy of the authors and of Steinkopff Publishers Darmstadt, FRG

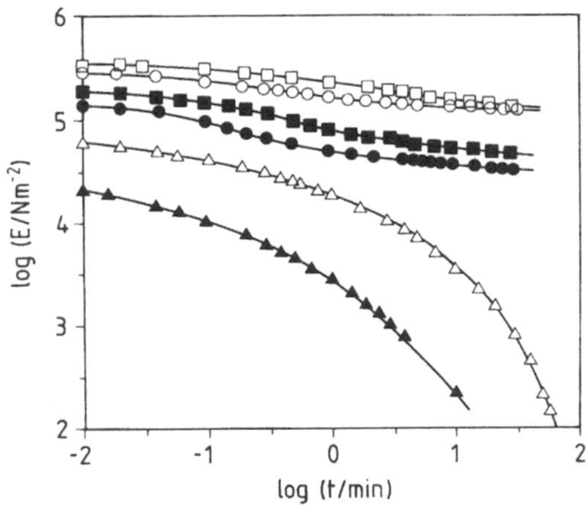

Fig. 204. Stress relaxation at 25 °C of 3% and 5% aqueous solutions of agarose specimens that were treated with 10%, 3% or 2% sodium hydroxide solutions; (□) 10% NaOH treated; (○) 3% NaOH treated; (△) 2% NaOH treated; *open symbols* and *filled symbols*: 5% and 3% agarose solutions, respectively (constructed after data presented in [522])

alkali treatment is dependent on sodium hydroxide concentration, as long as the concentration is lower than 3%; differences between 3 and 10% are marginal only.

11.2.3 Methoxy Groups Content

Agaroses obtained after purification of various agar agars may have different contents of methoxy groups, present as 6-O-methyl-galactose. The presence of these groups disturb the regularity of the agarose chains. Melting temperatures of 1.5% aqueous agarose gels have been reported for a considerable number of agarose samples [530]. From Fig. 205, it appears that the gels in general are more stable if the methoxy content is higher. On the other hand, excess methylation with methyl sulfate results in agarose samples with much lower melting temperatures, even for those samples having a lower methoxy content than shown in Fig. 205. Apparently, methylation with methyl sulfate attacks the hydroxyl groups randomly, so that the regularity of the agarose chains is disturbed to a much larger extent.

11.2.4 Alkali Metal Ions

The influence of alkali metal ions on the storage modulus E′ of agarose gels measured at 2.5 Hz is marginal as long as the temperature is lower than 55 °C. Above that temperature, the effect is strong; however, at those temperatures one cannot strictly speak of a real gel [530]. So the metal ions do not disturb the gelation process, but they have a clear influence on the rheological properties of the non-gelled solutions. Since agarose forms stable micelles, the addition of

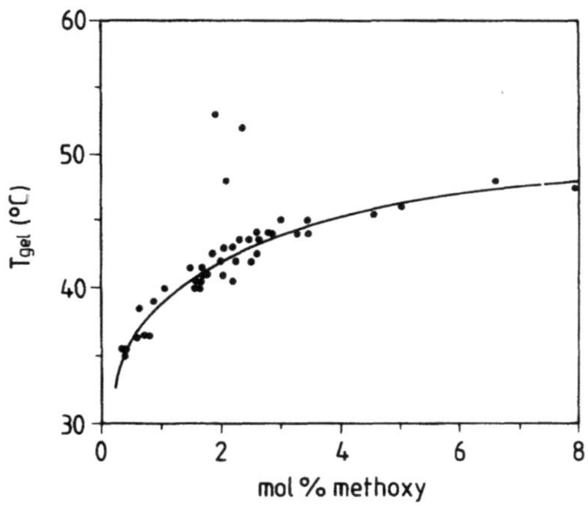

Fig. 205. Relationship between gelation temperature of 1.5% aqueous agarose solutions and the methoxy content of the agarose specimens. Reproduced from Carbohydr Res [Ref. 530] by the courtesy of Elsevier Science Ltd, The Boulevard, Langford Lane, Kidlington 0X5 1GB, UK

alkali metal ions does not change its molecular conformation and thus E′ of gels once formed is not greatly affected by the presence of alkali metal ions [530].

11.2.5 Concentration Dependence

Agarose forms aqueous gels at room temperature, even at very low concentrations. Clark et al. [487, 531] showed that the critical concentration is only about 0.2%. The concentration dependence is not a simple quadratic dependence: at low concentrations the exponent is much larger than 2 and at high concentrations smaller than 2; only at intermediate concentrations (around 10 wt%) might a region of quadratic dependence be present (see, e.g., Fig. 206), depending on the kind of agarose. Similar results were obtained by Nishinari for agar agar gels [520], whereas Tokita and Hikichi [532] observed a fourth power concentration dependence in the concentration range around 1 kg/m^3 (0.02–0.2 vol%, with shear moduli varying from 0.02 to 100 N/m^2). These authors also showed the presence of an equilibrium storage modulus for a 0.16 wt% aqueous gel at 20 °C, which was independent of frequency over approximately three decades.

11.2.6 Molecular Weight Dependence

Molecular weight dependence of the mechanical properties of gels is shown on samples varying in intrinsic viscosity (in 0.01 mol/l aqueous sodium salicylate) from 0.20 to 0.54 m^3/kg. Results composed of literature data [523] are shown in Figs. 207 and 208. Figure 207 shows the strong influence of molecular weight on the gelling behaviour at 25 °C: sample D ([η] = 0.54 m^3/kg) shows rubberlike behaviour (with only little stress relaxation), whereas the stress relaxation of sample A ([η] = 0.20 m^3/kg) is quite marked. The molecular weight dependence

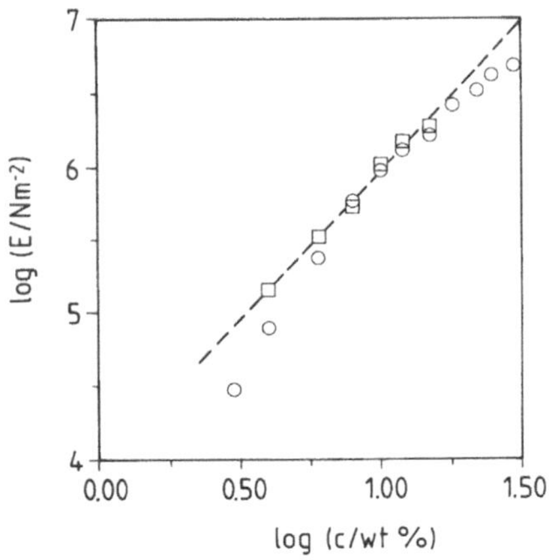

Fig. 206. Concentration dependence of the storage Young modulus of aqueous low molecular weight agarose gels ([η] = 0.17 m^3/kg); (□) [526]; (○) [527]; (------) slope = 2

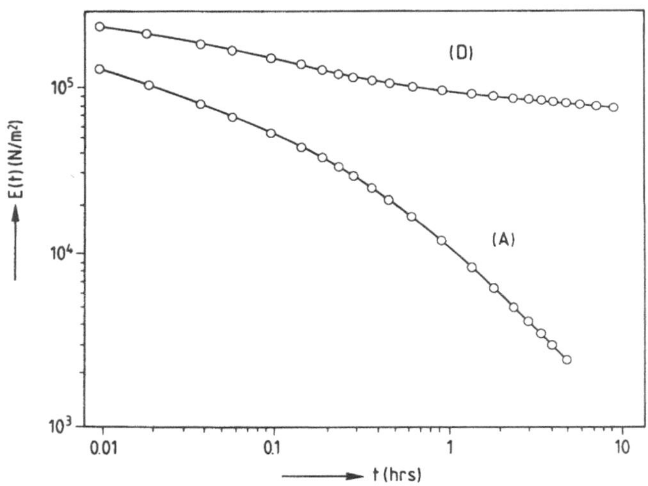

Fig. 207. Relaxation Young's modulus of 2.4% aqueous gels of agarose of various molecular weights measured at 25 °C; (A): [η] = 0.20 m^3/kg and (D): [η] = 0.54 m^3/kg (from data presented in [523])

is strong for samples with [η] < 0.33 m^3/kg (in fact results at 25 °C for 2.4% gels of agarose specimens with [η] = 0.54, 0.39 and 0.33 respectively show marginal differences only) [523]. Analogous results are shown in Fig. 208, where the temperature dependence of the storage Young modulus is shown for agarose samples, varying in number average molecular weight from 13 to 204 kg/mol: the molecular weight dependence is strong for the agaroses with $\bar{M}_n$ of 13–156 kg/mol; above that molecular weight the dependence is marginal [529].

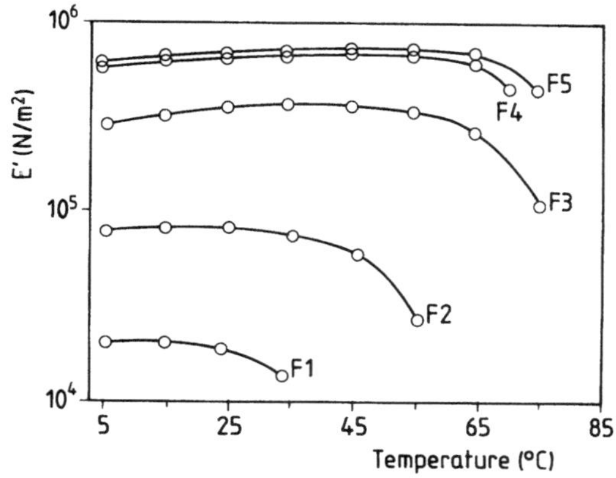

Fig. 208. Temperature dependence of the storage Young modulus of 3% aqueous gels of agarose of various molecular weight distributions; $\bar{M}_n$ and $\bar{M}_w$ in kg/mol: F1: 13 and 34; F2: 44.5 and 97.7; F3: 78.4 and 181; F4: 156 and 352; F5: 204 and 485, respectively ($\bar{M}_w/\bar{M}_n = 2.3 \pm 0.1$). Reproduced, with permission of the authors and publishers, from [529]

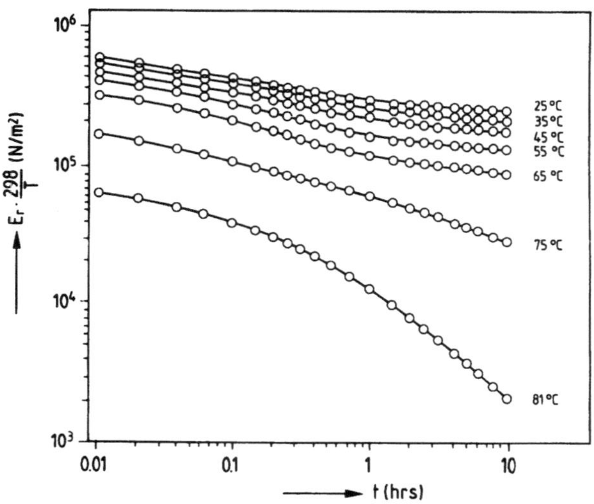

Fig. 209. Reduced relaxation Young's modulus (reduced to 298 K) of 4.4% aqueous agarose ($[\eta] = 0.54\ m^3/kg$) gels measured at various temperatures, as indicated. Reproduced from Rheol Acta [Ref. 523] by the courtesy of the authors and of Steinkopff Publishers Darmstadt, FRG

11.2.7 Temperature Dependence

Temperature dependence was already shown in Fig. 208, whereas in Fig. 209 for 4.4 wt% aqueous agarose gels [523] (i.e. the same agarose as in Fig. 207) the temperature dependence of the (time dependent) Young modulus is shown: above 65 °C the behaviour quickly becomes liquidlike. Agarose gels that were first set at 5 °C and subsequently heated [525] show rubber elasticity (modulus proportional to temperature) up to about 25 °C; from that temperature on, the increase slows down and a maximum value is obtained at about 35 °C (independent of concentration and molecular weight); the modulus decreases to zero at about 65 °C. This is in agreement with results of DSC measurements:

upon heating, endothermic peaks are obtained at 70 and 85 °C for gels of agarose samples with $[\eta] = 0.20$ and $0.54\ m^3/kg$, respectively, nearly independent of concentration. An exothermic peak is observed at about 30 °C, which is shifted to higher temperatures with increasing concentration, slightly lower than the temperature at which the storage modulus E′ shows a maximum. Eldridge-Ferry plots have curvatures concave to the $1/T_m$ axis. Nevertheless, an average ΔH_m value was estimated to be about 1300 kJ/mol for agarose gels, independent of the molecular weight. This rather large value is the consequence of the stiffness of the agarose helices.

11.2.8 Solvent Dependence

From their DSC and dynamic viscoelastic measurements, Watase and Nishinari [204] concluded that the gelling tendency of aqueous agarose solutions increases by the addition of dimethyl sulfoxide to water as a solvent, up to a mole fraction of 0.277 of DMSO. This is attributed to a strong interaction between the DMSO and water molecules: $(CH_3)_2SO.3H_2O$ and $(CH_3)_2SO.2H_2O$ (mole fractions of DMSO 0.25 and 0.33, respectively) are the complexes. Beyond this maximum, free DMSO acts as a good solvent for agarose, so that the formation of helical crosslinks is inhibited. It is interesting to note that the same phenomenon was observed in poly(vinyl alcohol) solutions (see Section 3).

11.3 Conclusions

The viscoelastic properties of aqueous agarose gels depend strongly on the degree of desulfation of its native polysaccharide agar agar: the propensity to form gels increases with increasing desulfation. The junction zones in agarose gels are the result of specific intermolecular hydrogen bonding to form double-stranded helices that will subsequently aggregate, whereas intramolecular hydrogen bonding contributes to the rigidity of the polymer chains.

Papers concerning viscoelastic properties are relatively scarce, although the dependences on concentration, molecular weight, temperature, methoxy group content and the presence of alkali metal ions were reported. Alkali metal ions have a positive influence on gelation behaviour. However, they affect the rheological properties of non-gelling solutions negatively, due to shrinkage of the polymer chain. This resembles the alkali metal dependence of the behaviour of aqueous gellan gum systems (see Section 13). Decrease of the solubility in water, obtained by desulfation or by methoxylation, has an increasing effect on the gelling propensity. It is evident that much more work has to be done in a systematical way in order to elucidate the network formation process. The large number of parameters that affect the network formation process makes it difficult to get a complete picture of the gelation process of aqueous agarose gels. Syneresis complicates the measurement of the viscoelastic behaviour of agarose gels.

12 Carrageenans

12.1 Introduction

Carrageenans are linear, sulfated polysaccharides of the type $(AB)_n$ consisting of β-1,3-linked D-galactose and α-1,4-linked 3,6-anhydro-D-galactose. In iota carrageenan both rings are sulfated, whilst in kappa carrageenan only the A-ring is sulfated (see Fig. 210, e.g. [1,533]). In general, the structure is less ideal: some residues are replaced by residues B′, so that kinks in the structure exist. Hence, the helix formation is less ideal and the result is gelation of low concentration solutions instead of crystallisation and phase separation. Aqueous, thermoreversible gels are formed at concentrations as low as 1%; kappa carrageenan forms brittle gels, whereas iota carrageenan forms soft gels. In the presence of alkali ions (e.g. K^+, Rb^+, Cs^+ and Ca^{2+}), the critical concentrations are considerably lower.

In marine algae (the source of carrageenans) other carrageenans are present, e.g. λ and μ carrageenans, with less regular, kinked structures; consequently, these carrageenans have no gelling properties.

The conformational changes in solutions of carrageenans and/or of their segments have been a subject of many studies, e.g. optical rotation [534–547], calorimetry [525, 548–551], conductivity [552] and spectroscopy [544, 545, 547, 553–556]. They all show changes in properties in the temperature region where a sol-gel transition is present. An example is given in Fig. 211, where the percentage of order of kappa carrageenan in solution (in the potassium form, concentration = 5.04 kg/m^3) is shown as a function of temperature, as calculated from optical rotation and conductivity measurements [552]. In Fig. 211, the conductivity itself is also shown as a function of temperature: the percentage of molecular order is calculated as

$$\text{Order} = \frac{\kappa_{\text{coil}} - \kappa_{\text{observed}}}{\kappa_{\text{coil}} - \kappa_{\text{helix}}}. \tag{53}$$

In general, the change is extremely rapid; only if the total ionic concentration (polyelectrolyte + added electrolyte) is low may the conformational transition induced by temperature be measured as a function of time; this time dependence was investigated with conductivity measurements by Rochas and Landry [552]. The time dependence of the coil-helix transition in a solution of kappa carrageenan (potassium form, concentration 1.4 kg/m^3) is shown in Figs. 212 and 213 (linear and logarithmic time scales, respectively). From the curve with the linear time scale one gets the impression that an equilibrium state is reached within 1 h (with 85% of helix order). However, from the curve with the logarithmic time scale it seems that it takes more time to reach an equilibrium value: from 0.25 h the curve becomes linear. It can be concluded that maximum order cannot be reached if the cooling rate is fast and/or if the temperature is low. The optical rotation is strongly dependent on temperature in the region where

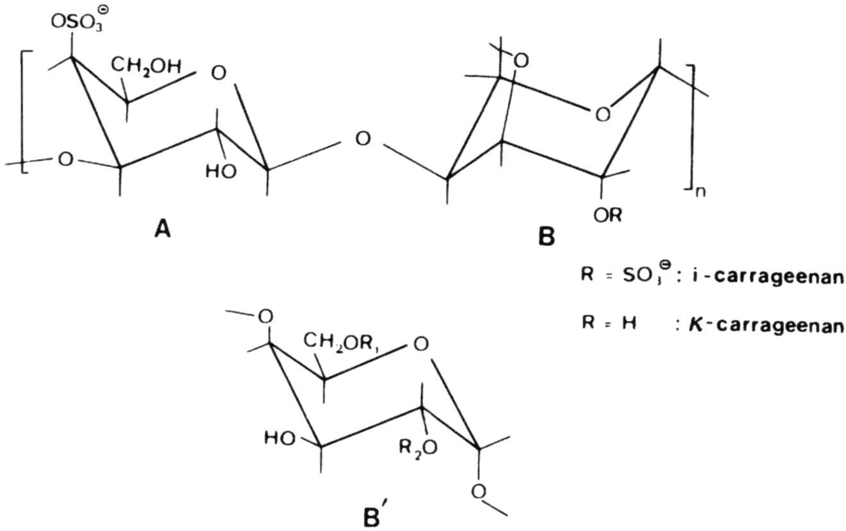

Fig. 210. Idealised AB repeat unit of iota and kappa carrageenan polymers based on 1,3-linked β-D-galactose residue (A) and 1,4-linked 3,6-anhydro-α-D-galactose residue (B). The sequence is broken occasionally by residues of general type B′. Reproduced from Adv Polym Sci [Ref. 1] by the courtesy of the authors and of Steinkopff Publishers Darmstadt, FRG

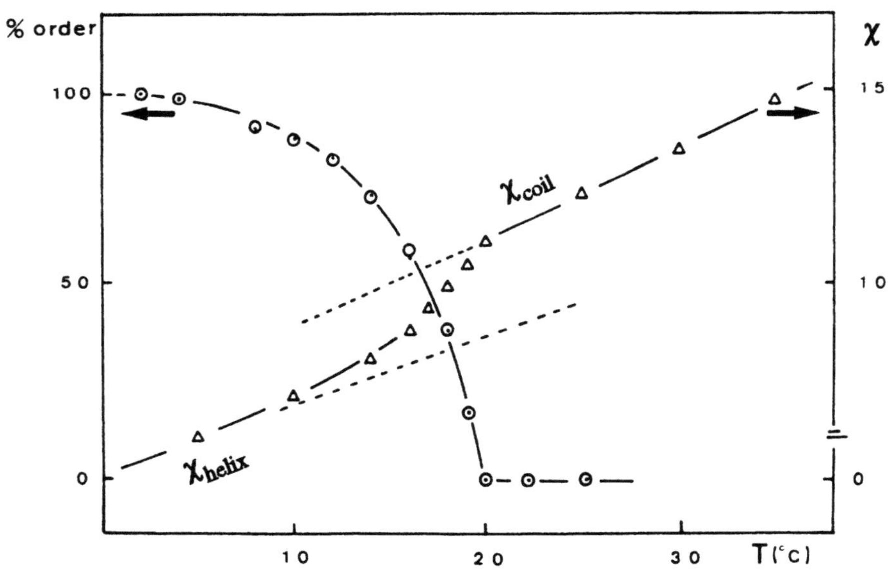

Fig. 211. Percentage of order of kappa carrageenan in solution (potassium form, 5.04 kg/m^3) obtained from optical rotation (○) and conductivity (●). Also shown are the measured values for the conductivity (Δ), expressed in mS. Reproduced from Carbohydr Polym [Ref. 552] by the courtesy of Elsevier Science Ltd, The Boulevard, Langford Lane, Kidlington 0X5 1GB, UK

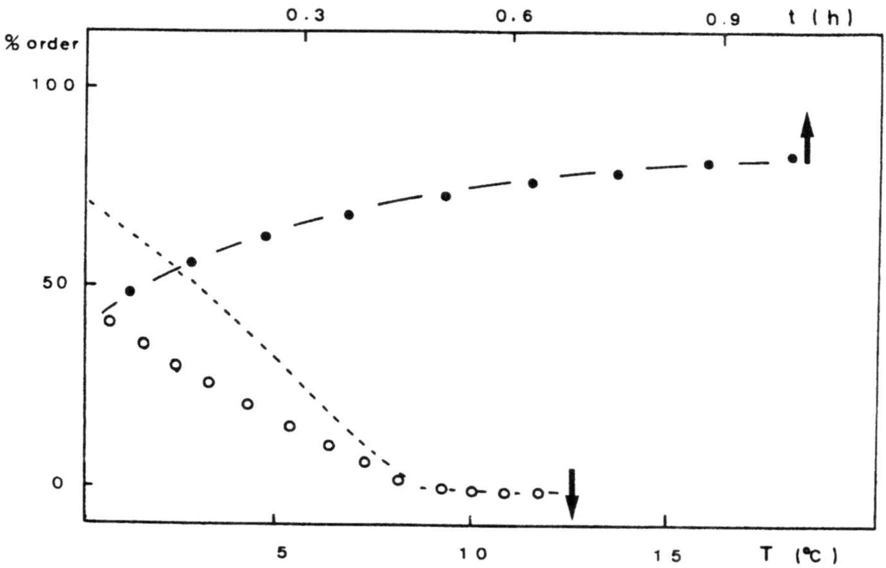

Fig. 212. Development of percentage of order of kappa carrageenan (potassium form, 1.4 kg/m^3) at 0 °C, determined by conductivity, plotted as a function of time (●). The measurement of time dependence was started immediately after cooling the solution to 0 °C with a cooling rate of 25 °C/h (○); by cooling with the lower rate of 5 °C/h a higher state of order is reached (*broken line*). Reproduced from Carbohydr Polym [Ref. 552] by the courtesy of Elsevier Science Ltd, The Boulevard, Langford Lane, Kidlington 0X5 1GB, UK

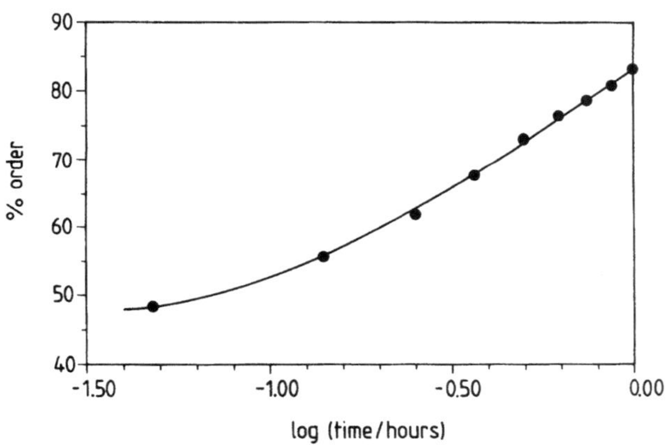

Fig. 213. Development of percentage of order of kappa carrageenan (potassium form, 1.4 kg/m^3) at 0 °C as shown in Fig. 212, plotted as a function of log time

a sol-gel transition is present. For iota carrageenan the optical rotation is reversible [536], whereas for kappa carrageenan a pronounced hysteresis is observed [534]. Investigations by Rochas et al. [544] on carrageenan segments demonstrate the existence of a coil-helix transition in 2.5 mol/l KCl at

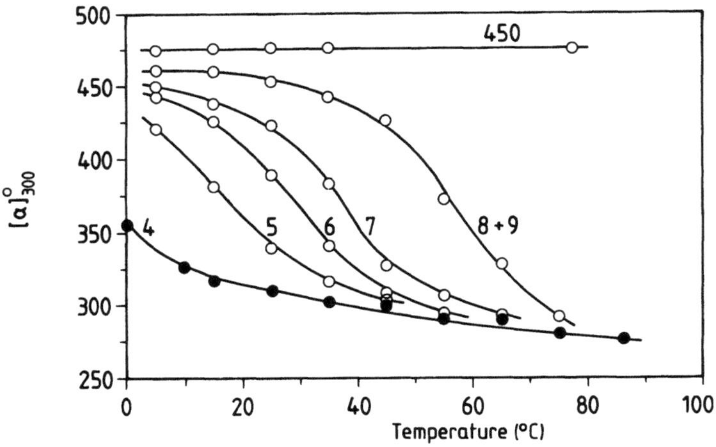

Fig. 214. Temperature dependence of optical rotation of carrageenan oligomers and polymer in the presence of 2.5 mol/l KCl. The parameters at the curves are the degrees of polymerisation. Reproduced, with permission of the authors and publishers, from [544]

temperatures between 0 and 75 °C, depending on the degree of polymerisation (4 to 9, respectively), whereas segments with a degree of polymerisation of 450 maintain their helical structure at least up to 75 °C (see Fig. 214). In NaCl solution, no transition to helices was observed.

X-ray investigations by Jones et al. [557] on kappa and iota carrageenan suggest that double-helices are also present in the solid state.

The gelation process of carrageenan solutions has been described by two mechanisms. In the first mechanism, suggested by Morris ER et al. [558], crosslinks are formed by segments of a double helix. These segments are then aggregated by ions as K^+ (see Fig. 215a). In the other mechanism, proposed by Smidsrød and Grasdalen [559, 560], monohelices are formed which are subsequently aggregated by K^+ ions to dimers, trimers etc. (see Fig. 215b). For an extensive discussion of the gelation mechanism, the reader is referred to [1] (pp 116–121).

12.2 Crosslink Structure

The efficiency of alkali ions in promoting the intermolecular association (crosslink formation) in iota carrageenan increases in the direction Na^+, K^+, Ca^{2+}, as is obvious from measurements reported by Morris VJ et al. [553, 555, 561, 562] (see Fig. 216) and Tako et al. [563, 564]. The increase in the Young modulus with concentration is much faster for Ca-carrageenan solutions than for K-carrageenan solutions, which in turn is much faster than for Na-carrageenan solutions. For this reason, it is said that Ca^{2+} ions favour the gelation of iota

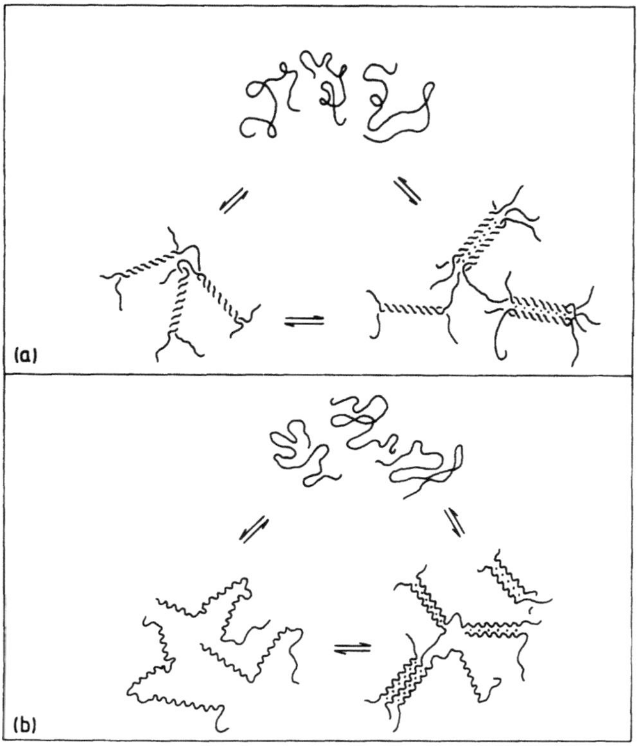

Fig. 215a, b. Schematic models for gel formation in kappa carrageenan; **a** by Morris ER et al. [558]; **b** by Smidsrød and Gasdalen [560]. K^+counterions are indicated by *solid circles* (●). Reproduced from Adv Polym Sci [Ref. 1] by the courtesy of the authors and of Steinkopff Publishers Darmstadt, FRG

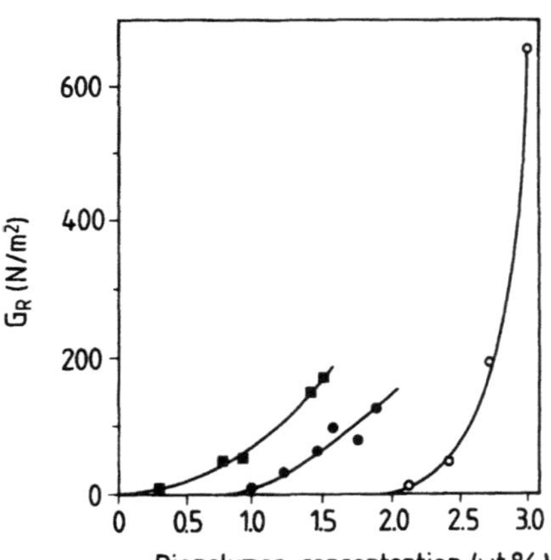

Fig. 216. Shear modulus vs biopolymer concentration for calcium (■), potassium (●) and sodium (○) iota carrageenans. Reproduced from J Chem Soc, Chem Comm [Ref. 553] by the courtesy of the authors and The Royal Society of Chemistry, Cambridge, UK

carrageenan. On the other hand, according to Tako et al. [563, 564], K^+ ions favour the gelation of kappa carrageenan. Although measurements by Morris VJ et al. [554, 561, 566] seem to contradict these findings for kappa carrageenan (these authors found that Ca^{2+} ions favour the gelation of kappa carrageenan more than K^+ ions), Tako et al. [563, 564] proposed a gelation mechanism for kappa carrageenan (favoured by K^+ ions) as well as for iota carrageenan (favoured by Ca^{2+} ions).

In kappa carrageenan, intramolecular bridges are formed in the presence of K^+ ions (or Rb^+ and Cs^+ ions), a bridge being formed first by an ionic bond between K^+ and the sulfate group of the one D-galactose residue and second by an electrostatic bond between the K^+ and the anhydro-O-3,6-ring of the other D-galactose residue (see Fig. 217). Because of the presence of the intramolecular bridge, the flexibility of the AB unit decreases. Many intramolecular cation bridges serve to keep the polymer chain rigid, which results in an intermolecular association. The univalent Na^+ and Li^+ ions are too small to be able to form a bridge: they are only able to be bonded ionically to the sulfate group without forming an electrostatic bonding with the anhydro-O-atoms.

In the more ionic iota carrageenan, the gelation mechanism is different: Ca^{2+} ions are more effective than K^+ ions. Intramolecular Ca^{2+} bridging may take place between the sulfate groups of adjacent anhydro-D-galactose and D-galactose residues (see Fig. 218). With univalent cations, the intramolecular bridging may take place in part between the anhydro-bridge oxygen atom and the sulfate group linked at C-4 of the D-galactose residues as in the K salt of kappa carrageenan. Upon cooling, intermolecular Ca^{2+} bridging may also take place between different molecules: the oxygen group of the sulfate groups at C-2 of the anhydro-D-galactose residues may contribute to the intermolecular Ca^{2+} bridges (see Fig. 219). The intermolecular Ca^{2+} bridges may gradually dissociate up to the transition temperature (45 or 50 °C) and the intramolecular Ca^{2+} bridges may dissociate rapidly above that temperature.

The model corresponds to a double-stranded helix, although sulfate groups may be present on the inside of the helix. A tertiary structure of the Ca salt of the

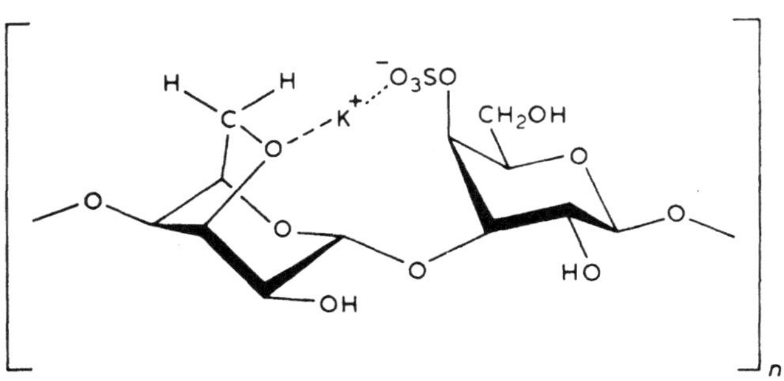

Fig. 217. Possible mode of intramolecular, cation selective bridge in the kappa carrageenan molecule in aqueous solution at low temperature: (····) ionic bonding and (----) electrostatic force of attraction. Reproduced from Carbohydr Res [Ref. 563] by the courtesy of the authors and Elsevier Science Ltd, The Boulevard, Langford Lane, Kidlington 0X5 1GB, UK

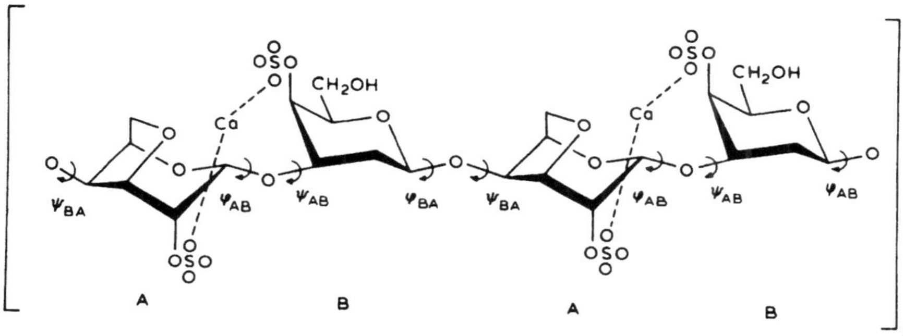

Fig. 218. Possible mode of intramolecular Ca^{2+} bridge of iota carrageenan. The *dashed lines* refer to the ionic force. The conformational changes are expressed in terms of four angles of rotation ϕ and ψ. Reproduced from Carbohydr Res [Ref. 564] by the courtesy of the authors and Elsevier Science Ltd, The Boulevard, Langford Lane, Kidlington 0X5 1GB, UK

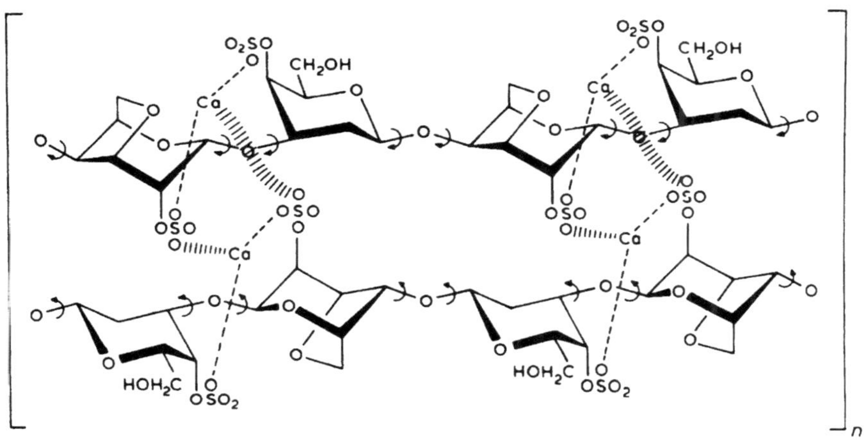

Fig. 219. Possible mode of intermolecular Ca^{2+} bridge of the Ca salt of iota carrageenan: (····) ionic bonding and (----) electrostatic forces of attraction. The *arrows* refer to the orientation of the conformational angles. Reproduced from Carbohydr Res [Ref. 564] by the courtesy of the authors and Elsevier Science Ltd, The Boulevard, Langford Lane, Kidlington 0X5 1GB, UK

iota carrageenan molecules may consist of two identical, right-handed, 3-fold helices in aqueous solution, as in the solid state. In the K-salt also intramolecular bridges are formed, less strong than in the Ca salt, however. In Na and K carrageenan the iota carrageenan molecule may form a single rod-like structure at room temperature. However, when the counter ions were exchanged by Ca^{2+} ions upon addition of $CaCl_2$, the intermolecular Ca bridging of the various molecules could take place by ionic forces. $CaCl_2$ addition to Ca carrageenan does not cause a strong increase in the mechanical properties; apparently one equivalent of Ca^{2+} suffices to form the necessary bridges.

12.3 Viscoelastic Behaviour

12.3.1 Alkali Treatment

The gelling ability of carrageenans is influenced by alkali treatment. Just as with agarose, deesterification of the sulfate esters will result in an enhanced tendency to form gels in aqueous solutions. By the formation of the 3, 6-anhydro bond the solubility in water is decreased. Moreover, sulfate groups hinder the formation of single and double helices, because they are bulky and repulse each other electrostatically. By the deesterification, the double-helix structure is stabilised by hydrogen bonds (see e.g. Sect. 11). The gelling behaviour of kappa carrageenans is especially affected by the alkali treatment [565]. The result of such treatment with 6% potassium hydroxide is shown in Fig. 220a: the increased gelation tendency of alkali treated kappa carrageenan is clearly visible: e.g. a 6% solution of pre-treated kappa carrageenan shows the same behaviour as a 10% solution of non-treated kappa carrageenan: not only is the modulus increased by the alkali treatment, but also the temperature where the modulus decreases rapidly is higher after the treatment. The melting enthalpies, obtained from an Eldridge-Ferry plot [160], of both kappa carrageenans are much the same: $\Delta H_m = 60$ kJ/(mol of crosslinks) as against $\Delta H_m = 33$ kJ/(mol of crosslinks) for iota carrageenan.

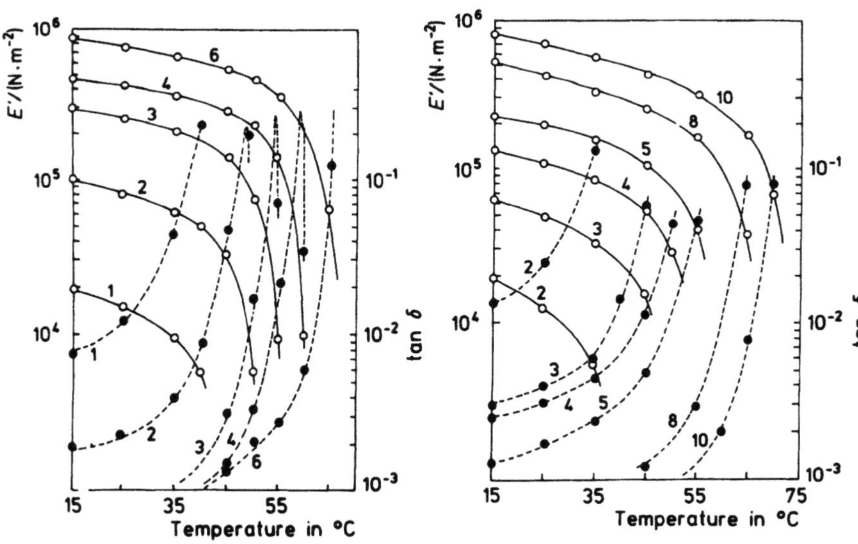

Fig. 220a, b. Temperature dependence of the storage Young modulus E' (*full lines*) and the loss tangent (*broken lines*) of aqueous kappa carrageenan gels of various concentrations, as indicated in wt%: **a** kappa carrageenan pre-treated with 6% KOH; **b** non-treated kappa carrageenan. Reproduced from Makromol Chem [Ref. 565] by the courtesy of the authors and of Hüthig & Wepf Verlag Publishers, Zug, Switzerland

12.3.2 Dependence of Electrolyte Concentration

The dependence of the gelation process of carrageenans on electrolyte concentration has been the subject of many investigations e.g. [522, 543–546, 553–555, 561, 562, 566–568]. In general, it was concluded that the alkali metal ions are able to increase the gelation tendencies of both iota and kappa carrageenans. The cation dependence of the gel modulus follows the Hofmeister series $Cs^+ > Rb^+ > K^+ \gg Na^+ > Li^+$, which is in agreement with what has been said before of the dependence of the crosslink structure on the alkali metal ions (see e.g. Fig. 216).

Rochas and Rinaudo [569] constructed a phase diagram, where the logarithm of the total ionic concentration is plotted vs the reciprocal transition temperature (see Fig. 221). It characterizes the conformation and gelation of kappa carrageenan in KCl solutions. In domain I (relatively low concentration and high temperature), only polymer coils are present; in domain II (relatively low concentration and low temperature), the kappa carrageenan molecules are in the ordered, helical conformation, whereas in domain III (relatively high concentration and low temperature), the system is in the (helical) gel state; the boundary between domains II and III is not sharply known and for that reason is indicated by a wavy line.

Below a critical concentration ($C^* = 7 \times 10^{-3}$ mol/l and $T_m = 20\,°C$), no gel formation is possible; the line corresponds to the transition between coils and helices; above C* gel formation is possible and a difference between a heating and a cooling process is perceptible (hysteresis). The phase diagrams of iota and kappa carrageenan differ, as shown in Fig. 222. It turns out that gels are more

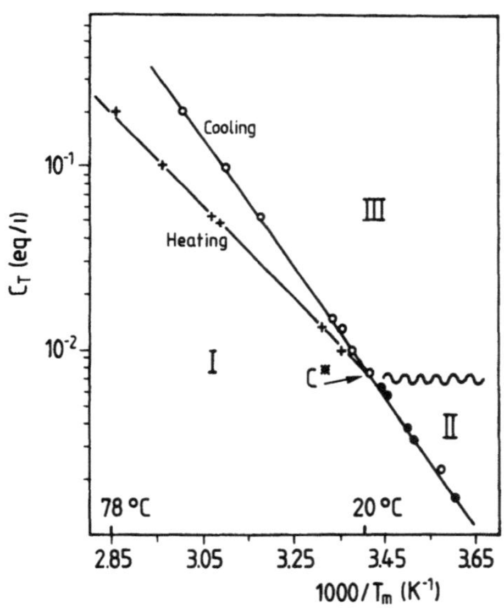

Fig. 221. Phase diagram of kappa carrageenan based on the melting temperature, T_m, and the total free potassium concentration, C_T. Reproduced from Biopolymers [Ref. 569] by the courtesy of the authors and of John Wiley & Sons, Inc

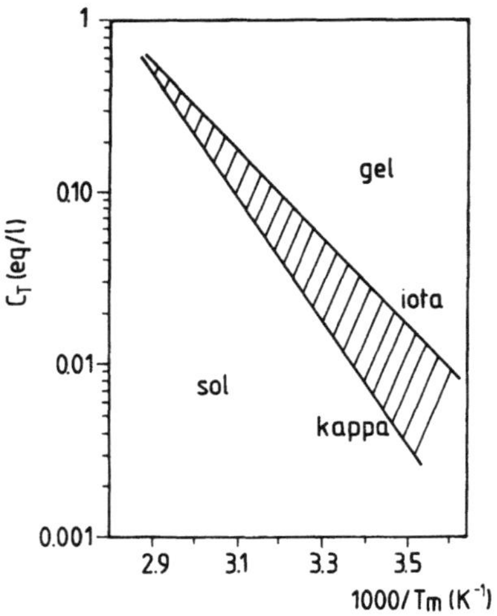

Fig. 222. Variation of the gelation temperature with concentration of free potassium counterion for kappa and iota carrageenan. Reproduced from Carbohydr Polym [Ref. 545] by the courtesy of Elsevier Science Ltd, The Boulevard, Langford Lane, Kidlington 0X5 1GB, UK

easily formed in kappa carrageenan solutions than in iota carrageenan solutions [545]: there is a domain where kappa carrageenan has a helical conformation (the gel state), but where iota carrageenan has a disordered conformation (the sol phase).

12.3.3 Molecular Weight Dependence

The dependence of the mechanical properties on the molecular weight of kappa carrageenan was investigated by Rochas et al. [546]. They found that the modulus depends on molecular weight up to some critical value of $\bar{M}_w \approx 200$ kg/mol, almost independent of the KCl concentration (see Fig. 223). A rough calculation, assuming Gaussian chains between small crosslinks, reveals a molecular weight between crosslinks of the order of 2 kg/mol. If one assumes that the fraction of dangling chain ends plays a role, if $\bar{M}_c$ is not less than one-tenth of the number average molecular weight one would expect an independence of the molecular weight for $\bar{M}_w > 40$ kg/mol. This means that the crosslinks are very extended along the polymer chain, in accordance with high values of order (85% or more) and also in agreement with the findings of Morris ER et al. [558], that at least ten polysaccharide segments are present in a crosslink. The conclusion might be that the crosslinks are extended to such a degree along the polymer chains that they play an important role in the value of the equilibrium shear modulus: 85% of relatively stiff material as against 15% of Langevin [525] chains.

Recently, Clark [570] calculated the molecular weight between crosslinks for the results shown in Fig. 223 by making use of the cascade theory. He concluded that molecular weights between crosslinks are 2.35–3.55 kg/mol, which is much

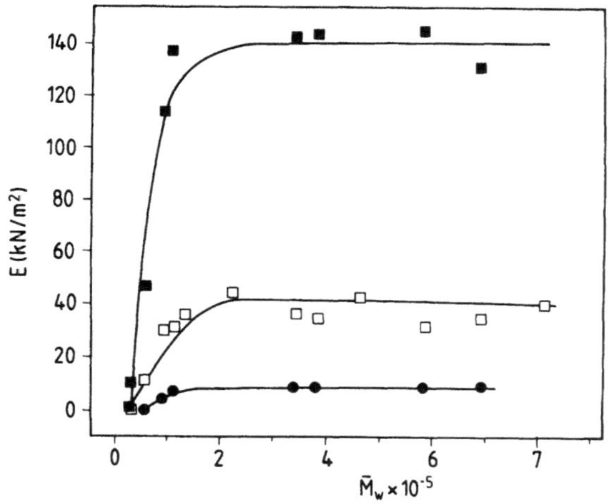

Fig. 223. Young's modulus E vs molecular weight of different fractions of kappa carrageenan in 0.1 mol/l KCl: (●), (□) and (■) polymer concentration 5, 10 and 20 kg/m^3, respectively. Reproduced from Carbohydr Polym [Ref. 546] by the courtesy of Elsevier Science Ltd, The Boulevard, Langford Lane, Kidlington 0X5 1GB, UK

higher than the value of 2 kg/mol mentioned above. This is in agreement with the relatively high value of the factor aRT, a measure of the average contribution to rubber elasticity per mole of elastically effective network chains (EANCs): the parameter a, for rubber elasticity being ≤ 1, varied from 14.4 to 42.7.

According to Clark and Ross-Murphy [1] the factor a appears to be substantially greater than 1 for biopolymer gels, especially at high concentrations. Under these conditions, the contribution of the EANCs is rather enthalpic than entropic.

12.3.4 *Concentration Dependence*

The concentration dependence of Young's modulus of kappa carrageenan at room temperature was investigated by Rochas and Landry [571] and found to be almost quadratic (slopes 2.02, 1.98 and 1.95 for KCl concentrations of 0.05, 0.1 and 0.5 mol/l respectively). Moreover, the Young modulus increases with the KCl concentration (see Fig. 224). However, clear deviations from this quadratic behaviour appear at high polymer concentrations, especially at the highest KCl concentration. By comparison with Fig. 223, it turns out that the curves must be independent of the carrageenan molecular weight, as long as the molecular weight exceeds a value of about 100 kg/mol.

On the other hand, Oakenfull and Scott [567] investigated the concentration dependence of the shear modulus of kappa and iota carrageenan solutions close to the gel temperature: their results are shown in Fig. 225 on double logarithmic scales. It appears that the slopes are greater than 2 (2.8 and 3.9 for kappa and iota carrageenan, respectively). Plashchina et al. [572] found this slope to be 2.4 ± 0.1 for kappa carrageenan solutions.

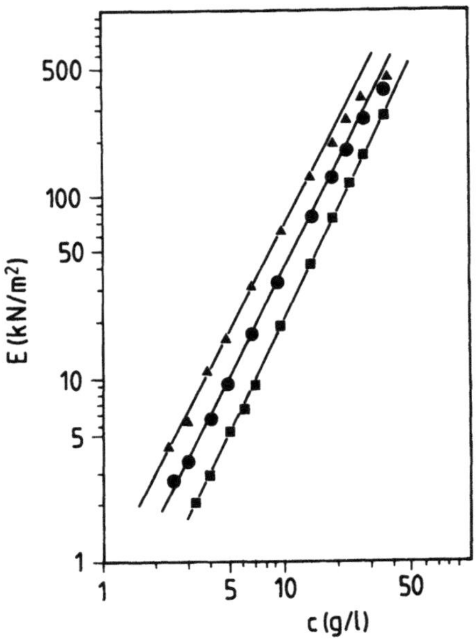

Fig. 224. Young's modulus E of kappa carrageenan gels as a function of the polymer concentration, for various KCl concentrations: (▲), (●) and (■) 0.25, 0.1 and 0.05 mol/l, respectively. Reproduced from Gums and Stabilisers for the Food Industry, Vol 4 [Ref. 571] by the courtesy of the authors and of Oxford University Press

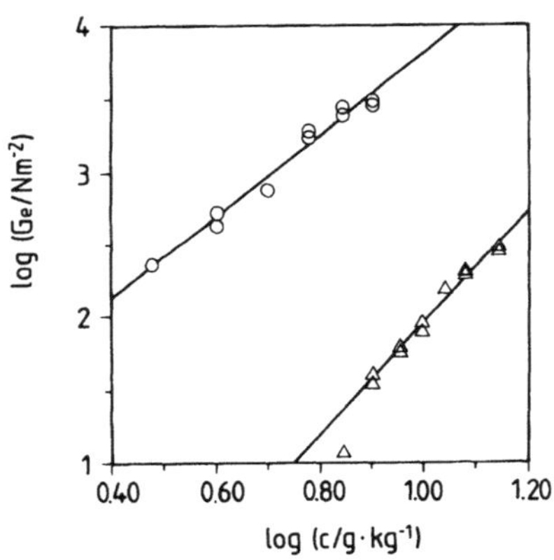

Fig. 225. Double logarithmic plot of shear modulus vs concentration for potassium salts of kappa and iota carrageenans at 25 °C. Reproduced from Gums and Stabilisers for the Food Industry, Vol 4 [Ref. 567] by the courtesy of the authors and of Oxford University Press

Oakenfull and Scott [567] also measured the gel time of kappa and iota carrageenan solutions as a function of polymer concentration. Its reciprocal value, a kind of gelation rate, was plotted vs concentration on double logarithmic scales. Straight lines were obtained with slopes of 12.5 ± 0.9 and 4.5 ± 0.6 for kappa and iota carrageenans, respectively (see Fig. 226). Their conclusion was, that in gels of kappa carrageenan, the average junction zone is an aggregate of about six double-helices; in gels of iota carrageenan the junction zones are considerably less complex, with two to three associated double-helices. From Figs. 225 and 226 it is clear that kappa carrageenan solutions form gels much easier, with a higher shear modulus, than iota carrageenan solutions.

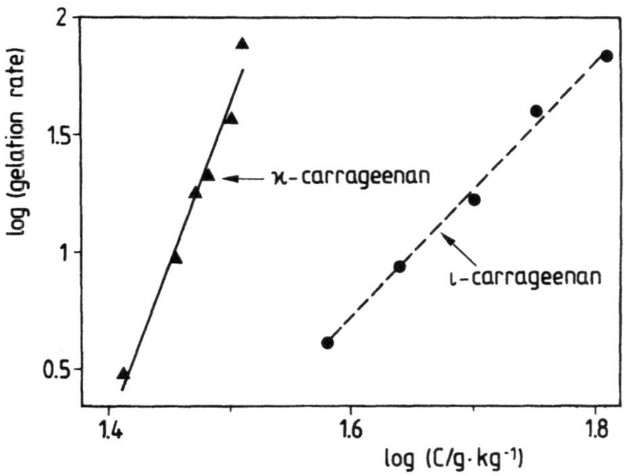

Fig. 226. Double logarithmic plot of "gelation rate" (100/t, t in minutes) vs weight concentration (g/kg) of kappa and iota carrageenan in aqeous solutions. Reproduced from Gums and Stabilisers for the Food Industry, Vol 4 [Ref. 567] by the courtesy of the authors and of Oxford University Press

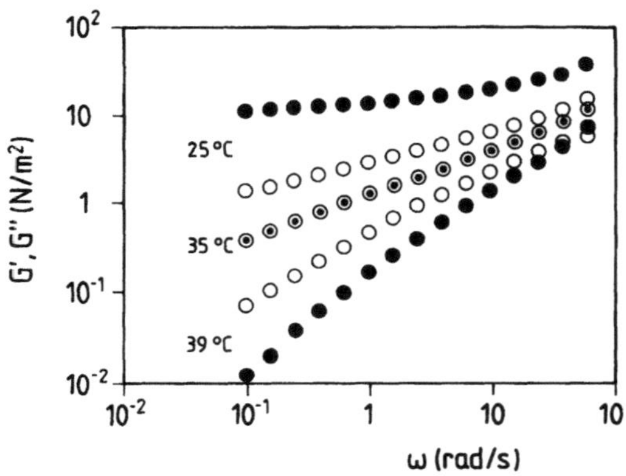

Fig. 227. Dynamic moduli of an aqueous 1% iota carrageenan solution vs angular frequency measured at 25, 35 and 39 °C; the system is in its critical state at 35 °C; (●) G′; (○) G″. Reproduced from Gums and Stabilisers for the Food Industry, Vol 5 [Ref. 573] by the courtesy of the authors and of Oxford University Press

12.3.5 Temperature Dependence, Critical Gelation

Cuvelier et al. [573] determined the frequency dependence of the dynamic moduli, G′ and G″, as a function of temperature for aqueous solutions of iota carrageenan. Results are shown in Fig. 227. At 39 °C the system shows liquidlike behaviour with G″ > G′, both moduli being strongly dependent on frequency, whereas at 25 °C the system behaves more rubberlike, with G″ < G′ and G′ only slightly depending on frequency. A real rubber plateau has not yet been reached, however. According to the Winter-Chambon model, the system is almost in its critical state at 35 °C: G′ and G″ are equal over nearly the whole frequency range measured, and the dependence is almost linear on the double logarithmic scales, with a slope $n \approx 0.5$. These results ($G' \approx G''$ and $n \approx 0.5$) are in agreement, as, according to the Winter-Chambon model, in its critical state the system should satisfy the relation already met in Eq. (6):

$$\frac{G''}{G'} = \tan(\delta) = \tan(\tfrac{1}{2}\pi n)$$

which, for n = 0.5, yields $G''/G' = \tan \pi/4 = 1$.

12.4 *Comparison of Viscoelastic Behaviour of Aqueous Agarose and Kappa Carrageenan Solutions*

Although the chemical structures of agarose and kappa carrageenan are very alike, they nevertheless differ strongly in their gelation behaviour. They both form helices by hydrogen bonding between OH groups. However, since stable micelles are formed naturally in aqueous agarose solutions, the addition of alkali metal ions does not change this molecular conformation. Hence, the Young's modulus of aqueous agarose gels is not greatly affected by the addition of alkali metal ions. On the other hand, the electrostatic repulsion of the sulfate groups present at the exterior part of the helical cylinder in kappa carrageenan hinders the formation of aggregates. By the addition of alkali metal ions the electrostatic repulsion is shielded, resulting in stabilization of micelle structures and formation of densely packed aggregates [522, 574]. This addition of alkali metal salts results in a strong increase in Young's modulus. Due to these different responses to the addition of alkali metal salts, there will also be a difference in swelling behaviour of aqueous gels after immersion in alkali metal salt solutions. The kappa carrageenan gels deswelled by a certain amount, whereas Young's modulus increased to a much greater extent than can be attributed to the increase of the biopolymer concentration in the gel. Immersion in CsCl and KCl solutions had much more influence than immersion in NaCl and LiCl solutions. The effect of immersion of agarose gels in these solutions turned out to be negligible [574].

A comparison between the temperature dependence of the viscoelastic properties of agarose and kappa carrageenan gels has been made by Watase and Nishinari [525]. Gels were obtained by cooling hot solutions to 5 °C, where they were allowed to maturate for one week. Dynamic mechanical measurements at 2.5 Hz (E′) and calorimetric (DSC) measurements were taken. Results for E′ are shown in Figs. 228 and 229. Remarkable differences in temperature dependence for both gels are obvious from these figures, as follows.

a) Agarose gels show a larger elastic modulus and a higher melting temperature than kappa carrageenan gels.
b) Agarose gels appear to be rubberlike up to about 35 °C, i.e. proportional to temperature: the dotted lines are calculated as

$$E'(^\circ C) = E'(5\,^\circ C)\,\frac{T(K)}{278}. \tag{54}$$

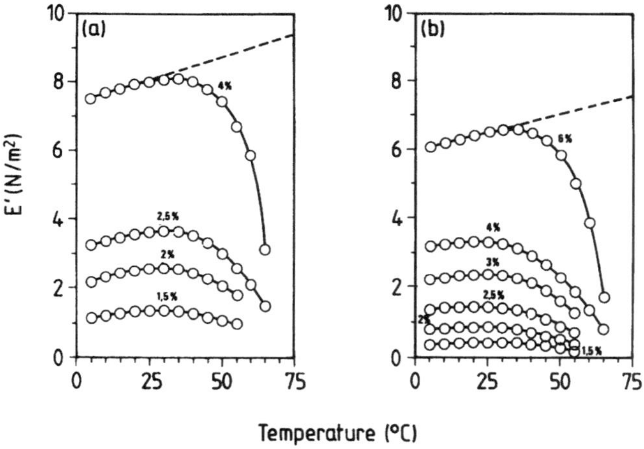

Fig. 228a, b. Temperature dependence of the dynamic Young modulus E′ for agarose gels of various concentrations: **a** $[\eta] = 0.54\ m^3/kg$ (at 35 °C in 10 mol/m^3 NaSCN); **b** $[\eta] = 0.20\ m^3/kg$ (under the same conditions). The *numbers at the curves* are the polymer concentrations in wt%. Reproduced from Polym J [Ref. 525] by the courtesy of the authors and The Society of Polymer Science, Japan

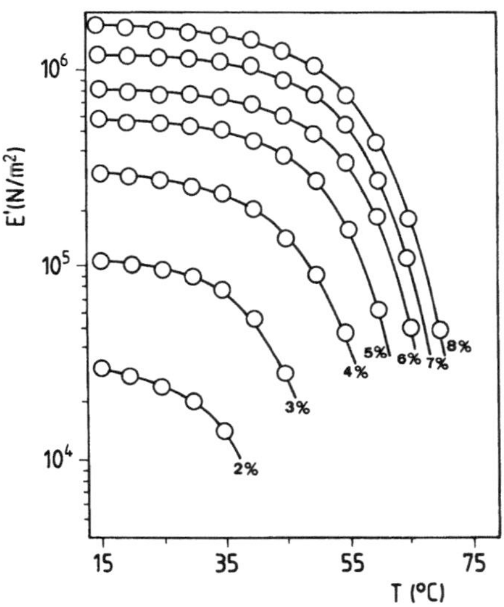

Fig. 229. Temperature dependence of the dynamic Young modulus E′ for kappa carrageenan gels of various concentrations. The *numbers at the curves* are the polymer concentrations in wt%. Reproduced from Polym J [Ref. 525] by the courtesy of the authors and The Society of Polymer Science, Japan

On the other hand, E′ of the kappa carrageenan gels gradually decreases with temperature in the same range. However, recently Richardson and Goycoolea [575] reported a rather sharp transition for 1 wt% potassium kappa carrageenan in 0.1 mol/l KCl around 55 °C: the storage modulus was increased from 1 to 1000 N/m^2 by lowering the temperature from 58 to 52 °C. (Moreover, these authors claim that with their sensitive home made rheometer provided with perforated coaxial cylinders, better results are obtained in the measurement of gel systems that are liable to slippage, such as, e.g., kappa carrageenan gels and gellan gum gels).

c) The temperature where E′ of the agarose gels starts to decrease rapidly (~35 °C) is almost independent of concentration and molecular weight: this is just the temperature in the DSC curves where the melting process starts. For the carrageenan gels this temperature of rapid decrease of E′ is shifted to higher temperatures with increasing concentration.

d) A rough extrapolation to $E' = 0$ shows that the gel-sol transitions of the agarose gels takes place at about 70 °C, independent of concentration and molecular weight; for the carrageenan gels this transition temperature is strongly dependent on concentration.

e) Because of the relatively low dependence of the gel-sol transition temperature on concentration, the melting enthalpy of the agarose gels as calculated from the Ferry Eldridge equation [160], is high: $-\Delta H_m \approx 1300$ kJ/mol; for the carrageenan gels the melting enthalpy is much smaller, in agreement with the stronger concentration dependence of the gel sol transition on temperature: $-\Delta H_m \approx 38$ kJ/mol. However, the plots of ln c vs $1/T_m$ are not really straight lines, so that there is some concentration dependence. This means that the crosslink structure of the agarose gels is much more complex than that of the carrageenan gels.

12.5 Conclusions

Gelation properties of carrageenans depend on the kind of carrageenan. Mostly used are iota and kappa carrageenans. As iota carrageenan contains more sulfate groups, its propensity to form gels is less than that of kappa carrageenan. The junction zones in aqueous carrageenan gels are mainly the result of intermolecular electrostatic forces, especially in the presence of alkali metal ions. The rigidity of the carrageenan chains is increased by intramolecular bridges. Two important models of network formation have been proposed in the literature: a) the formation of double-stranded helices, which may associate under the action of alkali metal ions and b) the formation of monohelices that aggregate under the action of alkali metal ions.

Viscoelastic behaviour is affected by alkali treatment (resulting in deesterification), electrolyte concentration, molecular weight, concentration and temperature. Many scientists have worked on the gelation properties of carrageenans and viscoelastic properties were also reported by many of them. Nevertheless, much work has still to be done in order to complete the picture of network junctions and network structure. However, the large number of parameters that affect the gelation properties makes it difficult to fill in the whole area of viscoelastic behaviour.

Due to difference in sulfate content between agarose and carrageenans, the viscoelastic properties of aqueous solutions and gels of these bipolymers are quite different.

13 Gellan Gum

13.1 Introduction

Gellan gum is a linear, anionic polysaccharide, composed of tetrasaccharide repeating units (β-D-glucose, β-D-glucuronic acid, β-D-glucose, α-L-rhamnose, i.e. residues A, B, C and D respectively) [576, 577]. It contains one carboxyl group (glucuronic acid) per repeating unit. Its chemical structure [578] is shown in Fig. 230. It is an extremely effective gelling agent and it can form gels at low pH and in the presence of salts. For these reasons it is used as a gelling agent in the food industry [579]. Its native form is an extracellular microbial polysaccharide, produced by the bacterium *Pseudomonas elodea* [580]. This can be obtained in large quantities and in an extremely pure form, due to today's biotechnology. It belongs to the gellan family, a novel group of five related polysaccharides that have the acidic tetrasaccharide repeating unit in common. They differ in substitution or branching to varying degrees (native gellan, gellan, welan, rhamsan and S-567). The native gellan as secreted by the bacterium, contains almost 100% L-glycerate and approximately 50% acetate substitution per tetrasaccharide unit, the former on the 6-position and the latter on the 2-position of the β-D-glucose A [581]. Its chemical structure [582] is shown in Fig. 231. This polysaccharide forms weak and rubbery aqueous gels.

After deesterification by heat treatment at high pH, gellan is obtained [583], which forms strong and brittle gels. Produced in a potassium salt buffer solution, the gellan becomes of the potassium type. The level of deesterification determines the properties of the gels that are formed [584, 585]. The carboxylic groups of the glucuronic acid unit induce electrostatic repulsion in solution and inhibit gelation. The gelation is promoted by the introduction of cations, due to their ability to shield the electrostatic repulsion of the carboxylic groups [586–592, 625]. The amount needed depends on the type and ionic strength of the cations used: divalent ions like Ca^{2+} and Mg^{2+} are much more effective than the monovalent K^+ and Na^+ ions.

According to Crescenzi et al. [584, 593], gellan gum may be made to undergo a reversible coil-helix transition in dilute solution by changing the temperature. Chapmann et al. [594] showed that Zimm plots are always affected by aggregation phenomena. On the other hand Zimm plots obtained by Dentini et al. [595] only showed formation of double-helices. Gunning and Morris VJ [596, 597] attribute these discrepancies to the filtration procedure: filtration of a stock solution through a 3 μm pore filter gave $\bar{M}_w = 4500$ kg/mol, whereas filtration through a 0.45 μm pore filter gave $\bar{M}_w = 106$ kg/mol. This might be the result of removal or of break up of the aggregates during filtration. Gunning and Morris prefer the latter explanation, because during filtration through a 0.45 μm pore filter there is only 4% loss of material.

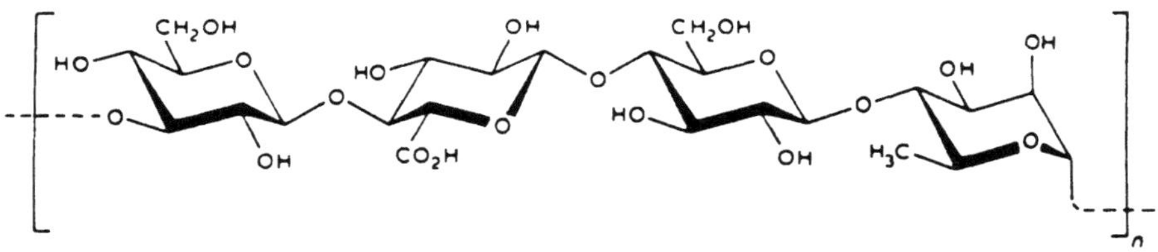

Fig. 230. Chemical structure of gellan. Reproduced from Gums and Stabilisers for the Food Industry, Vol 4 [Ref. 578] by the courtesy of the authors and of Oxford University Press

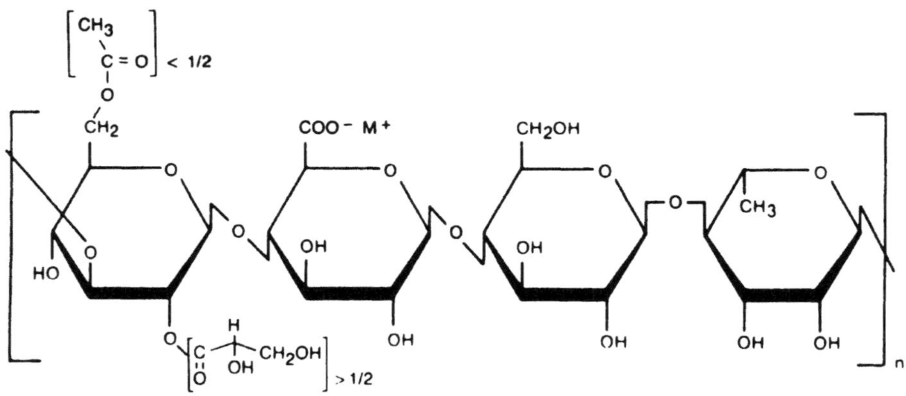

Fig. 231. Chemical structure of native gellan. Reproduced from Food Gels [Ref. 582] by the courtesy of the author and of Chapman & Hall

13.2 Optical Rotation

Evidence of a conformational transition comes mainly from optical rotation studies [585, 587, 598–601]. At room temperature the gellan chains are in low concentrated aqueous solutions (<0.1%) in the coiled state. Addition of 1 : 1 electrolytes up to approximately 0.08 mol/l brings about a conformational disorder → order transition [584, 593, 598, 599, 602], which becomes clear from the change in optical rotation (see Fig. 232a). The rate of the transition is very high. At 25 °C the transition takes place around 0.01 mol/l of 1: 1 electrolyte and is accompanied by an enthalpy of mixing [598] (see Fig. 232b). The transition point depends on the nature of the cations and of the anions as well, as is shown in Fig. 233. The transition temperature is slightly dependent on the salt concentration. If the salt concentration is sufficiently low, gelation does not take place. However, at the highest NaCl concentrations (50 and 65 mmol/l) the tendency of gellan to aggregate is discernable through the upward curvature at low temperatures. From this figure it is also clear that the transition is thermoreversible without any thermal hysteresis, because the cooling curves completely coincide

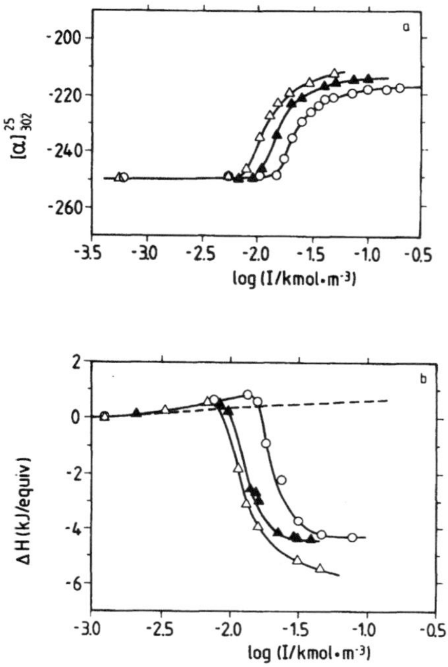

Fig. 232. a Gellan optical activity (302 nm, 25 °C) in dilute solution vs total ionic strength. **b** Enthalpy of mixing gellan solutions with 1:1 salt solutions vs total ionic strength at 25 °C. (△) NaCl; (▲) Me_4NCl; (○) Me_4NBr. Gellan concentration: 1.2 equiv/m^3 (0.09% w/v). Reproduced, with permission of the authors and publishers, from [598]

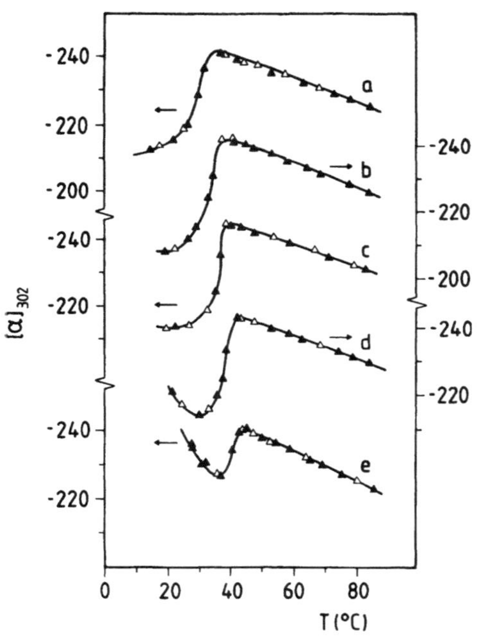

Fig. 233. Optical activity (302 nm) of sodium type gellan for different NaCl concentrations vs temperature. Salt molarity (in mmol/l): (a) 20; (b) 30; (c) 42; (d) 50; (e) 65; (▲) heating; (△) cooling; gellan concentration 1.2 equiv/m^3 (0.09% w/v). Reproduced, with permission of the authors and publishers, from [598]

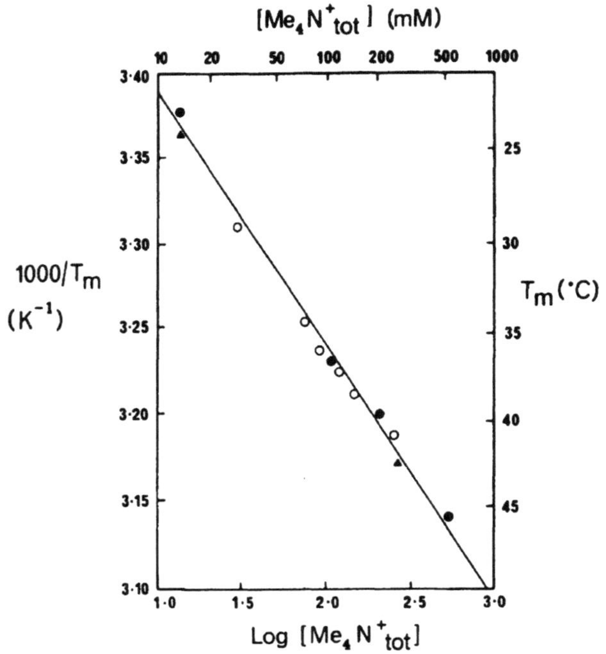

Fig. 234. Midpoint temperature T_m for the conformational transition of Me_4N^+ gellan in solution vs cation concentration $[Me_4N]^+_{tot}$ (counter-ions to the polymer + added Me_4NCl). (○) Data by Crescenzi et al. using optical rotation at 302 nm; gellan concentration 0.08 wt%; (●) and (▲) Data obtained by Robinson et al. using optical rotation at 436 nm and DSC; gellan concentration 1.0 wt%. Reproduced from Food Polymers, Gels and Colloids [Ref. 603] by the courtesy of the authors and The Royal Society of Chemistry, Cambridge, UK

with the heating curves. In Fig. 234 the transition temperature T_m (i.e. the mid point temperature of the transition), as obtained by Robinson et al. [603] with the aid of optical rotation and DSC experiments, is plotted vs the total cation concentration (i.e. counter ions to the polymer + added salt). They are in close agreement. It is remarkable that data obtained by Crescenzi et al. [584, 593], that are also plotted in this figure, are in excellent agreement with the results by Robinson et al., despite the difference in gellan concentration (0.08% and 1% respectively). As one might expect, the transition temperature is also a function of concentration, so it is not clear why the results agree so well. Approximately the same I-T_m dependence was reported by Cesàro et al. [604], who combined literature data of the transition temperature for gellan-Me_4NCl systems on the basis of measurements of optical rotation, calorimetry and intrinsic viscosity [584, 587, 593, 598]. Transition temperatures of gellan-NaCl systems were shifted to higher temperatures over approximately 3 °C. It is not really clear what the meaning is of these results. For co-operative conformational changes of linear polyelectrolytes the slope of these plots (log I vs $1/T_m$) is, according to Paoletti et al. [605], theoretically related to changes of the non-ionic enthalpy contribution and of the linear charge density. This would mean that these contributions are constant, independent of salt concentration and temperature, and equal for both NaCl-gellan and Me_4NCl-gellan systems.

From their studies on optical rotation, light scattering, intrinsic viscosity and electrical conductivity, Milas et al. [587] concluded that the conformational transition, controlled by temperature and ionic strength, corresponds to a single chain-double-helix transition:

$$2 \text{ single coils} \rightleftharpoons 1 \text{ double-helix.}$$

Aggregation of double-helices, which is promoted by the addition of salts, leads to gel formation. The influence of salts is rather complicated: they promote the aggregation of helices at lower temperature, whilst they reduce only the coil dimensions at higher temperature. The selectivity only exists in the gelation regime and depends strongly on the nature of the cation [587]:

$$(H_3C)_4N^+ < Li^+ < Na^+ < K^+.$$

This is different from the ion selectivity in kappa carrageenan, where it also exists in the disordered conformation [543], but similar to aqueous agarose systems.

13.3 Effect of Esterification

It is known that the gelation of native gellan is hindered by the presence of acetyl groups. Upon deesterification stronger gels are obtained. Crescenzi et al. [601] esterified the carboxyl groups of gellan gum with ethyl or propyl iodide. In Figs. 235 and 236 results of optical rotation as functions of added tetramethyl ammonium chloride and temperature are shown for gellans of various degrees of esterification [601]. With increasing degrees of esterification, higher ionic strengths are needed to induce the conformational changes and for de = 0.17 the conformational change is hardly detectable. The thermal stability is also reduced by esterification. From Fig. 236 it appears that the transition is thermoreversible without thermal hysteresis. Similar results were obtained with the propyl esters: for de ≥ 0.17 no transition was perceptible.

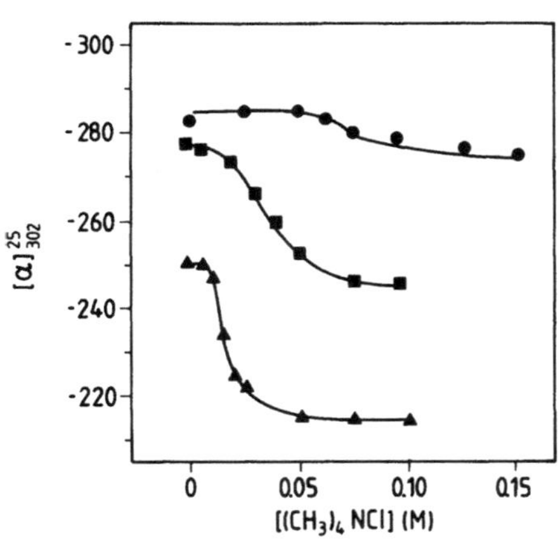

Fig. 235. Dependence of $[\alpha]^{25}_{302}$ of 0.1% gellan ethyl esters on Me_4NCl concentration, for different degrees of esterification: (●) de = 0.17; (■) de = 0.08; (▲) de = 0 (i.e. gellan). Reproduced from Carbohydr Res [Ref. 601] by the courtesy of the authors and Elsevier Science Ltd, The Boulevard, Langford Lane, Kidlington 0X5 1GB, UK

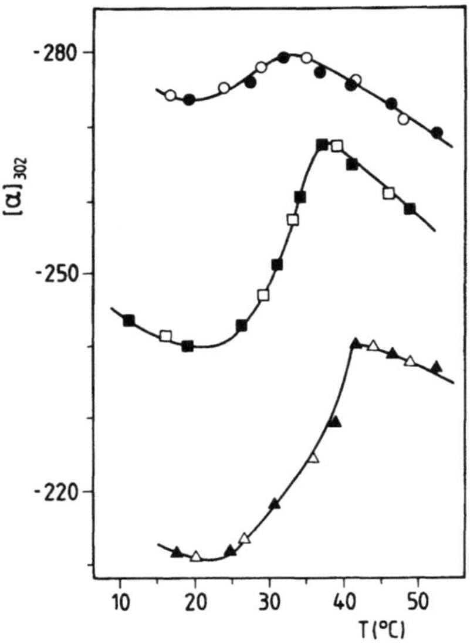

Fig. 236. Temperature dependence of $[\alpha]_{302}$ of 0.1% gellan ethyl esters in 100 mmol/l Me_4NCl, for different degrees of esterification: (●) and (○) de = 0.17; (■) and (□) de = 0.08; (▲) and (△) de = 0 (i.e. gellan); *full symbols*: heating; *open symbols*: cooling. Reproduced from Carbohydr Res [Ref. 601] by the courtesy of the authors and Elsevier Science Ltd, The Boulevard, Langford Lane, Kidlington 0X5 1GB, UK

13.4 Network Model

Many authors suggest that gelation occurs by the formation of double-stranded helices, followed by association [586, 587, 596, 597, 603, 606–610]. The junction zones seem to result from parallel alignment of the helical zones. Based on their studies on the conformational and physical properties of gellan solutions (optical rotation, viscoelasticity, NMR, DSC and intrinsic viscosity) and their dependence on ionic strength, Robinson et al. [603] suggested a gelation mechanism in aqueous solutions with or without added salt. The process is affected by molecular weight, polymer concentration, the origin of the gellan and by nature and concentration of added salt. Important results, that led the authors to this mechanism, are the calorimetric data shown in Fig. 237, where it becomes clear that in the presence of a relatively large amount of salt (NaCl) the endotherm in the heating process is split into two peaks: the low temperature peak is attributed to the uncoiling process of double-helices and the high temperature peak to the uncoiling process of cation mediated aggregates. If not enough salt is present to stabilize the aggregation of helices, then a weak gel is formed with weak, saltless aggregates of double-helices (see Fig. 238).

In a cooling process of gellan solutions, in the presence of salt, double-helices and cation mediated aggregates of double-helices are formed, whereas in the absence of (a sufficient amount of) salt only non-stabilized aggregates of double-helices are formed.

A practically similar model was suggested by Morris VJ and Gunning [596, 597] on the basis of their light scattering studies of gellan in the presence of tetramethyl ammonium chloride.

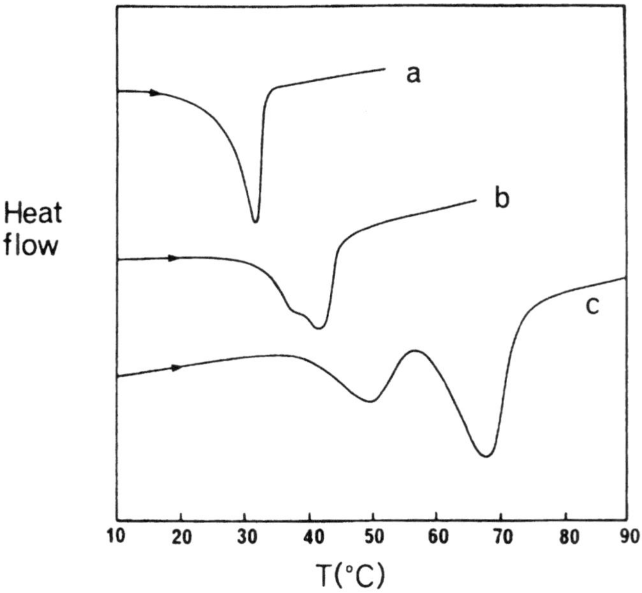

Fig. 237. DSC heating scans (0.1 K/min) for Na^+-gellan (1.0 wt%), showing the progressive development of biphasic melting behaviour with increasing total concentration of Na^+ (counter-ions to the polymer, 15 mmol/l + added NaCl): (a) 25 mmol/l; (b) 67 mmol/l; (c) 115 mmol/l. Reproduced from Food Polymers, Gels and Colloids [Ref. 603] by the courtesy of the authors and The Royal Society of Chemistry, Cambridge, UK

13.5 X-ray Diffraction

From X-ray diffraction [586, 606, 607, 610–620] of stretched fibres prepared from native gellan gum, it is evident that alignment of helical molecules is good but that the crystallinity is poor. On the other hand, deesterification causes increased crystallinity without a detectable change in the type or dimensions of the gellan helix.

X-ray studies on the lithium salt of gellan itself [612] revealed that the molecules exist as double-helices composed of two polysaccharide chains of the same polarity, having left handed, 3-fold helix symmetry and a pitch of 5.64 nm, intertwined about and staggered by 2.82 nm from each other along a common helix axis. The gellan mono-helix is stabilized by three intramolecular hydrogen bonds per tetrasaccharide unit, all of them involving the D-glucuronic acid residue and the D-glucose residues. By means of the interchain hydrogen bonds between the D-glucose (C) and the D-glucuronate residues, involving the COOH group, the double-helix is strengthened [586, 612, 613, 620].

Later on it was shown that this also occurs in gellan types of other monovalent cations (e.g. Na^+, K^+ and Rb^+) [620]. In native gellan the acetate groups are positioned on the exterior of the double-helix so they do not affect detectably the main chain conformations. This is not the case with the L-glycerate groups,

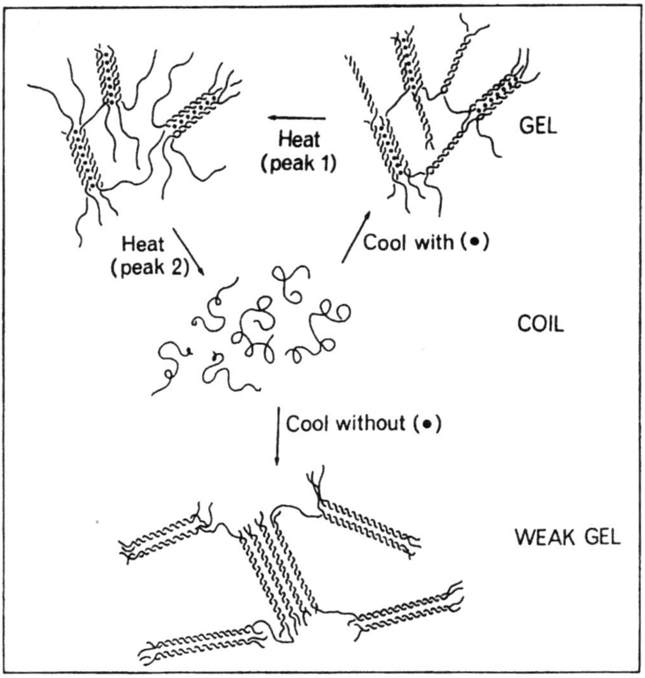

Fig. 238. Schematic representation of the model proposed by Robinson et al. for the conformation and functional interactions of gellan, showing (*top*) the processes envisaged on cooling and reheating in the presence of cations (●) that promote gel formation and (*bottom*) the "weak gel" network developed in absence of those cations (or in insufficient amount present to stabilize the association of double-helices). Reproduced from Food Polymers, Gels and Colloids [Ref. 603] by the courtesy of the authors and The Royal Society of Chemistry, Cambridge, UK

which are bigger and positioned in a crowded environment at the inner side of the double-helix, close to the COOH groups [581, 618]. On the other hand the acetate groups will hinder the aggregation of the double-helices. Hence, deesterification will affect the gelation tendency positively.

Aggregates are formed by double helix-water-double helix and through (double-helix)-K^+-water-K^+-(double-helix) complexes; Ca^{2+} salts show strong gelation promoting behaviour, due to its ability to (double-helix)-Ca^{2+}-(double-helix) interactions [586, 612–614, 618, 619].

The reason that KCl influences the viscoelastic properties of gellan more effectively than NaCl might be their differences in structure ordering properties of water [609]. Sodium ions belong to a group of structure ordering cations for water, and potassium to a group of structure disordering cations [621]. Hence, K^+ ions can shield the carboxyl groups of gellan more directly than Na^+ ions do. This is similar to aqueous kappa carrageenan gels [522, 574, 623].

13.6 Viscoelastic Behaviour

13.6.1 General Behaviour

A general view of viscoelastic behaviour of gellan solutions at low salt concentrations is reported by Robinson et al. [603] and shown in Fig. 239. At high temperature (30 °C) and low Me_4NCl concentration (100 mmol/l) the system behaves like a shear thinning polymer solution, typical for polysaccharide random coils. Also the Cox-Merz-rule (i.e. the complex viscosity $\eta^*(\omega) = G^*/\omega$ is equal to the steady shear viscosity $\eta(\dot{\gamma})$ for $\omega = \dot{\gamma}$) holds under these circumstances. At temperatures below the transition temperature, the solutions show weak-gel properties: $G' > G''$, both depending only slightly on frequency (slope in a log-log-plot approximately 0.2). The complex viscosity, η^*, is substantially larger than the steady shear viscosity, $\eta(\dot{\gamma})$, because in steady shear experiments the gel will be disrupted. Similar data were reported by Cesàro et al. [604].

13.6.2 Temperature and Concentration Dependence

The dynamic moduli of aqueous gellan solutions of various concentrations, as measured during cooling experiments [622], are shown in Fig. 240. G′ increases rather steeply at a certain temperature, which depends on concentration. The increase becomes less steep with decreasing concentration and shifts to lower temperatures. The gelation temperature was defined as the temperature, T_s, of the inflection point in the G′-temperature curve. According to the authors, an

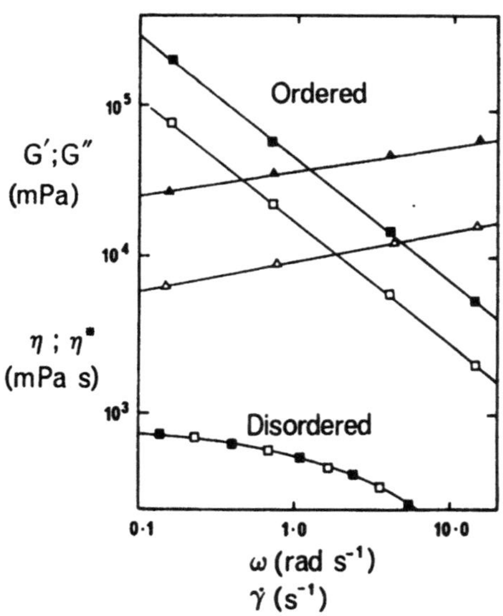

Fig. 239. Viscoelastic functions of Me_4N^+-gellan (1 wt%) in the disordered coil form (100 mmol/l Me_4NCl; 30 °C) and in the ordered state (250 mmol/l Me_4NCl; 5 °C); (▲) G′; (△) G″; (■) η* and (□) η (both viscosities in the ordered and disordered state). Reproduced from Food Polymers, Gels and Colloids [Ref. 603] by the courtesy of the authors and The Royal Society of Chemistry, Cambridge, UK

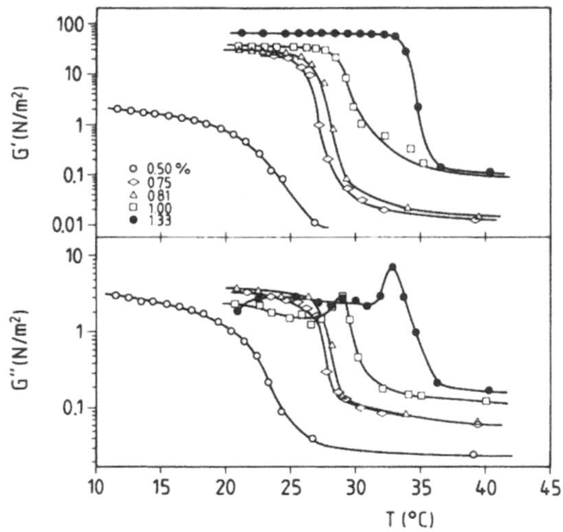

Fig. 240. Temperature dependence of the storage moduli G′ and the loss moduli G″ upon cooling gellan solutions of various concentrations; T_s is the temperature at the inflection of the cooling curves for G′; frequency 1 Hz; cooling rate 0.27 °C/min. Reproduced from Food Hydrocoll [Ref. 622] by the courtesy of the authors and of Oxford University Press

Eldridge-Ferry [160] plot (ln c vs $1/T_s$) yielded a heat of reaction of − 64 kJ/mol.

13.6.3 Time Dependence

The time dependence of the development of the dynamic moduli, after quenching to a certain temperature, shows a quite normal gradual increase of the dynamic moduli for low concentrations (0.5 and 0.75%). At higher concentrations (0.81 and 1%) a maximum in G′ is observed. This anomalous behaviour is also observed upon quenching a 0.8% solution to various temperatures: the development of G′ with time looks quite normal at temperatures of 28 °C and higher. On the other hand, at lower temperatures (15 and 20 °C) the development is much slower and seems to reach a maximum value. One could imagine that this anomalous behaviour is the result of slippage in the rheometer, due to syneresis of the gellan network, or of local network fracture as the authors suggest.

13.6.4 Dependence on Electrolyte Concentration

Many results of the dependence of viscoelastic and calorimetric behaviour on electrolyte concentration are reported by Nishinari et al. [579, 588–592]. These authors defined [588] the sol-gel transition temperature, T_{gel}, of gellan solutions as the temperature where, in a cooling experiment, the storage modulus starts to deviate from the base line (see Fig. 241). In comparison with the method of the

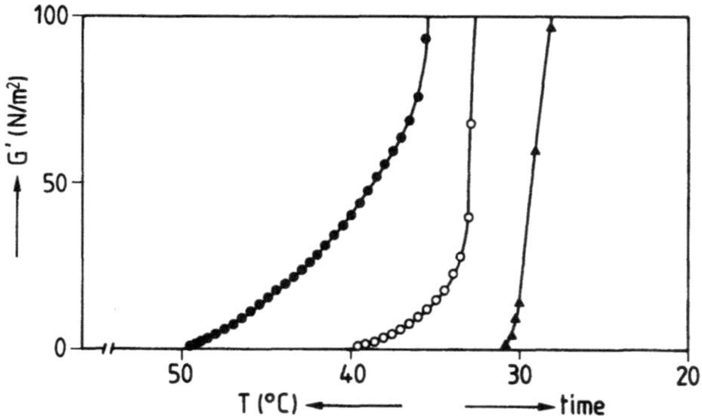

Fig. 241. Storage modulus vs temperature upon cooling gellan solutions of three concentrations: (▲) 0.99 wt%; (○) 1.28 wt%; (●) 1.48 wt%; small strain amplitude and frequency 2 Hz. Reproduced from Food Hydrocoll [Ref. 588] by the courtesy of the authors and of Oxford University Press

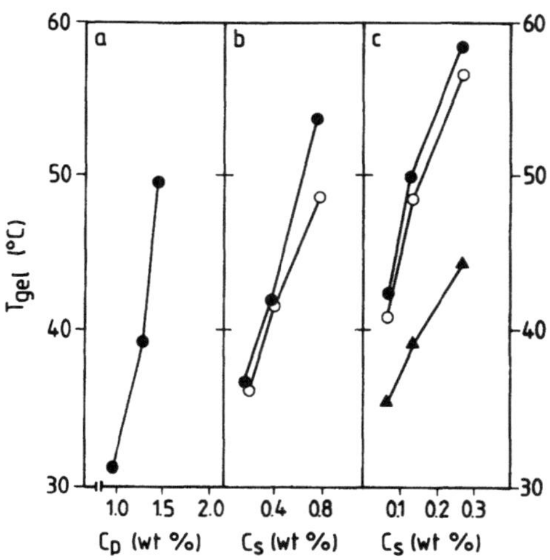

Fig. 242a–c. Sol-gel transition temperatures as functions of gellan concentration and salt concentration: **a** gellan without salt; **b** 0.4 wt% gellan containing NaCl (○) and KCl (●); **c** 0.2 wt% gellan containing $MgCl_2$ (○), $CaCl_2$ (●) and Ca-lactate (▲). Reproduced from Food Hydrocoll [Ref. 588] by the courtesy of the authors and of Oxford University Press

inflection temperature T_s, as used by Nakamura et al. [622], T_{gel} will be higher, especially at lower gellan concentrations. Hence, one would expect a more negative value for ΔH_{gel}. Results of T_{gel} as functions of polymer concentration and salt concentrations are shown in Fig. 242. Monovalent metal ions like K^+ and Na^+ have a rather strong influence on the gelation temperature and K^+ slightly more than Na^+, especially at the higher concentrations. The influence of divalent metal ions like Ca^{2+} and Mg^{2+} is much more pronounced. As was mentioned before, the divalent ions form strong, direct bonds between the double-helices. From Fig. 242c it is interesting to note that the influence of $Ca[CH_3CH(OH)COO]_2$, i.e. Ca-lactate, is less strong than that of both metal chlorides. Apparently it is not possible for the big lactate counter ions to find a position in the (double-helix)-Ca^{2+}-(double-helix) structure. As a result, it is also not easy for the Ca^{2+} ions to find a position in the structure without their counter ions. The phenomenon that at high salt concentrations, the equilibrium rubber modulus of the gels decreases again is remarkable. At this moment no explanation is available for this apparent discrepency, but it possibly originates from syneresis.

The gel melting temperature, as measured with the falling ball method, is much higher than the gelation temperature, as might be expected for gels with crystalline crosslinks. From Eldridge-Ferry [160] plots ($\ln c_{pol}$ vs $1/T_m$ and vs $1/T_{gel}$) (see Fig. 243), ΔH, the heat of reaction for crystallite formation of the gellan gels, was calculated to be $\Delta H_{gel} = -18$ kJ/mol. This is considerably less negative than -64 kJ/mol as reported by Nakamura et al. [622], notwithstanding the comment made before. The heat of melting of the crystalline crosslinks was calculated to be $\Delta H_m = 38$ kJ/mol. Apparently, in the maturing process of gellan gels, the crystallites grow, causing an increase in the melting enthalpy with respect to the gelation enthalpy.

Melting the gelation enthalpies appear to be strongly dependent on the sample used: Watase and Nishinari [589] also reported ΔH_m values of 76.6 kJ/mol for a potassium type of gellan (F1) and 132.2 kJ/mol for the same gellan in the presence of 100 mmol/l KCl. The gellan used in Fig. 243 is also of the potassium type. It is well-known that the molecular weight also plays an important role in influencing the melting enthalpy. Unfortunately, molecular weights were not reported.

Figure 244 shows the temperature dependences of the storage Young modulus, E′, and the loss tangent, tan δ, as functions of temperature for gels of various gellan concentrations and whether or not in the presence of 100 mmol/l KCl. The gels were matured at low temperatures. It has to be emphasized that the data of E′, measured at a frequency of 2.5 Hz, are not the equilibrium rubber moduli. Nevertheless, the increase with temperature is even more than might be

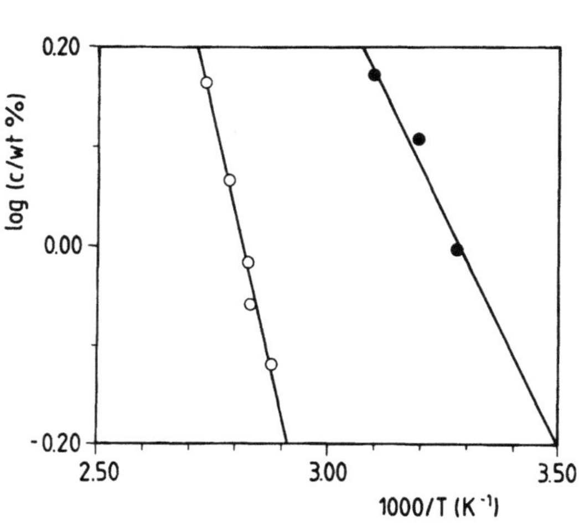

Fig. 243. Eldridge-Ferry plots for gellans gels: (○) melting temperature T_m and (●) gelation temperature T_{gel} (constructed from data presented in [588])

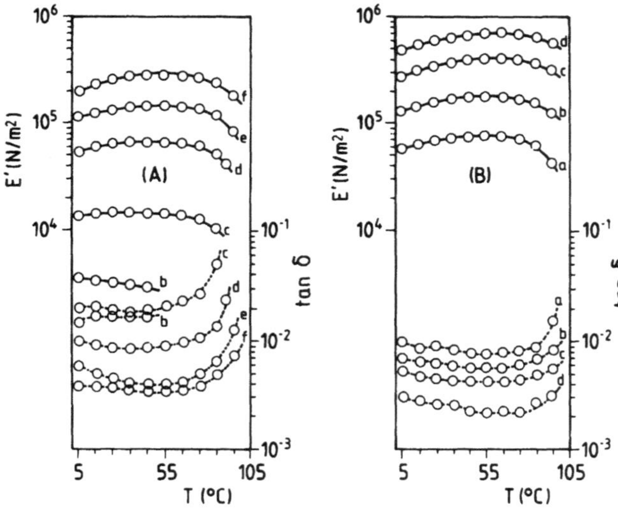

Fig. 244A, B. Temperature dependence of the storage Young modulus E′ and the loss tangent tan δ at 2.5 Hz and small strain amplitude: **A** for a potassium type gellan (called F1); **B** for the same gellan in the presence of 100 mmol/l KCl (called F3); gellan concentration in wt%: (a) 0.7; (b) 1.0; (c) 1.5; (d) 2.0; (e) 2.5 and (f) 3.0. Reproduced from Food Hydrocoll [Ref. 589] by the courtesy of the authors and of Oxford University Press

expected for Gaussian rubber elasticity. The authors suggest that the elasticity might be described by Langevin chains, just as was done for kappa carrageenan gels [525].

In general, the storage modulus increases gradually up to a maximum value and then levels off rapidly. The temperature of the maximum increases with increasing polymer concentration. From the figure it also becomes clear that the moduli of the gellan gels in the presence of an excess of KCl are higher by a factor of 10–30 with respect to the aqueous potassium type gellan gels. This factor decreases with increasing gellan concentration.

The exponent n in the concentration dependence of the storage modulus, $E = Ac^n$, is much larger than 2. However, the values of E′ measured at 2.5 Hz are in general not the equilibrium rubber plateau values. Hence, it is difficult to interpret these high values of n.

Very recently Nishinari et al. [590–592] reported their extensive studies into the influence of alkali metal ions on the viscoelastic and calorimetric behaviour of gellan gels. The gellan used was a sodium type gellan. Concentration (1, 2 and 3%) dependence of G′(ω) and G″(ω) was reported, as well as the influence of the addition of NaCl and KCl (5–100 mol/m^3) and of $MgCl_2$ and $CaCl_2$ (0.43–6.8 mol/m^3). Some results are shown in Figs. 245–247, where the dynamic moduli G′ and G″ of 1% gellan solutions are shown at temperatures of 0 and 25 °C and whether or not in the presence of 50 mmol/l NaCl. Some general results are as follows.

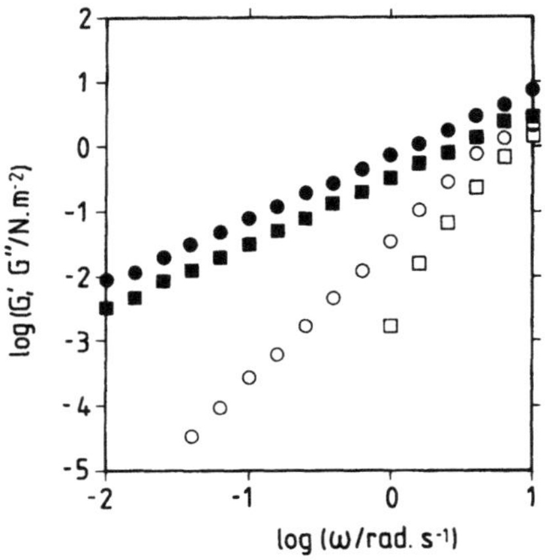

Fig. 245. Dynamic shear moduli vs angular frequency for a 1% gellan solution at 25 °C without and with 50 mmol/l NaCl; *circles*: without NaCl; *squares*: with 50 mmol/l NaCl; (○) and (□) G′; (●) and (■) G″ (constructed from data presented in [590])

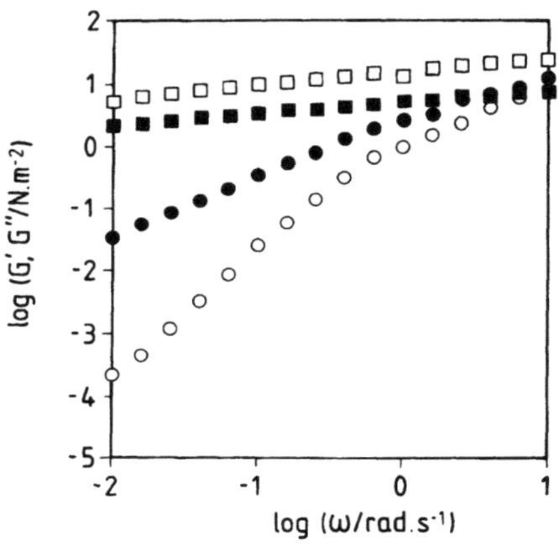

Fig. 246. Dynamic shear moduli vs angular frequency for a 1% gellan solution at 0 °C without and with 50 mmol/l NaCl; *circles*: without NaCl; *squares*: with 50 mmol/l NaCl; (○) and (□) G′; (●) and (■) G″ (constructed from data presented in [590])

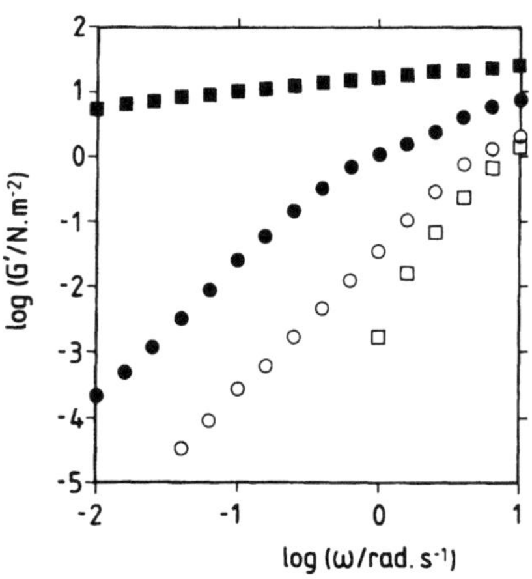

Fig. 247. Storage modulus vs angular frequency for a 1% gellan solution without and with 50 mmol/l NaCl; *open symbols* 25 °C; *filled symbols* 0 °C; (○) and (●) without NaCl; (□) and (■) with 50 mmol/l NaCl. Reproduced, with permission of the authors and publishers, from [592]

a) At low concentration (1%) the aqueous gellan system behaves like a dilute polymer solution: $G' < G''$ and both moduli strongly dependent on frequency; at a gellan concentration of 3% the system behaves gel-like at temperatures below 25 °C ($G' > G''$ and both moduli only slightly dependent on frequency, slope approximately 0.3 in a log-log-plot), due to intermolecular bondings with relatively low bonding energies.
b) At high temperature (25 °C) the system behaves like a polymer solution; the addition of NaCl has a decreasing effect on the dynamic moduli; at low temperature (0 °C) the aqueous system still behaves like a polymer solution, whereas by the addition of 50 mmol/l NaCl a gel-like system is obtained.
c) The addition of salt affects the viscoelastic behaviour in two possible ways: as long as the addition of salt does not initiate a gelation process, it only influences the gellan coil dimensions by a coil collapse, resulting in lower values of the dynamic moduli; on the other hand, if a gel is formed, then the presence of salt causes the moduli to increase and to become less frequency dependent: a horizontal plateau arises in the applied frequency range; this resembles the influence of salt in aqueous agarose systems.
d) Similar results were obtained in other salt solutions: approximately the same results for 1% gellan solutions were obtained in 22 mmol/l KCl, 2 mmol/l $MgCl_2$ and 1.5 mmol/l $CaCl_2$ solutions. Hence, the effects of the ions on the gelation process are: $Ca^{2+} > Mg^{2+} \gg K^+ > Na^+$; the increase in G' is more remarkable for K^+ than for Na^+ and for Ca^{2+} more than for Mg^{2+}. The role of divalent cations seems to be almost the same as that of monovalent cations in kappa carrageenan [588]. The shielding effect of the divalent cations is much larger than that of the monovalent cations. This is similar to aqueous kappa carrageenan gels and solutions, where the electrostatic repulsion of the sulfate groups is shielded more effectively by divalent than by monovalent cations [624].

DSC experiments revealed interesting results: cooling curves of 1% gellan solutions without salt showed an exothermic peak at $T_s = 29$ °C, whereas by subsequent heating from 5 °C an endothermic peak at 30.5 °C was observed. The difference of 1.5 °C was attributed to the scan rate: by extrapolation of data, measured at various scan rates, to zero scan rate, equal values for cooling and heating scans were obtained (28.6 °C). So thermal hysteresis is not present. With increasing concentration of NaCl or KCl, T_s increased up to 45 °C (100 mmol/l KCl), and the endothermic peak to about 80 °C. Hence, a pronounced thermal hysteresis is observed in aqueous gellan-salt systems. Upon cooling a solution in the presence of high concentrations of $MgCl_2$ or $CaCl_2$ (6.8 mmol/l) rather sharp exothermic peaks were observed, just as in NaCl and KCl solutions. However, the heating curves only showed a very broad endothermic peak, varying from 25 to 100 °C. This could be the result, as was suggested by the authors, of the presence of junction zones of varying stability. The thermal hysteresis in the presence of salt is in agreement with the differences in the temperature dependence of the storage modulus in

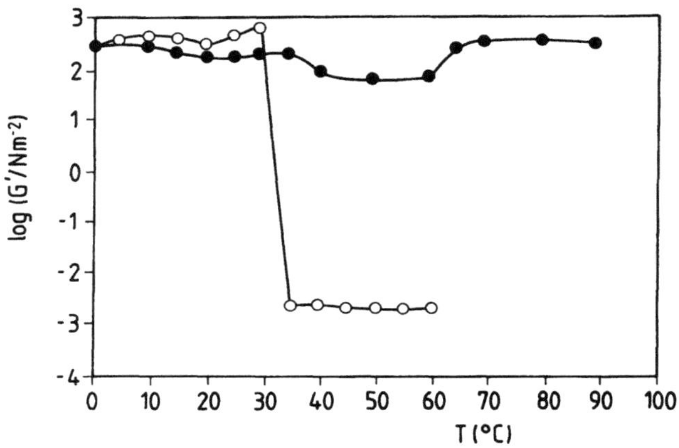

Fig. 248. Temperature dependence of the storage modulus during cooling (○) or heating (●) processes for a 1% gellan solution in the presence of 5.1 mmol/l $CaCl_2$; angular frequency 1.0 rad/s. Reproduced from Food Hydrocoll [Ref. 590] by the courtesy of the authors and of Oxford University Press

cooling and heating processes, as shown in Fig. 248 for a gellan solution in 5.1 mmol/l $CaCl_2$.

13.6.5 pH Dependence

Recently Moritaka et al. [579] investigated the pH dependence of the gelation process of gellan gums. DSC cooling curves for gellan solutions, whether or not in the presence of extra salt, did not show any peak at pH = 2. At pH = 4 an exothermic peak was observed, but at lower temperature than observed at pH values of 6–10. One would expect that at low pH gelation occurs, because in that case the carboxyl groups in gellan will not be ionized and will therefore not inhibit the gelation process. However, at pH = 2 the system shows phase separation and becomes turbid. Apparently ionized carboxylic groups play an important role in the gelation possibilities, because in that case interaction via cations (H^+, Li^+, Na^+, K^+ etc.) may result in the formation of junction zones.

13.7 Conclusions

Gellan gum was developed less than two decades ago. Despite its short existence, many studies concerning its gelation behaviour in aqueous solutions have been published. Many different techniques were used to characterize these systems. Networks are the result of the formation of junction zones consisting of double helices, which is controlled by temperature and by ionic strength of the solution, followed by aggregation, which is promoted by alkali metal salts. The gellan monohelix is stabilised by three intramolecular hydrogen bridges per tetrasaccharide repeating unit, whereas intermolecular hydrogen bonds strengthen the double-helix. The differences in the influences of alkali metal ions on the rheological behaviour of the solution and the gel state resemble their effects in aqueous agarose systems.

Viscoelastic behaviour has been studied relatively intensively for such a new polymer. It is affected by concentration, type of gellan (i.e. molecular weight), time, temperature, ionic strength and pH. Due to the pH dependence of the degree of ionisation of the carboxylic acid groups, repulsion of the polymer chains decreases with decreasing pH. Hence, by decreasing the pH, gelation will be promoted due to reduced solubility. At pH = 2 phase separation occurs and the gelation propensity is disturbed.

Despite the relatively large number of studies on the gelation behaviour of aqueous gellan gum solutions, much work, especially on their viscoelastic behaviour, has still to be done. The number of parameters that affect the gelation behaviour is even more than in aqueous agarose and carrageenan solutions, because the pH, which affects the gelation behaviour of gellan gum solutions, does not play a role in the degree of ionisation of the sulfate groups in agarose and carrageenan solutions.

A severe complication of the measurement of the viscoelastic behaviour of gellan gum gels is the occurrence of syneresis.

14 Concluding Remarks

The measurement of viscoelastic properties appears to be a powerful tool in the characterization of gel networks. In many cases viscoelastic properties are much more sensitive to relatively small changes in the network structure than other, independent techniques can detect.

In this review the viscoelastic properties of *thermoreversible* gels have been described. If one restricts oneself to this subject in the wide field of gel networks, the amount of available literature is rather limited. First, many physical gels are not thermoreversible and, second, the study of viscoelastic properties of those gels seems to be an disencouraging activity. In fact, incorrect or inadequate measurement programs are frequently used. In many cases the temperature dependence of the dynamic moduli of gels e.g. is measured at one frequency only and in the respective figures this temperature dependence is clearly shown. However, the presence of rubberlike behaviour cannot be concluded from these experiments and thus calculations on the basis of rubber elasticity to elucidate network parameters will not give reliable results. Determination of the frequency dependence of these moduli, a rather easy experiment during gradual temperature increase or decrease, would lead to a better understanding of the network behaviour.

The ageing, annealing or maturation behaviour of these systems depends in many cases on their temperature history. For that reason, it is important to be able to quench a solution to a desired temperature. In many instruments this is not possible in a proper way. Rheological measurement techniques are available in many different geometries. However, the most convenient way to study a fast ageing system would be to make use of a Couette system, where the sample is present between two coaxial cylinders. Thermostatting this sample is relatively easy by a thermostat mantle that is connected to a pumping thermostat. A temperature quench is easily reached by making use of several connected pumping thermostats. This method is in general not easily applicable for e.g. cone and plate systems.

Detailed viscoelastic studies have only been performed on a few thermoreversible gel systems, namely on gels of poly(vinyl chloride), poly(vinyl alcohol) and gelatin. On the other hand, investigations of gels of polymethyl methacrylates, polystyrenes, some block copolymers (in particular styrene-isoprene-styrene triblock copolymers and ethylene oxide-propylene oxide-ethylene oxide triblock copolymers) and of the biopolymers agarose, carrageenan and gellan gum, mostly studies of viscoelasticity, have been carried out in a more or less haphazard way, without a systematic approach. Finally, in the study of the gelation behaviour of solutions of polyacrylonitrile, polyethylenes and liquid crystalline polymers, the description of viscoelasticity is an underdeveloped area.

This great variety in the amount of literature on viscoelasticity made it impossible to compose a uniform review where every section has the same structure.

The measurement of viscoelastic properties can be an exciting study of the properties of gels, with many unexpected results. Examples mentioned in this review are:

a) the temperature and time dependence of the ageing process of poly(vinyl chloride)/plasticizer systems and in particular of special temperature histories, where it was even found that a double rubber plateau could be developed;
b) the same studies for aqueous gelatin, where it also appeared that relaxation takes place after a long time or at very low frequencies;
c) the strong dependence of the gelation ability of polyethylenes or poly(vinyl chloride) on chlorination and the wide difference in the effects of chlorination on these polymers, whether this chlorination took place in the solid phase or in the dissolved state;
d) the difference in gelation ability of poly(vinyl chloride) polymers obtained after extraction (i.e. the extracted and the residue poly(vinyl chlorides));
e) the strong dependence of the propensity to form gels on the previous dissolution temperature or the rejuvenating temperature of existing gels; in general this propensity depends on the disappearance of nuclei: just above the gel melting temperature crosslinks are still present, but they are too few to form a network; examples of these studies are gelatin and poly(methyl methacrylates);
f) the temporary character of aqueous poly(vinyl alcohol)/borate gels;
g) in many cases the Winter-Chambon method has been succesfully applied in recent years to the determination of the gel point of thermoreversible gels, although it never can be proven that the gel point obtained in this way really is the gel point of the system.

From these investigations, many conclusions can be drawn about network development. Although it appears that the study of viscoelastic properties is a powerful tool in the determination of the occurrence of changes in structure, it is in general not possible to fully characterize the network structure by this method only. For that purpose additional, independent techniques are needed, e.g. small-angle X-ray scattering, small-angle neutron scattering, optical rotation (dispersion), infrared spectroscopy, and nuclear magnetic resonance spectroscopy. By combination of results obtained from rheological investigation and at least one of these techniques, insight into the crosslinking process and the crosslink structure can be deepened tremendously.

Moreover, network models are needed to interpret the results obtained. In this review, fruitful use has been made of the model developed by the author for the process for crosslinking polydisperse polymer with multifunctional crosslinks. In fact, the model has been developed to be able to calculate network parameters of many of the gels described in this review. In particular, interesting results were obtained in the description of:

a) the crosslinking structure of PVC/plasticizer gels (in combination with SAXS);

b) aqueous gelatin gels (in combination with optical rotation);
c) the development of gelation kinetics and thermodynamics of aqueous gelatin gels;
d) aqueous PVA/borate gels (in combination with the protolytic constants of boric acid);
e) aqueous PVA/congo red gels (in crosslinking kinetics).

Numerous, less striking calculations were carried out to elucidate network parameters.

For many systems however, additional results to determine the network structure are not available. On the other hand, e.g. the gelation behaviour of aqueous PVA solutions was characterized by small-angle neutron scattering. In that way, it was possible to calculate the number of (multifunctional) crystalline crosslinks. Unfortunately equilibrium shear moduli were not reported, so that the functionality of the crystalline crosslinks, like in PVC, could not be calculated.

The general conclusion is that cooperation of polymer rheologists and polymer scientists working in other disciplines is very necessary. Only in that way can a synergistic result in the characterization of thermoreversible networks be obtained.

Acknowledgements. The author is much indebted to the following scientists for their valuable comments, remarks and advice preceeding the submission of this review: R. Addink, Z. Bashir, H. Berghmans, W. Burchard, G. Challa, S. Hvidt, P.F. Mijnlieff, K. Nishinari, S.B. Ross-Murphy, J. Spěváček and F.A. de Wolf, and in particular, to the editor Karel Dušek for his continuing interest and his unbelievable patience.

15 Appendix

For network formation with crosslinks of functionality f the overall equilibrium

$$1 \text{ crosslink} \rightleftharpoons \tfrac{1}{2} \text{ f potential crosslinking sites} \tag{A1}$$

is considered. The matching equilibrium constant K is given by

$$K = \frac{[\text{crosslinks}]}{[\text{potential junction sites}]^{0.5f}} \tag{A2}$$

If $\bar{\gamma}_w$ is the average number of crosslinks per weight average primary polymer molecule, or the number of junction zones (incorporated in crosslinks) per weight average primary polymer molecule, then the concentration on crosslinks is given by

$$[\text{crosslinks}] = \frac{\bar{\gamma}_w}{0.5f} \frac{c}{\bar{M}_w}. \tag{A3}$$

The number of potential junction zones per primary polymer molecule still present is equal to the maximum number per primary polymer molecule minus the number already present in crosslinks, hence equal to $\bar{\gamma}_{w,max} - \bar{\gamma}_w$, so that their concentration is equal to $(\bar{\gamma}_{w,max} - \bar{\gamma}_w) \times (c/\bar{M}_w)$. At the gel point, $\bar{\gamma}_w$ and c have to be replaced by $\bar{\gamma}_{w,g}$ and c_o respectively. On substituting these expressions into Eq. (A1) we obtain

$$K = \frac{\bar{\gamma}_w/0.5f}{[\bar{\gamma}_{w,max} - \bar{\gamma}_w]^{0.5f}\ [c/\bar{M}_w]^{0.5f-1}} \tag{A4}$$

and

$$K = \frac{\bar{\gamma}_w/0.5\,f}{[\bar{\gamma}_{w,max} - \bar{\gamma}_{w,g}]^{0.5f}\ [c_o/\bar{M}_w]^{0.5f-1}} w. \tag{A5}$$

Combination of both equations yields

$$\frac{\bar{\gamma}_w}{\bar{\gamma}_{w,g}} = \left[\frac{\bar{\gamma}_{w,max} - \bar{\gamma}_w}{\bar{\gamma}_{w,max} - \bar{\gamma}_{w,g}}\right]^{0.5f} \left[\frac{c}{c_o}\right]^{0.5f-1}. \tag{A6}$$

On substituting $\gamma_{w,g} = 2/(f-2)$ we obtain

$$\bar{\gamma}_w = \frac{2}{f-2}\left[\frac{\bar{\gamma}_{w,max} - \bar{\gamma}_w}{\bar{\gamma}_{w,max} - 2/(f-2)}\right]^{0.5f} \left[\frac{c}{c_o}\right]^{0.5f-1}. \tag{A7}$$

If we assume that $\bar{\gamma}_{w,max} \gg \bar{\gamma}_w$, then Eq. (A7) can be simplified to

$$\bar{\gamma}_w = \frac{2}{f-2}\left[\frac{c}{c_o}\right]^{0.5f-1}. \tag{A8}$$

For the crosslinking process of uniform polymer, $\bar{\gamma}_w$ is also given by

$$\bar{\gamma}_w = \bar{\gamma} = -\frac{\ln w_s}{1 - w_s^{0.5f}}. \tag{A9}$$

Combination of Eqs. (A8) and (A9) yields a relationship between the reduced concentration and the sol fraction in the equilibrium state. The reduced equilibrium shear modulus G_{er} in the equilibrium state of the crosslinking process of uniform primary polymer [39–44] can be expressed as:

$$G_{er} = \frac{G_e M}{cRT} = 2(1 - w_s)\left[\bar{\gamma}\,\frac{1 - w_s^{0.5f}}{1 - w_s}\,\frac{f-2}{f} - 1\right]. \tag{A10}$$

On substituting Eq. (A8) into Eq. (A10) we obtain

$$G_{er} = \frac{4}{f}\left[\frac{c}{c_o}\right]^{0.5f-1}(1 - w_s^{0.5f}) - 2(1 - w_s). \tag{A11}$$

For relatively high concentrations the sol fraction tends to zero in the equilibrium state. In that case the expression for the reduced equilibrium shear modulus will be

$$G_{er} = \frac{4}{f}\left[\frac{c}{c_o}\right]^{0.5f-1} - 2 \approx \frac{4}{f}\left[\frac{c}{c_o}\right]^{0.5f-1}. \tag{A12}$$

16 References

1. Clark AH, Ross-Murphy SB (1987) Adv Polym Sci 83: 57
2. Guenet J-M (1992) Thermoreversible gelation of polymers and biopolymers. Academic Press, London
3. Almdal K, Dyre J, Hvidt S, Kramer O (1993) Polymer Gels Networks 1: 5
4. Jordan Lloyd D (1926) In: Alexander J (ed) Colloid chemistry I. Chemical Catalog Company, New York, p 767
5. Hermans PH (1949) In: Kruyt HR (ed) Colloid science II. Elsevier, Amsterdam, p 483
6. Ferry JD (1962, 1970, 1980) Viscoelastic properties of polymers. John Wiley, New York, 1st edn p 391, 2nd edn p 557, 3rd edn p 529
7. Beltman H (1975) Doctoral Thesis, Wageningen, The Netherlands
8. Higgs PG, Ross-Murphy SB (1990) Int J Biol Macromol 12: 233
9. Ross-Murphy SB (1992) Polymer 33: 2622
10. Te Nijenhuis K, Dijkstra H (1975) Rheol Acta 14: 71
11. Te Nijenhuis K (1979) Doctoral Thesis Delft, The Netherlands
12. Te Nijenhuis K (1990) In Burchard W, Ross-Murphy SB (eds) Physical networks: polymers and gels, Elsevier Applied Sciences, London, Chapt 2
13. Dorrestijn A, Te Nijenhuis K (1990) Colloid Polym Sci 268: 895
14. Flory PJ (1974) Disc Farad Soc 57: 1
15. Gordon M, Hunter SC, Jove JA, Ward TC (1968) Nature (London) 217: 735
16. Kálal J, Gordon M, Devoy C (1972) Makromol Chem 152: 233
17. Peniche-Covas CAL, Dev SB, Gordon M, Judd M, Kajiwara K (1974) Farad Discuss Chem Soc 57: 165
18. Richardson RK, Ross-Murphy SB (1981) Int J Biol Macromol 3: 315
19. Takahashi A, Sakai M, Kato T (1980) Polym J 12: 335
20. Matsuda H, Kashiwagi R, Okabe M (1988) Polym J 20: 189
21. Adam M, Delsanti M, Pieransky P, Meyer R (1984) Rev Phys Appl 19: 253
22. Adam M, Delsanti M, Durand D (1985) Macromolecules 18: 2285
23. Te Nijenhuis K (1981) Colloid Polym Sci 259: 522
24. Te Nijenhuis K (1981) Colloid Polym Sci 259: 1017
25. Winter HH, Chambon F (1986) J Rheol 30: 367
26. Chambon F, Winter HH (1985) Polymer Bulletin 13: 499
27. Chambon F, Winter HH (1987) J Rheol 30: 683
28. Chambon F, Petrovic ZS, MacKnight W, Winter HH (1986) Macromolecules 19: 2146
29. Winter HH, Morganelli P, Chambon F (1988) Macromolecules 21: 532
30. Winter HH (1989) In: Kroschwitz JI, Bikalis N (eds) Supplement to encyclopaedia of polymer science and engineering. 2nd edn, John Wiley, New York, p 343
31. Te Nijenhuis K, Winter HH (1989) Macromolecules 22: 411
32. Dušek K (1984) Macromolecules 17: 716
33. Flory PJ (1941) J Am Chem Soc 63: 3096
34. Flory PJ (1947) J Am Chem Soc 69: 30
35. Flory PJ (1953) Principles of polymer chemistry. Cornell University Press, Ithaca, New York, Chapt IX
36. Stockmayer WH (1943) J Chem Phys 11: 45
37. Stockmayer WH (1944) J Chem Phys 12: 125
38. Morton SD, Ferry JD (1962) J Phys Chem 66: 1639
39. Te Nijenhuis K (1991) Makromol Chem 192: 603
40. Te Nijenhuis K (1991) Makromol Chem Macromol Symp 45: 117
41. Te Nijenhuis K (1993) Polym Gels Networks 1: 185
42. Te Nijenhuis K (1993) Polym Gels Networks 1: 199
43. Te Nijenhuis K (1991) In: Patsis AV (ed) Proceedings Vth international conference on crosslinked polymers. Luzern, Chapt 6
44. Te Nijenhuis K (1992) In: Kahoveč J (ed) Macromolecules 1992, Proceedings 34th IUPAC International Symposium on Macromolecules, Prague, 1992, VSP Interscience Publishers, Utrecht, Chapt 3
45. Charlesby A (1954) Proc R Soc (London) A222: 542

46. Charlesby A, Pinner SH (1959) Proc R Soc (London) A249: 367
47. Langley NR (1968) Macromolecules 1: 348
48. Langley NR, Ferry JD (1968) Macromolecules 1: 353
49. Langley NR, Polmanteer KE (1974) J Polym Sci, Polym Phys Ed 12: 1023
50. Pearson DS, Graessley WW (1978) Macromolecules 11: 528
51. Dossin LM, Graessley WW (1979) Macromolecules 12: 123
52. Graessley WW (1974) Adv Polym Sci 16: 1
53. Plotnikov J (1922) Zeitschr f wiss Photogr 21: 117
54. Flumiani G (1926) Zeitschr f Elektrochemie 32: 221
55. Flumiani G (1928) Kolloid Z 45: 152
56. Leaderman H (1943) Ind Eng Chem 35: 374
57. Reed MC (1943) Ind Eng Chem 35: 429
58. Doty P, Wagner H, Singer S (1947) J Phys Chem 51: 32
59. Hengstenberg J, Schuch E (1964) Makromol Chem 74: 55
60. Kratochvíl P (1964) Kolloid Z u Z Polymere 198: 95
61. Kratochvíl P (1965) Collection Czech Chem Commun 30: 1119
62. Kratochvíl P, Petrus V, Munk P, Bohdanečký, Šolc K (1967) J Polym Sci 16: 1257
63. Gezovich DM, Geil PH (1971) Int J Polym Mater 1: 3
64. Aiken W, Alfrey T, Janssen A, Mark H (1947) J Polym Sci 2: 178
65. Alfrey T, Wiederhorn N, Stein R, Tobolsky AV (1949) Ind Eng Chem 41: 107
66. Alfrey T, Wiederhorn N, Stein R, Tobolsky AV (1949) J Colloid Sci 4: 211
67. Bonart R (1966) Koloid Z u Z Polymere 213: 1
68. Singleton CJ, Stephenson T, Isner J, Geil PH, Collins EA (1977) J Macromol Sci B14: 29
69. Nakajima A, Hamada H, Hayashi S (1966) Makromol Chem 95: 40
70. Pezzin G, Gligo N (1966) J Appl Polym Sci 10: 1
71. Crugnola A, Danusso F (1968) J Polym Sci Part B 6: 535
72. Kratochvíl P, Bohdanečký M, Šolc K, Kolínský M, Ryska M, Lím D (1968) J Polym Sci, Part C 23: 9
73. Nakajima A, Hayashi S (1969) Kolloid Z u Z Polymere 229: 12
74. Harrison MA, Morgan PH, Park GS (1972) Eur Polym J 8: 1361
75. Abdel-Alim AH (1975) J Appl Polym Sci 19: 2179
76. Ceccorulli G, Pizzolli M, Pezzin G (1977) J Macromol Sci B14: 499
77. Gert C, Meyer HM (1978) Angew Makromol Chem 74: 81
78. Ehlers JF, Goldstein KR (1950) Kolloid Z 118: 151
79. Colborne RS (1970) J Appl Polym Sci 14: 127
80. Hellwege KH, Johnsen U, Kockott D (1964) Koloid Z u Z Polymere 194: 5
81. Crugnola A (1968) Chim Ind (Milan) 50: 880
82. Juijn J (1972) Doctoral thesis. Delft, The Netherlands
83. Juijn J, Gisolf JH, De Jong WA (1969) Kolloid Z u Z Polymere 235: 1157
84. Juijn J, Gisolf JH, De Jong WA (1973) Kolloid Z u Z Polymere 251: 456
85. Ohta S, Kajiyama T, Takayanagi M (1976) Polym Eng Sci 16: 456
86. Lemstra PJ, Keller A, Cudby M (1978) J Polym Sci Part A2 16: 1507
87. Wenig W (1978) J Polym Sci Part A2 16: 1635
88. Straff RS, Uhlmann DR (1976) J Polym Sci Part A2 14: 353
89. Keller A (1979) Polymer 20: 1371
90. Guerrero SJ, Keller A, Soni PL, Geil PH (1980) J Polym Sci Polym Phys Ed 18: 1533
91. Asahina M, Okada K (1960) Kobunshi Kagaku 17: 607
92. Dorrestijn A, Keijzers AEM, Te Nijenhuis K (1981) Polymer 22: 305
93. Guerrero SJ, Meader D, Keller A (1981) J Macromol Sci Phys B20: 185
94. Mammi M, Naidi V (1963) Nature 199: 247
95. Baker C, Maddams WF, Preedy JE (1977) J Polym Sci, Polym Phys Ed 15: 1041
96. Keller A (1983) In: Hiltner A (ed) Structure and property relationships of polymeric solids. Plenum, New York, Chapt 2
97. Yang YC, Geil PH (1983) J Macromol Sci Phys B22: 463
98. He X, Herz J, Guenet JM (1988) Macromolecules 21: 1757
99. Najeh M, Munch J-P, Guenet J-M (1992) Macromolecules 25: 7018
100. Mutin PH, Guenet JM (1986) Polymer 27: 1098
101. Candau SJ, Dormoy Y, Mutin PH, Debeauvais F, Guenet JM (1987) Polymer 28: 1334
102. Mutin PH, Guenet JM, Hirsch E, Candau SJ (1988) Polymer 29: 30

103. Spěváček J, Schneider B (1975) Makromol Chem 176: 3409
104. Spěváček J (1982) Makromol Chem, Rapid Commun 3: 629
105. Abied H, Brûlet A, Guenet JM (1990) Colloid Polym Sci 268: 403
106. Mutin PH, Guenet JM (1989) Macromolecules 22: 843
107. Kawanishi K, Takeda Y, Inoue T (1986) Polym J 18: 411
108. Komatsu M, Inoue T, Miyasaka K (1986) J Polym Sci, Polym Phys Ed 24: 303
109. Jackson RS, Bower DI, Maddams WF (1990) Polymer 31: 857
110. Walter AT (1954) J Polym Sci 13: 207
111. Stühlen F, Meier L (1971) Kunststoffe 61: 867
112. Pezzin G (1971) Pure Applied Chem 25: 241
113. Fitzgerald ER, Ferry JD (1953) J Colloid Sci 8: 1
114. Ferry JD, Fitzgerald ER (1953) J Colloid Sci 8: 224
115. Ferry JD, Plazek DJ, Heckler GE (1968) J Chim Phys 55: 152
116. Birnboim MH, Ferry JD (1961) J Applied Phys 32: 2305
117. Ferry JD (1980) Viscoelastic properties of polymers, 3rd Ed, Wiley, New York, Chapt 11
118. Tobolsky AV, McLoughlin JR (1955) J Phys Chem 59: 989
119. Shen MC, Tobolsky AV (1965) Adv Chem Ser 48: 118
120. Teplov BF, Gubanov EF, Chenborisova LYa, Ovchinnikov YuV, Teitel'baum BYa, Maklakov AI (1972) Polym Sci USSR 13: 2719
121. Anagstopoulos CE, Coran AY, Gamrath HR (1960) J Applied Polym Sci 4: 181
122. Lyngaee-Jørgensen J (1974) Polym Eng Sci 14: 342
123. Dorrestijn A, Lemstra PJ, Berghmans H (1983) Polym Commun 24: 226
124. Garcia AI, Muñoz ME, Peña JJ, Santamaría A (1990) Macromolecules 23: 5251
125. Gallego F, Muñoz ME, Peña JJ, Santamaría A (1988) J Polym Sci, Polym Phys Ed 26: 1871
126. Kawanishi K, Komatsu M, Inoue T (1987) Polymer 28: 980
127. Te Nijenhuis K (1996) Polym Gels Networks 4: 415
128. Finch CA (ed) (1992) Polyvinyl alcohol, developments. Wiley, Chichester
129. Bunn CW (1948) Nature 161: 159
130. Pezron E, Leibler L, Ricard A, Audebert R (1985) Macromolecules 21: 1126
131. Deuel H, Neukom H (1949) Makromol Chem 3: 13
132. Ahad E (1974) J Appl Polym Sci 18: 1587
133. Saito S, Okuyama H, Kishimoto H, Fujiyama Y (1955) Kolloid Z 144: 41
134. Morawetz H (1965) Macromolecules in solution. Interscience Publishers, New York, p 76
135. Rehage G (1963) Kunststoffe 53: 605
136. Halboth H, Rehage G (1974) Angew Makromol Chem 38: 111
137. Rehage G (1975) Progr Colloid Polym Sci 57: 7
138. Prins W, Rimai L, Chompff AJ (1972) Macromolecules 5: 104
139. Pines E, Prins W (1973) Macromolecules 6: 888
140. Feke GT, Prins W (1974) Macromolecules 7: 527
141. Wu W, Shibayama M, Roy S, Kurokawa H, Coyne LD, Nomura S, Stein RS (1990) Macromolecules 23: 2245
142. Wu W, Kurokawa H, Roy S, Stein RS (1991) Macromolecules 24: 4328
143. Pritchard JG (1970) Polyvinyl alcohol. Gordon & Breach Science Publishers, London
144. Maeda H, Kawai T, Kashiwagi R (1956) Chem High Polymers (Japan) 13: 193
145. Prokopová E, Štern P, Quadrat O (1985) Colloid Polym Sci 263: 899
146. Štern P, Prokopová E, Quadrat O (1987) Colloid Polym Sci 265: 234
147. Prokopová E, Štern P, Quadrat O (1987) Colloid Polym Sci 265: 903
148. Mrkvičková L, Prokopová E, Quadrat O (1987) Colloid Polym Sci 265: 978
149. Prokopová E, Biroš J, Lednický F, Sundelöf LO (1988) Eur Polym J 24: 61
150. Prokopová E, Schmidt P (1989) Eur Polym J 25: 1319
151. Standt UD (1983) J Macromol Sci, Rev Macromol Chem Phys C23: 376
152. Braun D, Walter E (1980) Colloid Polym Sci 258: 376
153. Shibatani K (1970) Polym J 1: 348
154. Takahashi A, Hiramitsu S (1974) Polym J 6: 103
155. Ogasawara K, Nakajima T, Yamaura K, Matsuzawa S (1975) Progr Colloid Polym Sci 58: 145
156. Ogasawara K, Nakajima T, Yamaura K, Matsuzawa S (1976) Colloid Polym Sci 254: 456
157. Ogasawara K, Nakajima T, Yamaura K, Matsuzawa S (1976) Colloid Polym Sci 254: 553
158. Ogasawara K, Nakajima T, Yamaura K, Matsuzawa S (1976) Colloid Polym Sci 254: 982
159. Matsuzawa S, Yamaura K, Maeda R, Ogasawara K (1979) Makromol Chem 180: 229

160. Eldridge JE, Ferry JD (1954) J Phys Chem 58: 992
161. Yamaura K, Takahashi T, Tanigami T, Matsuzawa S (1987) J Appl Polym Sci 33: 1983
162. Nishinari K, Watase M (1983) Polym Commun 24: 270
163. Nishinari K, Watase M (1983) Polym Commun 24: 345
164. Watase M, Nishinari K, Nambu M (1983) Cryo-Letters 4: 197
165. Watase M, Nishinari K (1985) J Polym Sci, Polym Phys Ed 23: 1803
166. Watase M, Nishinari K (1985) Makromol Chem 186: 1081
167. Watase M, Nishinari K (1988) Makromol Chem 189: 871
168. Watase M, Nishinari K (1989) Makromol Chem 190: 155
169. Watase M, Nishinari K (1989) Polym J 21: 567
170. Watase M, Nishinari K (1989) Polym J 21: 597
171. Berghmans H, Stoks W (1986) In: Kleintjens LA, Lemstra PJ (eds) Integration of fundamental polymer science and technology. Elsevier, London, Vol. 1, p 218
172. Berghmans H (1988) In: Kleintjens LA, Lemstra PJ (eds) Integration of fundamental polymer science and technology. Elsevier, London, Vol. 2, p 296
173. Stoks W, Berghmans H, Moldenaers P, Mewis J (1988) Brit Polym J 20: 361
174. Stoks W, Berghmans H (1991) J Polym Sci, Polym Phys Ed B29: 609
175. Nishinari K, Watase M (1993) Polym J 25: 463
176. Hyon S-H, Cha W-I, Ikada Y (1989) Polym Bull 22: 119
177. Hyon S-H, Cha W-I, Graiver D, Ikada Y (1993) J Applied Polym Sci. 47: 339
178. Ohkura M, Kanaya T, Kaji K (1992) Polymer 33: 3686
179. Ohkura M, Kanaya T, Kaji K (1992) Polymer 33: 5044
180. Kanaya T, Ohkura M, Kaji K, Furusaka M, Muthukumar M (1994) Macromolecules 27: 5609
181. Kanaya T, Ohkura M, Takeshita H, Kaji K, Furusaka M, Yamaoka H, Wignall GD (1995) Macromolecules 28: 3168
182. Tanigami T, Murase K, Yamaura K, Matsuzawa S (1994) Polymer 35: 2573
183. Okada S, Hoshino H, Urakawa H, Kajiwara T, Ito T (1988) Polym Prep Jpn, V37, No 9: 3064
184. Bolewski K, Rychły B (1968) Kolloid ZuZ Polymere 228: 48
185. Schultz RK, Myers RR (1969) Macromolecules 2: 281
186. Koike A, Nemoto N, Inoue T, Osaki K (1995) Macromolecules 28: 2339
187. Ochiai H, Shimizu S, Tadokoro Y, Murakami I (1981) Polymer 22: 1456
188. Matsuzawa S, Yamaura K, Tanigami T, Somura T, Nakata M (1987) Polym Commun 28: 105
189. Keita G, Ricard A (1990) Polymer Bull 24: 627
190. Keita G, Ricard A (1990) Polymer Bull 24: 633
191. Sinton SW (1987) Macromolecules 20: 2430
192. Shibayama M, Sato M, Kimura Y, Fujiwara H, Nomura S (1988) Polymer 29: 336
193. Shibayama M, Yoshizawa H, Kurokawa H, Fujiwara H, Nomura S (1988) Polymer 29: 2066
194. Shibayama M, Kurokawa H, Nomura S, Muthukumar M, Stein RS, Roy S (1992) Polymer 33: 2883
195. Cheng ATY, Rodriguez R (1981) J Appl Polym Sci 26: 3895
196. Kurokawa H, Shibayama M, Ishimaru T, Nomura S, Wu W (1992) Polymer 33: 2182
197. Maerker JM, Sinton SW (1986) J Rheol 30: 77
198. Dittmar C, Priest WJ (1955) J Polym Sci 18: 275
199. Okada N, Sakurada I (1951) Bull Inst Chem Res, Kyoto Univ 26: 95; [CA 47: 1423d]
200. Okada N, Sakurada I (1951) Bull Inst Chem Res, Kyoto Univ 26: 94; [CA 47: 1423h]
201. Shibayama M, Ikkai F, Moriwaki R, Nomura S (1994) Macromolecules 27: 1738
202. Shibayama M, Ikkai F, Nomura S (1994) Macromolecules 27: 6383
203. Beltman H, Lyklema J (1974) Discuss Faraday Chem Soc 57: 92
204. Watase M, Nishinari K (1988) Polym J 20: 1125
205. Borchard W, Pyrlik M, Rehage G (1971) Makromol Chem 145: 169
206. Pyrlik M, Borchard W, Rehage G, Uerpmann EP (1974) Angew Makromol Chem 36: 133
207. Pyrlik M, Rehage G (1975) Rheol Acta 14: 303
208. Pyrlik M, Rehage G (1976) Colloid Polym Sci 254: 329
209. Könnecke K, Rehage G (1981) Colloid Polym Sci 259: 1062
210. Rehage G, Wagner D (1982) Polym Prep (Am Chem Soc, Div Polym Chem) 23: 29
211. Könnecke K, Rehage G (1983) Makromol Chem 184: 2679
212. Spěváček J, Schneider B (1987) Adv Colloid Interface Sci 27: 81
213. Spěváček J (1990) Makromol Chem, Macromol Symp 39: 71
214. Spěváček J, Schneider B (1974) Makromol Chem 175: 2939

215. Biroš J, Máša Z, Pouchlý J (1974) Eur Polym J 10: 629
216. Rehage G, Wagner D (1985) In: Dubin P (ed) Microdomains in polymer solutions. Plenum Press, New York, p 87
217. Katime I, Quitana JR (1988) Makromol Chem 189: 1373
218. Schomaker E (1988) Doctoral thesis. Groningen, The Netherlands
219. Ten Brinke G, Schomaker E, Challa G (1985) Macromolecules 18: 1925
220. Schomaker E, Ten Brinke G, Challa G (1985) Macromolecules 18: 1930
221. Brinkhuis RHG (1991) Doctoral thesis. Groningen, The Netherlands
222. Watanabe WH, Ryan CF, Fleischer PC, Garrett BS (1961) J Phys Chem 65: 896
223. Glusker DL, Gallucio RA, Evans RA (1964) J Am Chem Soc 86: 187
224. Ryan CF, Fleischer PC (1965) J Phys Chem 69: 3384
225. Fujishige S, Goeldi P, Elias H-G (1971) J Macromol Sci Chem A5: 1011
226. Dybal J, Spěváček J (1988) Makromol Chem 189: 2099
227. De Boer A, Challa G (1976) Polymer 17: 633
228. Berghmans H, Donkers A, Frenay L, Stoks W, De Schrijver FE, Moldenaers P, Mewis J (1987) Polymer 28: 97
229. Spěváček J, Schneider B (1980) Polym Bull 2: 227
230. Spěváček J, Schneider B, Baldrian J, Dybal J, Štokr J (1983) Polym Bull 9: 495
231. Mrkvičková L, Stejskal S, Spěváček J, Horská J, Quadrat O (1983) Polymer 24: 700
232. Sedláček B, Spěváček J, Mrkvičková L, Stejskal S, Horská J, Baldrian J, Quadrat O (1984) Macromolecules 17: 825
233. Spěváček J, Schneider B, Straka J (1990) Macromolecules 23: 3042
234. Spěváček J (1978) J Polym Sci, Phys Ed 16: 523
235. Spěváček J, Schneider B, Bohdanecký M, Sikora A (1982) J Polym Sci, Polym Phys Ed 20: 1623
236. Spěváček J, Schneider B, Dybal J, Štokr J, Baldrian J, Pelzbauer J (1984) J Polym Sci, Polym Phys Ed 22: 617
237. Dybal J, Spěváček J, Schneider B (1986) J Polym Sci, Polym Phys Ed 24: 657
238. Borchard W, Kalawrytino G, Mohadjer M, Pyrlik M, Rehage G (1973) Angew Makromol Chem 29/30: 471
239. Spěváček J, Fernandez-Pierola I (1987) Makromol Chem 188: 861
240. Feitsma EL, De Boer A, Challa G (1975) Polymer 16: 515
241. Lohmeyer JGHM, Kransen G, Tan YY, Challa G (1975) J Polym Sci, Polym Letters Ed 13: 725
242. Challa G, De Boer A, Tan YY (1976) Int J Polym Mater 4: 239
243. Lohmeyer JGHM, Tan YY, Lako P, Challa G (1978) Polymer 19: 1171
244. Vorenkamp EJ, Bosscher F, Challa G (1979) Polymer 20: 59
245. Bosscher F, Keekstra D, Challa G (1981) Polymer 22: 124
246. Vorenkamp EJ, Challa G (1981) Polymer 22: 1705
247. Bosscher F, Ten Brinke G, Challa G (1982) Macromolecules 15: 1442
248. Spěváček J, Schneider B (1978) Polymer 19: 63
249. Spěváček J, Schneider B (1974) J Polym Sci, Polym Letters Ed 12: 349
250. Spěváček J, Schneider B (1975) Makromol Chem 176: 729
251. Spěváček J, Fernandez-Pierola I, Pleštil J (1991) Polym Prep 32(3): 423
252. Spěváček J, Straka J (1993) Makromol Chem, Macromol Symp 72: 201
253. Spěváček J, Schneider B (1980) Colloid Polym Sci 250: 621
254. Dybal J, Štokr J, Schneider B (1983) Polymer 24: 971
255. Spěváček J (1985) J Mol Struct 129: 175
256. Liquori AM, Anzuino G, Coiro VM, d'Alagni M, De Santis P, Savino M (1965) Nature 206: 358
257. Chiang R, Burke JJ, Threlkeld JO, Orofino TA (1966) J Phys Chem 70: 3591
258. Liu HZ, Liu KJ (1968) Macromolecules 1: 157
259. Kock HJ, Rehage G (1984) Colloid Polym Sci 262: 182
260. Coiro VM, De Santis P, Liquori AM, Mazzarella L (1969) J Polym Sci, Part C 16: 4591
261. Kusanagi H, Tadokoro H, Chatani Y (1976) Macromolecules 9: 531
262. Kusuyama H, Takase M, Higashihata Y, Tseng HT, Chatani Y, Tadokoro H (1982) Polymer 23: 1256
263. Berghmans M, Thijs S, Cornette M, Berghmans H, De Schryver FC, Moldenaers P, Mewis J (1994) Macromolecules 27: 7669
264. Fazel N, Fazel Z, Brûlet A, Guenet J-M (1992) J Phys II France 2: 1617

265. Fazel Z, Fazel N, Guenet J-M (1992) J Phys II France 2: 1745
266. Benoit H, Picot C (1966) Pure Applied Chem 12: 545
267. Guenet JM, Klein M (1990) Makromol Chem, Macromol Symp 39: 85
268. Guenet JM, Willmott NF, Elsmore PA (1983) Polym Commun 24: 230
269. Boyer RF, Baer E, Hiltner A (1985) Macromolecules 18: 27
270. Lobanov AM, Frenkel SYa (1980) Polym Sci USSR 22: 1150
271. Boyer RF (1980) J Macromol Sci, Phys B18: 461
272. Arnauts J, Berghmans H (1987) Polym Commun 28: 66
273. Arnauts J, Berghmans H (1990) In: Burchard W, Ross-Murphy SB (eds) Physical networks. Elsevier London, Chapt 3
274. Van den Broecke PH, Berghmans H (1990) Makromol Chem, Macromol Symp 39: 59
275. Hikmet RM, Callister S, Keller A (1988) Polymer 29: 1378
276. Callister S, Keller A, Hikmet RM (1990) Makromol Chem, Macromol Symp 39: 19
277. Hiltner A (1985) In: Keinath SE, Miller RL, Rieke JK (eds) Order in the amorphous "state" of polymers. Plenum Press, New York
278. Ferry JD, Grandine LD (1953) J Appl Phys 24: 679
279. Wellinghoff ST, Shaw J, Baer E (1979) Macromolecules 12: 932
280. Clark J, Wellinghoff ST, Miller WG (1983) Polym Preprints Am Chem Soc, Div Polym Chem 24(2) : 86
281. Holly HH, Venkataraman SK, Chambon F, Winter HH (1988) J Non-Newt Fluid Mech 27: 17
282. Koltisko B, Keller A, Litt M, Baer E, Hiltner A (1986) Macromolecules 19: 1207
283. Xie XM, Tanioka A, Miyasaka K (1990) Polymer 31: 281
284. Tan HM, Moet A, Hiltner A, Baer E (1983) Macromolecules 16: 28
285. Vandeweert P, Berghmans H, Tervoort Y (1991) Macromolecules 24: 3547
286. Arnauts J, Berghmans H, Koningsveld R (1993) Makromol Chem 194: 77
287. Hikmet RM, Callister S, Keller A (1988) In: Kleintjens LA, Lemstra PJ (eds) Integration of fundamental polymer science and technology. Vol 2, Elsevier, London, p 306
288. Frank FC, Keller A (1988) Polym Commun 29: 186
289. Gan YS, François J, Guenet JM (1986) Macromolecules 19: 173
290. François J, Gan YS, Guenet JM (1986) Macromolecules 19: 2755
291. François J, Gan YS, Sarazin D, Guenet JM (1988) Polymer 29: 898
292. Guenet JM, Klein M, Menelle A (1989) Macromolecules 22: 493
293. Gan YS, François J, Guenet JM, Gautier-Manuel B, Allain C (1985) Makromol Chem, Rapid Commun 6: 225
294. Lehsaini N, François J, Weill G (1993) Macromolecules 26: 7333
295. Lehsaini N, Boudenne N, Zilliox JG, Sarazin D (1994) Macromolecules 27: 4353
296. Bisschops J (1954) J Polym Sci 12: 583
297. Bisschops J (1955) J Polym Sci 17: 81
298. Bisschops J (1955) J Polym Sci 17: 89
299. Łabudzińska A, Wasiak A, Ziabicky A (1967) J Polym Sci, Part C 16: 2835
300. Łabudzińska A, Ziabicky A (1971) Koloid Z u Z Polymere 243: 21
301. Paul DR (1967) J Appl Polym Sci 11: 439
302. Bashir Z (1992) Polymer 33: 4304
303. Bashir Z (1992) J Polym Sci, Part B (Physics) 30: 1299
304. Bashir Z, Atureliya SK, Church SP (1993) J Mater Sci 28: 2721
305. Herbert IR, Tipping A, Bashir Z (1993) J Polym Sci, Part B (Physics) 31: 1459
306. Klement JJ, Geil PH (1968) J Polym Sci, Polym Phys Ed 6: 138
307. Bashir Z (1994) J Polym Sci, Polym Phys Ed 32: 1115
308. Bashir Z, Church SP, Waldron D (1994) Polymer 35: 967
309. Beckmann J, Zenke D (1993) Colloid Polym Sci 271: 436
310. Beevers RB (1968) Macromolecular Reviews 3: 113
311. Lindstedt B, Flodin P (1987) In: Rånby B (ed) Physical chemistry of colloids and macromolecules. Blackwell Scientific Publications, Oxford, p 127
312. Flodin P (1988) Makromol Chem, Macromol Symp 22: 253
313. Finke V, Zachmann HG (1990) Makromolekulares Symposium, Hamburg
314. Chiang R (1965) J Polym Sci (Part A) 3: 2019
315. Holland VF, Mitchell SB, Hunter WL, Lindenmeyer PH (1962) J Polym Sci 62: 145
316. Yamazaki H, Kajita S, Kamide K (1987) Polym J 19: 995
317. Kumamaru F, Kajiyama T, Takayanagi M (1980) J Cryst Growth 48: 202

318. Bashir Z, Church SP, Price DM (1993) Acta Polym 44: 211
319. Autureliya SK, Bashir Z (1993) Polymer 34: 5116
320. Ko MB, Kwon IH, Jo WH, Son TW (1994) J Polym Sci (Physics) 32: 945
321. Pfragner J, Schurz J (1979) Rheol Acta 18: 717
322. Tanaka F (1989) Macromolecules 22: 1988
323. Frushour BG (1982) Polym Bull 7: 1
324. Lemstra PJ, Smith P (1980) Br Polym J 12: 212
325. Smith P, Lemstra PJ, Pijpers JPL, Kiel AM (1981) Colloid Polym Sci 259: 1070
326. Smith P, Lemstra PJ, Booij HC (1981) J Polym Sci, Polym Phys Ed 19: 877
327. Aerts L (1992) Doctoral thesis. Leuven, Belgium
328. Aerts L, Kunz M, Berghmans H, Koningsveld R (1993) Makromol Chem 194: 2697
329. Pennings AJ, Van der Mark JMMA, Booij HC (1970) Kolloid Z u Z Polymere 236: 99
330. Smith P, Lemstra PJ (1980) J Mater Sci 15: 505
331. Pennings AJ (1977) J Polym Sci, Polym Symp 59: 55
332. Narh KA, Barham PJ, Keller A (1982) Macromolecules 15: 464
333. Coombes AG, Keller A (1979) J Polym Sci, Polym Phys Ed 17: 1637
334. Barham PJ, Hill MJ, Keller A (1980) Colloid Polym Sci 258: 899
335. Edwards CO, Mandelkern L (1982) J Polym Sci, Polym Lett Ed 20: 355
336. Mandelkern L (1986) ACS Polym Prep 27(1) : 206
337. Mandelkern L, Edwards CO, Domszy RC, Davidson MW (1985) In: Dubin P (ed) Microdomains in polymer solution. Plenum Press, New York, p 121
338. Domszy RC, Alamo R, Edwards CO, Mandelkern L (1986) Macromolecules 19: 310
339. Mandelkern L (1987) ACS Polym Prep 28(1): 130
340. Tan HM, Chang BH, Baer E, Hiltner A (1983) Eur Polym J 19: 1021
341. Okabe M, Isayama M, Matsuda H (1985) Polym J 17: 369
342. Matsuda M, Araki K, Fujimatsu H, Kuroiwa S (1981) Polym J 13: 587
343. Matsuda H, Fujimatsu H, Kuroiwa S (1981) Polym J 13: 807
344. Takahashi A, Nakamura T, Kagawa I (1972) Polym J 3: 207
345. Matsuda H, Imaizumi M, Fujimatsu H, Kuroiwa S, Okabe M (1984) Polym J 16: 151
346. Jo WH, Kwon IH, Seoul C (1989) Polym Eng Sci 29: 1569
347. Li Z, Mark JE, Chan EKM, Mandelkern L (1989) Macromolecules 22: 4273
348. Chung B, Zachariades AE (1986) ACS Polym Prep 27(1) : 208
349. Chung B, Zachariades AE (1987) ACS Symp Ser 350: 22
350. Chung B, Zachariades AE (1986) in Russo PS (ed) Reversible polymer gels and related systems, ACS, New York, p 22
351. Tuzar Z, Kratochvil P (1976) Adv Colloid Interface Sci 6: 201
352. Zhou Z, Chu B (1987) Macromolecules 20: 3089
353. Raspaud ED, Lairez D, Adam M, Carton J-P (1994) Macromolecules 27: 2956
354. Mortensen K, Brown W, Jørgensen EB (1994) Macromolecules 27: 5654
355. Glotzer S, Bansil R, Gallagher PD, Gyure MF, Sciortino F, Stanley HE (1993) Physica A 201: 482
356. Bates FS (1991) Science 251: 898
357. Spitteler PHJ (1994) Doctoral thesis. Twente University, The Netherlands
358. Spitteler PHJ, Mijnlieff PF, paper in preparation
359. Visscher K (1993) Doctoral thesis. Twente University, The Netherlands
360. Visscher K, Mijnlieff PF (1991) Rheol Acta 30: 559
361. Visscher K, Mijnlieff PF (1990) Makromol Chem, Macromol Symp 39: 99
362. Nguyen-Misra M, Mattice WL (1995) Macromolecules 28: 1444
363. Stacey CJ, Kraus G (1977) Polym Eng Sci 17: 627
364. Canham PA, Lally TP, Price C, Stubbersfield RB (1980) J Chem Soc, Faraday Trans I 76: 1857
365. Żuchowska D, Meissner W (1985) Makromol Chem 186: 2355
366. Baloch MK, Rauf A (1988) Makromol Chem 189: 1517
367. Küçükyavuz Z (1990) Makromol Chem 191: 2205
368. Pico ER, Williams MC (1977) Polym Eng Sci 17: 573
369. Watanabe H, Kuwahara S, Kotaka T (1984) J Rheol 28: 393
370. Elworthy PH, Macfarlane CB (1964) J Chem Soc: 311
371. Xu R, Winnik MA, Hallett FR, Riess G, Croucher MD (1991) Macromolecules 24: 87
372. Polik WF, Burchard W (1983) Macromolecules 16: 978
373. Napper DH (1983) Polymeric stabilization of colloidal dispersions. Academic Press, London

374. Bahadur P, Li P, Almgren M, Brown W (1992) Langmuir 8: 1903
375. Zhou Z, Chu B (1988) J Colloid Interface Sci 126: 171
376. Zhou Z, Chu B (1988) Macromolecules 21: 2548
377. Mortensen K, Pedersen JK (1993) Macromolecules 26: 805
378. Brown W, Schillén K, Almgren M, Hvidt S, Bahadur P (1991) J Phys Chem 95: 1850
379. Wanka G, Hoffmann H, Ulbricht W (1990) Colloid Polym Sci 268: 101
380. Brown W, Schillén K, Hvidt S (1992) J Phys Chem 96: 6038
381. Hvidt S, Jørgensen EB, Brown W, Schillén K (1994) J Phys Chem 98: 12320
382. Mortensen K, Brown W, Nordén B (1992) Phys Rev Lett 68: 2340
383. Mortensen K (1993) Progr Colloid Polym Sci 93: 72
384. Linse P, Malmsten M (1992) Macromolecules 25: 5434
385. Malmsten M, Lindman B (1992) Macromolecules 25: 5440
386. Malmsten M, Lindman B (1992) Macromolecules 25: 5446
387. Atkins PW (1982) Physical Chemistry. Oxford University Press, 2nd edn
388. Mortensen K (1992) Europhys Lett 19: 599
389. Rassing J, Attwood D (1983) Int J Pharma 24: 57
390. Rassing J, McKenna WP, Bandyopadhyay S, Eyring EM (1994) J Mol Liq 27: 165
391. Wang P, Johnston TP (1991) J Appl Polym Sci 43: 283
392. Glatter O, Scherf G, Schillén K, Brown W (1994) Macromolecules 27: 6046
393. Jørgensen EB, Jensen JH, Hvidt S (1994) J Non-Cryst Solids 172–174: 972
394. Bahadur P, Pandya K (1992) Langmuir 8: 2666
395. Hvidt S, private communication
396. He X, Herz J, Guenet J-M (1987) Macromolecules 20: 2003
397. He XW, Fillon B, Herz J, Guenet J-M (1987) Polym Bull 17: 45
398. He X, Herz J, Guenet J-M (1989) Macromolecules 22: 1390
399. Shukla P (1992) Polymer 33: 365
400. Onsager L (1949) Ann NY Acad Sci 51: 627
401. Flory PJ (1956) Proc Roy Soc London Ser A 234: 60
402. Flory PJ (1956) Proc Roy Soc London Ser A 234: 73
403. Wee EL, Miller WG (1971) J Chem Phys 75: 1446
404. Miller WG, Wu CC, Wee EL, Santee GL, Rai JH, Goebel KG (1974) Pure Appl Chem 38: 37
405. Miller WG, Rai JH, Wee EL (1974) In: Johnson JF, Porter RS (eds) Liquid crystals and ordered fluids. Vol. 2, Plenum, New York, p 243
406. Miller WG, Kou L, Tohyama K, Voltaggio V (1978) J Polym Sci, Polym Symp 65: 91
407. Miller WG (1978) Ann Rev Phys Chem 29: 519
408. Tohyama K, Miller WG (1981) Nature 289: 813
409. Koningsveld R, Stockmayer WH, Nies E (1990) Makromol Chem, Macromol Symp 39: 1
410. Shukla P, Muthukumar M (1988) Polym Eng Sci 28: 1304
411. Shukla P, Muthukumar M (1991) J Polym Sci, Part B Polym Phys 29: 1373
412. Murthy AK, Muthukumar M (1987) Macromolecules 20: 564
413. Shukla P, Muthukumar M, Langley KH (1992) J Applied Polym Sci 44: 2115
414. De Gennes PG (1979) Scaling concepts in polymer physics. Cornell University Press, Ithaca, NY
415. Doty P, Bradbury JH, Holtzer AM (1956) J Am Chem Soc 76: 947
416. Sasaki S, Shiraki C, Uematsu I (1982) Polym J 14: 205
417. Sasaki S, Toluma K, Uematsu I (1983) Polymer Bull 10: 539
418. Hermans J (1962) J Colloid Sci 17: 638
419. Kiss G, Porter RS (1978) J Polym Sci, Polym Symp 65: 193
420. Guezala S, Muñoz Mª E, Pana JJ, Santamaria A (1990) Polymer 31: 651
421. Greiner A, Rochefort WE, Greiner K, Schmidt H-W, Pearson DS (1992) Makromol Chem, Rapid Commun 13: 25
422. Hellendoorn EW, De Boer R (1965) Conserva 14: 100
423. Veis A (1964) The macromolecular chemistry of gelatin. Academic Press, New York
424. Balian G, Bowes JH (1977) In: Ward AG, Courts A (eds) The science and technology of gelatin. Academic Press, New York, Chapt 1
425. Josse J, Harrington WF (1964) J Mol Biol 9: 269
426. Harrington WF (1964) J Mol Biol 9: 613
427. Privalov PL, Tiktopulo EI (1970) Biopolymers 9: 127
428. Eastoe JE, Leach AA (1977) In: Ward AG, Courts A (eds) The science and technology of gelatin. Academic Press, New York, Chapt 1

429. Johns P, Courts A (1977) In: Ward AG, Courts A (eds) The science and technology of gelatin. Academic Press, New York, Chapt 5
430. Privalov PL (1982) Adv Protein Chem 35: 1
431. Heidemann E, Roth W (1982) Adv Polym Sci 43: 143
432. Heidemann E (1987) Das Leder 38: 81
433. Heidemann E (1991) Das Leder 42: 21
434. Dölz R, Heidemann E (1986) Biopolymers 25: 1069
435. McClain PE, Wiley EG (1972) J Biol Chem 247: 692
436. Privalov PL, Serdyuk IN, Tiktopulo EI (1971) Biopolymers 10: 1777
437. Berendsen HJC, Migchelsen C (1965) Ann NY Acad Sci (1965) 125: 365
438. Migchelsen C, Berendsen HJC (1973) J Chem Phys 59: 296
439. Piez KA, Sherman MR (1970) Biochemistry 9: 4134
440. Thakur S, Vadolas D, Germann H-P, Heidemann E (1986) Biopolymers 25: 1081
441. Bella J, Brodsky B, Berman HM (1994) Science 266: 75
442. Engel J, Chen H-T, Prockop DJ (1977) Biopolymers 16: 601
443. Venugopal MG, Ramshaw JAM, Braswell E, Zhu D, Brodsky B (1994) Biochemistry 33: 7948
444. Long CW, Thomas M, Brodsky B (1995) Biopolymers 35: 621
445. Anachi BR, Siegel DL, Baum J, Brodsky B (1995) FEBS Letters 368: 551
446. Cowan PM, North ACT, Randall JT (1953) Randall JT (ed) The nature and structure of collagen. Butterworths, London, p 241
447. Ramachandran GN, Kartha G (1954) Nature, Lond 174: 269
448. Ramachandran GN, Kartha G (1955) Nature, Lond 176: 593
449. Rich A, Crick FHC (1955) Nature, Lond 176: 915
450. Cowan PM, McGavin S, North ACT (1955) Nature, Lond 176: 1062
451. Ref. 2, p 10
452. Rich A, Crick FHC (1958) In: Stainsby G (ed) Recent advances in gelatin and glue research. Pergamon, London, p 20
453. Brodsky B, Shah NK (1995) FASEB J, Serial Rev 9: 1537
454. Brodsky B, Li M-H, Long CG, Apigo J, Braun J (1992) Biopolymers 32: 447
455. Ramachandran GN, Chandrasekharan R (1968) Biopolymers 6: 1649
456. Eyre DR (1980) Science 207: 1317
457. Brown FR, Hopfinger AJ, Blout ER (1972) J Mol Biol 63: 101
458. Djabourov M, Leblond J, Papon P (1988) J Phys France 49: 319
459. Todd A (1961) Nature 191: 567
460. Harrington WF, Rao NV (1970) Biochemistry 9: 3714
461. Leick A (1904) Ann Phys 14: 139
462. Arisz L (1915) In: Ostwald Wo (ed) Kolloidchemische Beihefte. Band VII, Steinkopf Verlag, Dresden, p 1
463. Poole HJ (1925) Trans Farad Soc 21: 114
464. Hatschek E (1932) J Phys Chem 36: 2994
465. Sheppard SE, Sweet SS (1921) J Am Chem Soc 43: 539
466. Ferry JD (1948) Adv Protein Chem 4: 1
467. Ross-Murphy SB (1991) In: Dickinson E (ed) Food polymers, gels and colloids. Royal Soc Chem, Special Publication No 82, RSC London, p 357
468. Kramer O (1992) private communication
469. Djabourov M, Leblond J, Papon P (1988) J Phys France 49: 333
470. Michon C, Cuvelier G, Launay B (1993) Rheol Acta 32: 94
471. Amis EJ, Hodgson DF, Yu Q (1991) Polym Prep 32 (3) : 447
472. Hodgson DF, Yu Q, Amis EJ (1992) In: Aharoni SM (ed) Synthesis, characterization and theory of polymeric networks and gels. p 243
473. Hsu, S-h, Jamieson AM (1993) Polymer 34: 2602
474. Pouradier J (1971) J Chim Phys 10: 1547
475. Stainsby G (1977) In: Wards AG, Courts A (eds) The science and technology of gelatin. Academic Press, London, p 179
476. Ledward DA (1990) In: Phillips GO, Williams PA, Wedlock DJ (eds) Gums and stabilisers for the food industry. IRL Press, London, Vol 5, p 145
477. Eagland D, Pilling G, Wheeler RG (1974) Farad Discuss Chem Soc 57: 181
478. Busnell JP, Clegg SM, Morris ER (1988) In: Phillips GO, Williams PA, Wedlock DJ (eds) Gums and stabilisers for the food industry. IRL Press, London, Vol 4, p 105

479. De Wolf FA, Keller RCA (1996) Progr Colloid Sci 102: 9
480. Cumper CWN, Alexander AE (1952) Austr J Res A5: 153
481. Ferry JD (1948) J Am Chem Soc 70: 2244
482. Narayanamurti D (1956) Kolloid Z 145: 80
483. Szczesniak AS, MacAllister RV (1964) J Appl Polym Sci 8: 1391
484. Croome RJ (1976) In: Cox RJ (ed) Photographic gelatin II. Academic Press, London, p 101
485. Ferry JD, Eldridge JE (1949) J Phys Colloid Chem 53: 184
486. Saunders PR, Ward AG (1953) In: Harrison VGW (ed) Proc 2nd Intern Congr Rheol. Oxford, Butterworths, London, p 197
487. Clark AH, Ross-Murphy SB (1985) Brit Polym J 17: 64
488. Hermans J (1965) J Polymer Sci, Part A, 3: 1859
489. Gordon M, Ross-Murphy SB (1975) Pure Appl Chem 43: 1
490. Godard P, Biebuyck JJ, Daumerie M, Naveau H, Mercier JP (1978) J Polym Sci Polym Phys Ed 16: 1817
491. Tarr I (1974) Disc Faraday Soc 57: 83
492. Flory PJ, Garrett RR (1958) J Am Chem Soc 80: 4836
493. Oth JFM, Dumitri ET, Spurr Jr OK, Flory PJ (1957) J Am Chem Soc 79: 3288
494. Stauffer D, Coniglo A, Adam M (1982) Adv Polym Sci 44: 103
495. Oakenfull DG, Scott A (1986) In: Phillips GO, Wedlock DJ, Williams PA (eds) Gums and stabilisers for the food industry. Elsevier Applied Science, London, Vol 3, p 465
496. Oakenfull DG, Morris VJ (1987) Chem Ind 201
497. Oakenfull DG, Scott A (1988) In: Phillips GO, Wedlock DJ, Williams PA (eds) Gums and stabilisers for the food industry. IRL Press, Oxford, Vol 4, p 127
498. Ross-Murphy SB (1991) Rheol Acta 30: 401
499. Ross-Murphy SB (1990) Carbohydr Polym 14: 281
500. Pouradier J (1967) J Chim Phys 64: 1616
501. Smith CR (1919) J Am Chem Soc 41: 135
502. Flory PJ, Weaver ES (1960) J Am Chem Soc 82: 4518
503. Harrington WF, Von Hippel PH (1961) Arch Biochem Biophys 92: 100
504. Harrington WF, Karr GM (1970) Biochemistry 9: 3725
505. Hauschka PV, Harrington WF (1970) Biochemistry 9: 3734, 3745 and 3754
506. Mascuga DD (1972) Biopolymers 11: 2521
507. Yuan L, Veis A (1973) Biophys Chem 1: 117
508. Djabourov M, Papon P (1983) Polymer 24: 537
509. Djabourov M, Maquet J, Theveneau H, Leblond J, Papon P (1985) Brit Polym J 17: 169
510. Djabourov M, Leblond J (1987) In: Russo P (ed) Reversible polymer gels and related systems. ACS Symp Ser 350: Chapt 14
511. Djabourov M, Lechaire J-P, Gaill F (1993) Biorheology 30: 191
512. Chatellier JY, Durand D, Emery JR (1985) Int J Biol Macromol 7: 311
513. Durand D, Emery JR, Chatellier JY (1985) Int J Biol Macromol 7: 315
514. Watase M, Arakawa K (1961) Bull Chem Soc Japan 34: 1233
515. Watase M, Arakawa K (1967) Bull Chem Soc Japan 40: 473
516. Watase M, Arakawa K (1968) Bull Chem Soc Japan 41: 1830
517. Tako M, Nakamura S (1988) Carbohydr Res 180: 277
518. Hayashi A, Kinoshita K, Yasueda S (1980) Polym J 12: 447
519. Arnott S, Fulmer A, Scott WF, Dea ICM, Moorehouse R, Rees DA (1974) J Mol Biol 90: 269
520. Nishinari K (1976) Japan J Appl Phys 15: 1263
521. Watase M, Nishinari K (1981) Rheol Acta 20: 155
522. Watase M, Nishinari K (1982) Rheol Acta 21: 318
523. Watase M, Nishinari K (1983) Rheol Acta 22: 580
524. Watase M, Nishinari K (1986) J Calorim, Anal Thermodyn Chim 17: 219
525. Watase M, Nishinari K (1986) Polym J 18: 1017
526. Watase M, Nishinari K (1987) Makromol Chem 188: 1177
527. Watase M, Nishinari K, Clark AH, Ross-Murphy SB (1989) Macromolecules 22: 1196
528. Nishinari K, Watase M, Kohyama K, Nishinari N, Oakenfull D, Koide S, Ogino K, Williams PA, Phillips GO (1992) Reports Progress Polym Phys Jpn 35: 169
529. Nishinari K, Watase M (1993) Reports Progress Polym Phys Jpn 36: 661
530. Guiseley KB (1970) Carbohydr Res 13: 247
531. Clark AH, Richardson RK, Ross-Murphy SB, Stubbs JM (1983) Macromolecules 16: 1367

532. Tokita M, Hikichi K (1987) Phys Rev A 35: 4329
533. Rees DA (1977) Adv Carbohydr Chem 24: 267
534. Dea ICM, McKinnon AA, Rees DA (1972) J Mol Biol 68: 153
535. Rees DA, Scott WE, Williamson FB (1970) Nature (London) 227: 390
536. McKinnon AA, Rees DA, Williamson FB (1969) J Chem Soc, Chem Commun 701
537. Norton IT, Goodall DM (1978) J Chem Soc, Chem Comm 515
538. Norton IT, Goodall DM, Morris ER, Rees DA (1979) J Chem Soc, Chem Comm 988
539. Norton IT, Goodall DM (1983) J Chem Soc, Faraday Trans I 79: 2475
540. Norton IT, Goodall DM, Morris ER, Rees DA (1983) J Chem Soc, Faraday Trans I 79: 2489
541. Norton IT, Goodall DM, Morris ER, Rees DA (1983) J Chem Soc, Faraday Trans I 79: 2501
542. Norton IT, Morris ER, Rees DA (1984) Carbohydr Res 134: 89
543. Rochas C, Rinaudo M (1980) Biopolymers 19: 1675
544. Rochas C, Rinaudo M, Vincendon M (1983) Int J Biol Macromol 5: 111
545. Rochas C, Rinaudo M, Landry S (1989) Carbohydr Polym 10: 115
546. Rochas C, Rinaudo M, Landry S (1990) Carbohydr Polym 12: 255
547. Bryce TA, McKinnon AA, Morris ER, Rees DA, Thom D (1974) Faraday Disc Chem Soc 57: 221
548. Smidsrød O, Andresen I-L, Grasdalen H, Larsen B, Painter T (1980) Carbohydr Res 80: C11
549. Rochas C, Rinaudo M (1982) Carbohydr Res 105: 227
550. Rochas C, Mazet J (1984) Biopolymers 23: 2825
551. Rochas C (1985) J Thermal Analysis 30: 1375
552. Rochas C, Landry S (1987) Carbohydr Polym 7: 435
553. Morris VJ, Belton PS (1980) J Chem Soc, Chem Comm 983
554. Morris VJ (1982) Int J Biol Macromol 4: 155
555. Morris VJ, Belton PS (1982) Progr Fd Nutr Sci 6: 55
556. Day DH, Phillips GO, Williams PA (1988) In: Phillips GO, Williams PA, Wedlock DJ (eds) Gums and stabilisers for the food industry 4. IRL Press, Oxford p 497
557. Jones RA, Staples EJ, Penman A (1973) J Chem Soc, Perkin Trans 2: 1608
558. Morris ER, Rees DA, Robinson C (1980) J Mol Biol 138: 349
559. Grasdalen H, Smidsrød O (1981) Macromolecules 14: 1845
560. Smidsrød O, Grasdalen H (1982) Carbohydr Polym 2: 270
561. Morris VJ (1982) Progr Fd Nutr Sci 6: 119
562. Morris VJ, Chilvers GR (1981) J Sci Food Agric 32: 1235
563. Tako M, Nakamura S (1986) Carbohydr Res 155: 200
564. Tako M, Nakamura S, Kohda Y (1987) Carbohydr Res 161: 247
565. Watase M, Nishinari K (1987) Makromol Chem 188: 2213
566. Morris VJ, Chilvers GR (1983) Carbohydr Polym 3: 129
567. Oakenfull D, Scott A (1988) In: Phillips GO, Williams PA, Wedlock DJ (eds) Gums and stabilisers for the food industry 4. IRL Press, Oxford, p 127
568. Oakenfull D, Scott A (1990) In: Phillips GO, Williams PA, Wedlock DJ (eds) Gums and stabilisers for the food industry 5. IRL Press, Oxford, p 507
569. Rochas C, Rinaudo M (1984) Biopolymers 23: 735
570. Clark AH (1994) Carbohydr Polym 23: 247
571. Rochas C, Landry S (1988) In: Phillips GO, Williams PA, Wedlock DJ (eds) Gums and stabilisers for the food industry 4. IRL Press, Oxford, p 445
572. Plashchina IG, Grinberg NV, Braudo EE, Tolstoguzov VB (1980) Colloid Polym Sci 258: 939
573. Cuvelier G, Peigny-Noury C, Launay B (1990) In: Phillips GO, Wedlock DJ, Williams PA (eds) Gums and stabilisers for the food industry 5. IRL Press, Oxford, p 549
574. Watase M, Nishinari K (1982) Colloid Polym Sci 260: 971
575. Richardson RK, Goycoolea FM (1994) Carbohydr Polym 24: 223
576. Jansson P, Lindberg B, Sandford PA (1983) Carbohydr Res 124: 135
577. O'Neill MA, Selvendran RR, Morris VJ (1983) Carbohydr Res 124: 123
578. Rinaudo M (1988) In: Phillips GO, Williams PA, Wedlock DJ (eds) Gums and stabilizers for the food industry 4. IRL Press, Oxford, p 119
579. Moritaka H, Nishinari K, Taki M, Fukuba H (1995) J Agric Food Chem 43: 1685
580. Kang KS, Veeder GS (1982) US Patent 4,326,053
581. Kuo MS, Mort AJ, Dell A (1986) Carbohydr Res 156: 173
582. Sanderson GR (1990) In: Harris P (ed) Food gels. Elsevier Applied Science Publishers, London, p 201

583. Moorehouse R, Colegrave GT, Sandford PA, Baird JK, Kang KS (1981) ACS Symp Ser 150: 111
584. Crescenzi V, Dentini M, Dea ICM (1987) Carbohydr Res 160: 283
585. Grasdalen H, Smidsrød O (1987) Carbohydr Polym 7: 371
586. Chandrasekaran R, Puigjaner LC, Joyce KL, Arnott S (1988) Carbohydr Res 181: 23
587. Milas M, Shi X, Rinaudo M (1990) Biopolymers 30: 451
588. Moritaka H, Fukuba H, Kumeno K, Nakahama N, Nishinari K (1991) Food Hydrocoll 4: 495
589. Watase M, Nishinari K (1993) Food Hydrocoll 7: 449
590. Miyoshi E, Takaya T, Nishinari K (1994) Food Hydrocoll 8: 505
591. Miyoshi E, Takaya T, Nishinari K (1994) Food Hydrocoll 8: 529
592. Miyoshi E, Takaya T, Nishinari K (1994) Rep Progr Polym Phys Japan 37: 751
593. Crescenzi V, Dentini M, Coviello T, Rizzo R (1986) Carbohydr Res 149: 425
594. Chapmann HD, Chilvers GR, Miles MJ, Morris VJ (1988) In: Phillips GO, Williams PA, Wedlock DJ (eds) Gums and stabilizers for the food industry 4. IRL Press, Oxford, p 147
595. Dentini M, Coviello T, Burchard W, Crescenzi V (1988) Macromolecules 21: 3312
596. Gunning AP, Morris VJ (1990) Int J Biol Macromol 12: 338
597. Morris VJ (1992) In: Chandrasekaran R (ed) Frontiers in carbohydrate research-2. Elsevier Applied Science Publishers, London, p 191
598. Crescenzi V, Dentini M, Coviello T (1992) In: Chandrasekaran R (ed) Frontiers in carbohydrate research-2. Elsevier Applied Science Publishers, London, p 100
599. Crescenzi V, Dentini M (1988) In: Phillips GO, Williams PA, Wedlock DJ (eds) Gums and stabilizers for the food industry 4. IRL Press, Oxford, p 63
600. Campana S, Ganter J, Milas M, Rinaudo M (1992) Carbohydr Res 231: 31
601. Crescenzi V, Dentini M, Segatori M, Tiblandi C, Callegaro L, Benedetti L (1992) Carbohydr Res 231: 73
602. Morris VJ, Brownsey GJ, Harris JE, Gunning AP, Stevens BJH, Johnson AWB (1989) Carbohydr Res 191: 315
603. Robinson G, Manning CE, Morris ER (1991) In: Dickinson E (ed) Food polymers, gels and colloids. The Royal Society of Chemistry, Cambridge, p 22
604. Cesàro A, Gemini A, Navarini L (1992) Polymer 33: 4001
605. Paoletti S, Smidsrød O, Grasdalen H (1984) Biopolymers 23: 1771
606. Carroll V, Miles MJ, Morris VJ (1982) Int J Biol Macromol 4: 432
607. Carroll V, Chilvers GR, Franklin D, Miles M, Morris VJ, Ring SG (1983) Carbohydr Res 114: 181
608. Pettitt DJ (1986) In: Phillips GO, Wedlock DJ, Williams PA (eds) Gums and stabilisers for the food industry 3. Elsevier Applied Science Publishers, London, p 451
609. Brownsey GJ, Chilvers GR, I'Anson KJ, Morris VJ (1984) Int J Biol Macromol 6: 211
610. Attwool PT, Atkins EDT, Upstill C, Miles MJ, Morris VJ (1986) Gums and stabilisers for the food industry 3. Elsevier Applied Science Publishers, London, p 135
611. Miles MJ, Morris VJ, O'Neill MA (1984) In: Phillips GO, Wedlock DJ, Williams PA (eds) Gums and stabilisers for the food industry 2. Pergamon Press, Oxford p 485
612. Chandrasekaran R, Millane RP, Arnott S, Atkins EDT (1988) Carbohydr Res 175: 1
613. Chandrasekaran R, Millane RP, Arnott S (1988) In: Phillips GO, Williams PA, Wedlock DJ (eds) Gums and stabilisers for the food industry 4. IRL Press, Oxford, p 183
614. Chandrasekaran R (1989) In: Millane RP, BeMiller JN, Chandrasekaran R (eds) Frontiers in carbohydrate research-1: food applications. Elsevier Applied Science Publishers, London, p 15
615. Chandrasekaran R, Thailambal VG (1990) Carbohydr Polym 12: 431
616. Lee EJ, Chandrasekaran R (1990) Carbohydr Res 214: 11
617. Chandrasekaran R, Thailambal VG (1990) ACS Symp Ser 430: 300
618. Chandrasekaran R (1991) In: Levine H, Slade L (eds) Adv Exp Med Biol, Vol 302: Water relationships in foods. Plenum Press, New York, p 773
619. Chandrasekaran R, Radha A, Thailambal VG (1992) Carbohydr Res 224: 1
620. Chandrasekaran R, Lee EJ, Radha A, Thailambal VG (1992) In: Chandrasekaran R (ed) Frontiers in carbohydrate research-2. Elsevier Applied Science Publishers, London, p 65
621. Hinton JF, Amisz ES (1971) Chem Rev 71: 627
622. Nakamura K, Harada K, Tanaka Y (1993) Food Hydrocoll 7: 435
623. Watase M, Nishinari K (1985) Colloid Polym Sci 263: 744
624. Watase M, Nishinari K (1986) In: Phillips GO, Wedlock DJ, Williams PA (eds) Gums and stabilisers for the food industry 3. Elsevier Applied Science Publishers, London, p 185
625. Miyoshi E (1996) Doctoral thesis, Osaka, Japan

Editor: K. Dušek
Received: July 1996

Author Index Volumes 101-130

Author Index Vols. 1-100 see Vol. 100

Index of Polymers

Index of Solvents

General Index

MIX
Papier aus verantwortungsvollen Quellen
Paper from responsible sources
FSC® C105338

If you have any concerns about our products,
you can contact us on
ProductSafety@springernature.com

In case Publisher is established outside the EU,
the EU authorized representative is:
Springer Nature Customer Service Center GmbH
Europaplatz 3, 69115 Heidelberg, Germany

Printed by Libri Plureos GmbH
in Hamburg, Germany